Universitext

For other titles in this series, go to
http://www.springer.com/series/223

Steven H. Weintraub

Galois Theory

Second Edition

Springer

Steven H. Weintraub
Department of Mathematics
Lehigh University
Bethlehem, PA, USA
shw2@lehigh.edu

ISBN: 978-0-387-87574-3 e-ISBN: 978-0-387-87575-0
DOI: 10.1007/978-0-387-87575-0

Library of Congress Control Number: 2008937989

Mathematics Subject Classification (2000): 12-01, 12F10, 11R32

Printed on acid-free paper

springer.com

To Judy, after 17 years, and to Blake

Contents

Preface to the First Edition

This is a textbook on Galois theory. Galois theory has a well-deserved reputation as one of the most beautiful subjects in mathematics. I was seduced by its beauty into writing this book. I hope you will be seduced by its beauty in reading it.

This book begins at the beginning. Indeed (and perhaps a little unusually for a mathematics text), it begins with an informal introductory chapter, Chapter 1. In this chapter we give a number of examples in Galois theory, even before our terms have been properly defined. (Needless to say, even though we proceed informally here, everything we say is absolutely correct.) These examples are sort of an airport beacon, shining a clear light at our destination as we navigate a course through the mathematical skies to get there.

Then we start with our proper development of the subject, in Chapter 2. We assume no prior knowledge of field theory on the part of the reader. We develop field theory, with our goal being the Fundamental Theorem of Galois Theory (the *FTGT*). On the way, we consider extension fields, and deal with the notions of normal, separable, and Galois extensions. Then, in the penultimate section of this chapter, we reach our main goal, the *FTGT*.

Roughly speaking, the content of the *FTGT* is as follows: To every Galois extension $\mathbf{E}$ of a field $\mathbf{F}$ we can associate its Galois group $G = \mathrm{Gal}(\mathbf{E}/\mathbf{F})$. By definition, G is the group of automorphisms of $\mathbf{E}$ that are the identity on $\mathbf{F}$. Then the *FTGT* establishes a one-to-one correspondence between fields $\mathbf{B}$ that are intermediate between $\mathbf{E}$ and $\mathbf{F}$, i.e., between fields $\mathbf{B}$ with $\mathbf{F} \subseteq \mathbf{B} \subseteq \mathbf{E}$, and subgroups of G. This connection allows us to use the techniques of group theory to answer questions about fields that would otherwise be intractable. (Indeed, historically Galois theory has been used to solve questions about fields that were outstanding for centuries, and even for millenia. We will treat some of these questions in this book.) In the final section of this chapter, we return to the informal examples with which we started the book, as well as treating more intricate and advanced ones that we can handle with our new-found knowledge.

In Chapter 3 we further develop and apply Galois theory. In this chapter we deal with a variety of different topics, some of which we mention here. In the first section we use Galois theory to investigate the field of symmetric functions and the ring of symmetric polynomials. Galois theory allows us to completely determine the structure of finite fields, and we do this in the third section. Two important properties of fields are the existence of primitive elements and normal bases, and we prove these in Sections 3.5 and 3.6. We can also say quite a bit about abelian extensions (i.e., extensions with abelian Galois groups), and we treat these in Section 3.7.

We develop Galois theory in complete generality, with careful consideration to the situation in positive characteristic as well as in characteristic 0. But we are especially interested in algebraic number fields, i.e., finite extensions of the field of rational numbers **Q**. We devote Chapter 4 to considering extensions of **Q**. Again we deal with a variety of different topics. We consider cyclotomic polynomials. We consider the question as to when equations are solvable by radicals (and prove Abel's theorem that the general equation of degree at least 5 is not). We show that the three classical geometric problems of Greek antiquity: trisecting the angle, duplicating the cube, and squaring the circle, are unsolvable with straightedge and compass. We deal with quadratic fields and their relation to cyclotomic fields, and we deal with radical polynomials, i.e., polynomials of the form $X^n - a$, which have a particular theory.

In Chapter 5 we consider more advanced topics in Galois theory. In particular, we prove that every field has an algebraic closure, and that the field of complex numbers **C** is algebraically closed, in Section 5.3, and we develop Galois theory for infinite algebraic extensions in Section 5.4.

Note that in Chapters 1 through 4, we assume that all field extensions are finite, except where explicitly stated otherwise. In Chapter 5 we allow extensions to be infinite.

There are three appendices. In the first we develop some necessary group theory. (We have put this in the appendix for logical reasons. The main text deals with field theory, and to develop the necessary group theory would lead to digressions in the main line of argument. Thus we collect these facts in an appendix in order to have a clear line of argument.) The second appendix revisits some material in the text from a more advanced point of view, which we do not want to presuppose of the reader, while the third appendix presents an elementary but tricky argument that enables us to avoid relying on Dirichlet's theorem about primes in an elementary progression at one point.

Our approach has been heavily influenced by Artin's classic 1944 text *Galois Theory*. Artin's approach emphasized linear algebra, and our approach has the same (and perhaps greater) emphasis. We have tried to have minimal prerequisites for this book, but, given this emphasis, the reader should have a sound knowledge of linear algebra. Beyond that the reader should know the basic facts about groups and rings, and especially about polynomial rings. As a source for this background material we naturally recommend our previous book, *Algebra: An Approach via Module Theory*, by William A. Adkins and Steven H. Weintraub, Springer-Verlag Graduate Texts in Mathematics No. 136. We refer to this text as [AW] when we have occasion to cite it. Also, the reader should be familiar with elementary number theory (the material contained in any standard undergraduate course in the subject will more than suffice). Finally, the Krull topology is the key to understanding infinite Galois extensions, so in the final section of this book (Section 5.4), and in this section alone, the reader must have a good knowledge of point-set topology.

There is, roughly speaking, enough material in this book for a year-long course in Galois theory. A one-semester course could consist of Chapters 1 and 2 (both of which can be easily covered in a semester) plus additional material from Chapters 3 and 4 as interest dictates and time permits.

We would like to mention that our previous book, [AW], treated groups, rings, modules, and linear algebra. This book treats field theory, so together these two books cover the topics of a standard one-year graduate algebra course. Also, given our particular attention to Galois theory over **Q**, we feel this book would be especially well-suited to students with an interest in algebraic number theory.

Our numbering system in this book is fairly standard. Theorem a.b.c refers to Theorem b.c in section b of Chapter a (or Appendix a). We denote the end of proofs by $\square$, as usual. In case a result is immediate, we simply append this symbol to its statement. Theorems, etc., are in italics, so are naturally set off from the remaining text. Definitions, etc., are in roman, and so are not. To delimit them, we end them with the symbol $\diamond$. Our notation is also fairly standard, but we call the reader's attention to the following conventions: We use $A \subseteq B$ to mean that A is a subset of B, while $A \subset B$ means that A is a proper subset of B. We denote fields by boldface letters and the integers by $\mathbb{Z}$. We will often be considering the situation where the field **E** is an extension of the field **F**, and in this situation we will use greek letters ($\alpha, \beta, \gamma, \ldots$) to denote elements of **E** and roman letters ($a, b, c, \ldots$) to denote elements of **F**. The greek letter ω will denote the primitive complex cube root of unity $\omega = (-1 + i\sqrt{3})/2$. The letter p will always denote a prime. Finally, id will denote the identity automorphism of whatever object is under consideration.

June 2005 *Steven H. Weintraub*

Preface to the Second Edition

I am pleased that there has been enough interest in this book to justify a second edition in such a short time since the first edition appeared.

The main change in this second edition is that I have added a new chapter, Chapter 6, that deals with transcendental extensions, thereby expanding the scope of the book. In addition, I have added smaller amounts of material: the statement and proof of Newton's Identities at the end of Section 3.1, material on linear disjointness at the end of Section 3.4, an expanded treatment of Example 2.9.7, and an additional corollary at the end of Section 4.7. Also, there is one consistent change, purely for esthetic reasons: Defined terms are in italics, instead of boldface, as they were previously. Beyond that, I have taken the opportunity to correct all (known) typos and minor errors in the first edition. (To my knowledge, there were no major errors.) It would be nice to hope that in adding around 25 pages to the text, I have not introduced any new errors; alas, that is probably too optimistic. Well, one can hope, anyway.

I have made it a point to preserve the numbering from the first edition intact. That is, in all cases Theorem a.b.c from the first edition remains Theorem a.b.c in the second edition, etc. Thus readers of the first edition should not be too disadvantaged.

June 2008 *Steven H. Weintraub*

1

Introduction to Galois Theory

1.1 Some Introductory Examples

In this section we will proceed informally, neither proving our claims nor even carefully defining our terms. Nevertheless, as you will see in the course of reading this book, everything we say here is absolutely correct. We proceed in this way to show in advance what our main goals are, and hence to motivate our development.

Example 1.1.1. Let $\mathbf{F} = \mathbf{Q}$, the field of rational numbers, and consider the polynomial $X^2 - 2 \in \mathbf{Q}[X]$. We let $\mathbf{E}$ be the "splitting field" of this polynomial, i.e., the "smallest" field which contains "all" the roots of this polynomial. Then $\mathbf{E} = \mathbf{Q}(\sqrt{2})$, the field obtained by "adjoining" $\sqrt{2}$ to $\mathbf{Q}$. Then $\mathbf{Q}(\sqrt{2}) = \{a + b\sqrt{2} \mid a, b \in \mathbf{Q}\}$, so as a vector space, $\dim_{\mathbf{F}} \mathbf{E} = 2$. We ask for all automorphisms of $\mathbf{E}$ that fix $\mathbf{F}$. If $\sigma : \mathbf{E} \to \mathbf{E}$ is such an automorphism, then it must take a root of $X^2 - 2$ to a root of $X^2 - 2$. Thus there are two possibilities: $\sigma_1(\sqrt{2}) = \sqrt{2}$, in which case $\sigma_1 : \mathbf{E} \to \mathbf{E}$ is the identity map, or $\sigma_2(\sqrt{2}) = -\sqrt{2}$, in which case $\sigma_2 : \mathbf{E} \to \mathbf{E}$ is given by $\sigma_2(a+b\sqrt{2}) = a - b\sqrt{2}$. Thus the group of automorphisms of $\mathbf{E}$ that fix $\mathbf{F}$, known as the Galois group $\mathrm{Gal}(\mathbf{E}/\mathbf{F})$, is $\{\sigma_1 = \mathrm{id}, \sigma_2\}$, isomorphic to $\mathbb{Z}/2\mathbb{Z}$, of order 2. Note that $|\mathrm{Gal}(\mathbf{E}/\mathbf{F})| = \dim_{\mathbf{F}} \mathbf{E}$. ◇

Example 1.1.2. Let $\mathbf{F} = \mathbf{Q}$ and consider the polynomial $(X^2 - 2)(X^2 - 3) \in \mathbf{Q}[X]$. The splitting field of this polynomial is $\mathbf{E} = \mathbf{Q}(\sqrt{2}, \sqrt{3})$ and $\mathbf{E} = \{a + b\sqrt{2} + c\sqrt{3} + d\sqrt{6} \mid a, b, c, d \in \mathbf{Q}\}$, so $\dim_{\mathbf{F}} \mathbf{E} = 4$. Again we ask for all automorphisms σ of $\mathbf{E}$ that fix $\mathbf{F}$. We must have that σ takes a root of $X^2 - 2$ to a root of $X^2 - 2$, so $\sigma(\sqrt{2}) = \pm\sqrt{2}$, and that σ takes a root of $X^3 - 3$ to $X^2 - 3$, so $\sigma(\sqrt{3}) = \pm\sqrt{3}$. Indeed, the values of σ on $\sqrt{2}$ and $\sqrt{3}$ can be

S.H. Weintraub, *Galois Theory*, DOI 10.1007/978-0-387-87575-0_1,

chosen independently, and these determine the values of σ on all elements of **E**, so we have the four automorphisms of **E** given by

$$\begin{aligned}
\sigma_1(a+b\sqrt{2}+c\sqrt{3}+d\sqrt{6}) &= a+b\sqrt{2}+c\sqrt{3}+d\sqrt{6},\\
\sigma_2(a+b\sqrt{2}+c\sqrt{3}+d\sqrt{6}) &= a-b\sqrt{2}+c\sqrt{3}-d\sqrt{6},\\
\sigma_3(a+b\sqrt{2}+c\sqrt{3}+d\sqrt{6}) &= a+b\sqrt{2}-c\sqrt{3}-d\sqrt{6},\\
\sigma_4(a+b\sqrt{2}+c\sqrt{3}+d\sqrt{6}) &= a-b\sqrt{2}-c\sqrt{3}+d\sqrt{6};
\end{aligned}$$

thus, $G = \mathrm{Gal}(\mathbf{E}/\mathbf{F}) = \{\sigma_1 = \mathrm{id}, \sigma_2, \sigma_3, \sigma_4\}$ is isomorphic to $\mathbb{Z}/2\mathbb{Z} \oplus \mathbb{Z}/2\mathbb{Z}$, and $|\,\mathrm{Gal}(\mathbf{E}/\mathbf{F})| = 4 = \dim_{\mathbf{F}} \mathbf{E}$.

Now we ask for all fields **B** "intermediate" between **E** and **F**, i.e., fields **B** containing **F** and contained in **E**. Of course $\mathbf{B}_1 = \mathbf{F}$ and $\mathbf{B}_5 = \mathbf{E}$ are certainly intermediate fields. Here are three others that easily appear:

$$\mathbf{B}_2 = \mathbf{Q}(\sqrt{2}),\ \mathbf{B}_3 = \mathbf{Q}(\sqrt{3}),\ \mathbf{B}_4 = \mathbf{Q}(\sqrt{6}).$$

Now let us look at subgroups H of G. It is easy to list all such H. They are $H_1 = G$, $H_5 = \{\mathrm{id}\}$ and also

$$H_2 = \{\mathrm{id}, \sigma_3\}, \quad H_3 = \{\mathrm{id}, \sigma_2\}, \quad H_4 = \{\mathrm{id}, \sigma_4\}.$$

Note we have a one-to-one correspondence between $\{H_i\}$ and $\{\mathbf{B}_i\}$ given by

$$\mathbf{B}_i = \mathrm{Fix}(H_i)$$

where $\mathrm{Fix}(H)$ is the subfield of **E** fixed by all elements of the subgroup H. In fact, the Fundamental Theorem of Galois Theory (*FTGT*) tells us there is a one-to-one correspondence between subgroups H of $\mathrm{Gal}(\mathbf{E}/\mathbf{F})$ and fields **B** intermediate between **E** and **F** given by

$$H \longleftrightarrow \mathbf{B} \text{ where } \mathbf{B} = \mathrm{Fix}(H).$$

It is easy to determine all the subgroups of G, as this is a finite group, so this gives us a way of determining all intermediate fields (a priori, a much more different task), and in our case this shows that $\mathbf{B}_1, \ldots, \mathbf{B}_5$ are indeed *all* the intermediate fields.

Let us make a further observation here, which the *FTGT* also tells us is true in general: If **B** is an intermediate field, then **B** is an **F**-vector space and **E** is a **B**-vector space, and

$$\dim_{\mathbf{F}} \mathbf{B} = [G : H], \quad \dim_{\mathbf{B}} \mathbf{E} = |H|.$$

One more observation: For each intermediate field $\mathbf{B}$, we may consider $\mathrm{Gal}(\mathbf{B}/\mathbf{F})$, the group of automorphisms of $\mathbf{B}$ fixing $\mathbf{F}$. If $\mathbf{B} = \mathbf{F}$, then $\mathrm{Gal}(\mathbf{B}/\mathbf{F}) = \mathrm{Gal}(\mathbf{F}/\mathbf{F})$ consists of the identity element alone, while if $\mathbf{B} = \mathbf{E}$, then $\mathrm{Gal}(\mathbf{B}/\mathbf{F}) = \mathrm{Gal}(\mathbf{E}/\mathbf{F})$ is the whole original Galois group. For $\mathbf{B} = \mathbf{B}_i$, $i = 2, 3, 4$, $\mathrm{Gal}(\mathbf{B}/\mathbf{F})$ is isomorphic to $\mathbb{Z}/2\mathbb{Z}$, of order 2. Thus in all five cases,

$$|\,\mathrm{Gal}(\mathbf{B}/\mathbf{F})| = \dim_{\mathbf{F}} \mathbf{B}. \qquad \diamond$$

Example 1.1.3. Let $\mathbf{F} = \mathbf{Q}$ and consider the polynomial $X^3 - 2 \in \mathbf{Q}[X]$. We let $\mathbf{E}$ be the splitting field of this polynomial. Then $\sqrt[3]{2} \in \mathbf{E}$ and, since $\mathbf{E}$ is a field, $(\sqrt[3]{2})^2 = \sqrt[3]{4} \in \mathbf{E}$. Now $\mathbf{E}$ contains *all* the roots of $x^3 - 2$, so if ω is a primitive cube root of 1, then $\omega\sqrt[3]{2} \in \mathbf{E}$, and since $\mathbf{E}$ is a field, $\omega\sqrt[3]{2}/\sqrt[3]{2} = \omega \in \mathbf{E}$. (Also, $\omega^2 \in \mathbf{E}$, but note that $\omega^3 = 1$, $\omega^3 - 1 = 0$, $(\omega - 1)(\omega^2 + \omega + 1) = 0$ and $\omega \neq 1$, so $\omega^2 + \omega + 1 = 0$ and $\omega^2 = -1 - \omega$; thus once we add ω to $\mathbf{F}$ we already have ω^2 as well.) Then $\mathbf{E} = \mathbf{Q}(\omega, \sqrt[3]{2})$ and $\mathbf{E} = \{a + b\sqrt[3]{2} + c\sqrt[3]{4} + d\omega + e\omega\sqrt[3]{2} + f\omega\sqrt[3]{4} \mid a, b, c, d, e, f \in \mathbf{Q}\}$ and $\dim_{\mathbf{F}} \mathbf{E} = 6$.

Now we wish to determine $\mathrm{Gal}(\mathbf{E}/\mathbf{F})$. Again, any automorphism of $\mathbf{E}$ must take a root of $X^3 - 2$ to another root of $X^3 - 2$, so $\sigma(\sqrt[3]{2}) = \sqrt[3]{2}$, $\omega\sqrt[3]{2}$, or $\omega^2\sqrt[3]{2}$. Also, as we just observed, ω is a root of $X^2 + X + 1$, whose other root is ω^2, so $\sigma(\omega) = \omega$ or ω^2. There are three choices for $\sigma(\sqrt[3]{2})$ and two choices for $\sigma(\omega)$. We may make these choices independently, and these choices determine σ on all of $\mathbf{E}$, so $|\,\mathrm{Gal}(\mathbf{E}/\mathbf{F})| = 6$.

In particular we have the following elements:

$$\varphi\colon \mathbf{E} \to \mathbf{E} \text{ with } \varphi(\sqrt[3]{2}) = \omega\sqrt[3]{2}, \quad \varphi(\omega) = \omega,$$
$$\psi\colon \mathbf{E} \to \mathbf{E} \text{ with } \psi(\sqrt[3]{2}) = \sqrt[3]{2}, \quad \psi(\omega) = \omega^2.$$

It is easy to check that $\varphi^3 = \mathrm{id}$ and $\psi^2 = \mathrm{id}$.

Let us compute further:

$$\varphi\psi(\sqrt[3]{2}) = \varphi(\psi(\sqrt[3]{2})) = \varphi(\sqrt[3]{2}) = \omega\sqrt[3]{2},$$
$$\varphi\psi(\omega) = \varphi(\psi(\omega)) = \varphi(\omega^2) = \omega^2,$$

while

$$\psi\varphi^2(\sqrt[3]{2}) = \varphi(\psi^2(\sqrt[3]{2})) = \psi(\omega^2\sqrt[3]{2}) = \omega\sqrt[3]{2},$$
$$\psi\varphi^2(\omega) = \psi(\varphi^2(\omega)) = \psi(\omega) = \omega^2.$$

Thus we see that $\varphi\psi = \psi\varphi^2$, and so

$$G = \mathrm{Gal}(\mathbf{E}/\mathbf{F}) = \langle \psi, \varphi \mid \varphi^3 = 1, \psi^2 = 1, \varphi\psi = \psi\varphi^{-1} \rangle$$

is isomorphic to the dihedral group of order 6.

Now it is easy to list all subgroups H of G; by the *FTGT* this gives us all intermediate fields $\mathbf{B}$. The subgroups H are

$$H_1 = G, \quad H_2 = \{\text{id}, \varphi, \varphi^2\}, \quad H_3 = \{\text{id}, \psi\},$$
$$H_4 = \{\text{id}, \varphi\psi\}, \quad H_5 = \{\text{id}, \varphi^2\psi\}, \quad H_6 = \{\text{id}\}.$$

Then we can compute that the corresponding subfields $\mathbf{B}_i = \text{Fix}(H_i)$ are given by

$$\begin{aligned}
\mathbf{B}_1 &= \mathbf{Q},\\
\mathbf{B}_2 &= \mathbf{Q}(\omega) = \{a + d\omega \mid a, d \in \mathbf{Q}\},\\
\mathbf{B}_3 &= \mathbf{Q}(\sqrt[3]{2}) = \{a + b\sqrt[3]{2} + c\sqrt[3]{4} \mid a, b, c \in \mathbf{Q}\},\\
\mathbf{B}_4 &= \mathbf{Q}(\omega\sqrt[3]{2}) = \{a + c\sqrt[3]{4} + e\omega\sqrt[3]{2} + c\omega\sqrt[3]{4} \mid a, c, e \in \mathbf{Q}\},\\
\mathbf{B}_5 &= \mathbf{Q}(\omega^2\sqrt[3]{2}) = \{a + b\sqrt[3]{2} + b\omega\sqrt[3]{2} + f\omega\sqrt[3]{4} \mid a, b, f \in \mathbf{Q}\},\\
\mathbf{B}_6 &= \mathbf{Q}(\omega, \sqrt[3]{2}).
\end{aligned}$$

Again we see that for each H and corresponding $\mathbf{B}$,

$$\dim_{\mathbf{F}} \mathbf{B} = [G : H], \quad \dim_{\mathbf{B}} \mathbf{E} = |H|.$$

Once again, let us compute $\text{Gal}(\mathbf{B}/\mathbf{F})$ for each intermediate field $\mathbf{B}$. For $\mathbf{B} = \mathbf{B}_1$, $\text{Gal}(\mathbf{B}/\mathbf{F}) = \{\text{id}\}$ and for $\mathbf{B} = \mathbf{B}_6$, $\text{Gal}(\mathbf{B}/\mathbf{F}) = G$. More interestingly, $\text{Gal}(\mathbf{B}_2/\mathbf{F})$ is isomorphic to $\mathbb{Z}/2\mathbb{Z}$, with the nontrivial element τ obtained as follows: ω is a root of $X^2 + X + 1$, so τ must take ω to the other root of $x^2 + x + 1$, which is $\omega^2 = -1 - \omega$. Thus

$$\tau(a + d\omega) = a + d(-1 - \omega) = (a - d) - d\omega.$$

(Actually, $\mathbf{B}_2$ and τ have an alternate description. By the quadratic formula, $\omega = (-1+i\sqrt{3})/2$ and $\omega^2 = \bar{\omega} = (-1-i\sqrt{3})/2$. Thus $\mathbf{B}_2 = \mathbf{Q}(i\sqrt{3}) = \{a' + b'i\sqrt{3} \mid a', b' \in \mathbf{Q}\}$ and $\tau(a' + b'i\sqrt{3}) = a' - b'i\sqrt{3}$, so τ is simply complex conjugation on $\mathbf{B}_2$). In each of these three cases we see that $|\text{Gal}(\mathbf{B}/\mathbf{F})| = \dim_{\mathbf{F}} \mathbf{B}$.

On the other hand, consider $\mathbf{B}_3$. An automorphism of $\mathbf{B}_3 = \mathbf{Q}(\sqrt[3]{2})$ fixing $\mathbf{Q}$ is determined by its effect on $\sqrt[3]{2}$. Since $\sqrt[3]{2}$ is a root of $X^3 - 2$, it must take $\sqrt[3]{2}$ to a root of this same polynomial. But $\sqrt[3]{2}$ is the *only* root of $X^3 - 2$ in $\mathbf{B}_3$, so any automorphism of $\mathbf{B}_3$ must fix $\sqrt[3]{2}$ and hence must be the identity. A similar argument works for $\mathbf{B}_4$ and $\mathbf{B}_5$, so we see that $1 = |\text{Gal}(\mathbf{B}_i/\mathbf{F})| < \dim_{\mathbf{F}} \mathbf{B}_i = 3$ for $i = 3, 4, 5$. Thus the situation here is different.

Why is it different? The answer is that the subgroups H_1, H_2, and H_6 of G are normal subgroups of G, while the subgroups H_3, H_4, and H_5 are not. As we will see from the *FTGT*, in case H is a normal subgroup of G and $\mathbf{B} = \mathrm{Fix}(H)$, then $|\,\mathrm{Gal}(\mathbf{B}/\mathbf{F})| = \dim_{\mathbf{F}} \mathbf{B}$ and, moreover, $\mathrm{Gal}(\mathbf{B}/\mathbf{F})$ is the quotient group G/H. You can verify that this is true here for H_1 and H_6 (which is immediate) and for H_2 (which is more interesting). Also, we'll remark that H_3, H_4, and H_5 are conjugate subgroups of G and that $\mathbf{B}_3$, $\mathbf{B}_4$ and $\mathbf{B}_5$ are "conjugate" intermediate fields, in that there are automorphisms of $\mathbf{E}$ fixing $\mathbf{F}$ and permuting these subfields; indeed $\varphi(\mathbf{B}_3) = \mathbf{B}_4$ and $\varphi^2(\mathbf{B}_3) = \mathbf{B}_5$.

Note that in Example 1.1.2 the Galois group G was abelian, so *every* subgroup was normal, and hence for *every* intermediate field $\mathbf{B}$, we had $|\,\mathrm{Gal}(\mathbf{B}/\mathbf{F})| = \dim_{\mathbf{F}} \mathbf{B}$ and in fact $\mathrm{Gal}(\mathbf{B}/\mathbf{F}) = G/H$. ◇

Example 1.1.4. Let $\mathbf{F} = \mathbf{Q}$ and consider the polynomial $X^2 - 5 \in \mathbf{Q}[X]$. The splitting field of this polynomial is $\mathbf{B} = \mathbf{Q}(\sqrt{5})$ and the Galois group $\mathrm{Gal}(\mathbf{B}/\mathbf{F}) = \{\mathrm{id}, \varphi\}$ where $\varphi(\sqrt{5}) = -\sqrt{5}$. Now consider the polynomial $\Phi_5(X) = (X^5 - 1)/(X - 1) = X^4 + X^3 + X^2 + X + 1$. This polynomial is irreducible. Let $\mathbf{E}$ be the splitting field of this polynomial. Then $\mathbf{E} = \mathbf{Q}(\zeta_5)$ where $\zeta_5 = \exp(2\pi i/5)$, a fifth root of 1, and $\dim_{\mathbf{F}} \mathbf{E} = 4$. The automorphism of $\mathbf{E}$ fixing $\mathbf{F}$ must take ζ_5 to some other root of $\Phi_5(X)$, and any root is possible, as $\Phi_5(X)$ is irreducible. Now $\Phi_5(X)$ has roots ζ_5, ζ_5^2, ζ_5^3, and ζ_5^4. Let $\sigma \in \mathrm{Gal}(\mathbf{E}/\mathbf{F})$ with $\sigma(\zeta_5) = \zeta_5^2$. Then $\sigma^2(\zeta_5) = \zeta_5^4$ and $\sigma^3(\zeta_5) = \zeta_5^8 = \zeta_5^3$. By the *FTGT*, $|\,\mathrm{Gal}(\mathbf{E}/\mathbf{F})| = 4$ so $\mathrm{Gal}(\mathbf{E}/\mathbf{F}) = \{\mathrm{id}, \sigma, \sigma^2, \sigma^3\}$ is cyclic of order 4.

Now $0 = \Phi_5(\zeta_5) = \zeta_5^4 + \zeta_5^3 + \zeta_5^2 + \zeta_5 + 1$, so $(\zeta_5 + \zeta_5^4)^2 = \zeta_5^2 + \zeta_5^3 + 2 = \zeta_5 + \zeta_5^4 + 1$, so $\theta = \zeta_5 + \zeta_5^4$ is a root of $X^2 - X - 1$, as is $\theta' = \zeta_5^2 + \zeta_5^3$. By the quadratic formula $\theta = (1 + \sqrt{5})/2$ and $\theta' = (1 - \sqrt{5})/2$. Thus we see that $\mathbf{E} \supset \mathbf{Q}(\theta) = \mathbf{Q}(\sqrt{5}) = \mathbf{B}$, and $\dim_{\mathbf{B}} \mathbf{E} = 2$ as $\theta = \zeta_5 + \zeta_5^4 = \zeta_5 + \zeta_5^{-1}$, so ζ_5 is a root of the quadratic equation $X^2 - \theta X + 1$, a polynomial in $\mathbf{B}[X]$. (Its other root is ζ_5^{-1}.)

By the *FTGT*, $\mathrm{Gal}(\mathbf{E}/\mathbf{B})$ has order 2, and is a subgroup of $\mathrm{Gal}(\mathbf{E}/\mathbf{F})$. Then it must be the subgroup $\{\mathrm{id}, \sigma^2\}$, and that checks as $\sigma(\theta) = \sigma(\zeta_5 + \zeta_5^4) = \zeta_5^2 + \zeta_5^3 = \theta'$ so σ^2 does not fix $\mathbf{B}$, but $\sigma^2(\theta) = \theta$ (and $\sigma^2(\theta') = \theta'$) so σ^2 does. Also, $\mathrm{Gal}(\mathbf{B}/\mathbf{Q}) = \mathrm{Gal}(\mathbf{E}/\mathbf{Q})/\mathrm{Gal}(\mathbf{E}/\mathbf{B}) = \{\mathrm{id}, \sigma, \sigma^2, \sigma^3\}/\{\mathrm{id}, \sigma^2\}$ has σ as a representative of the nontrivial element. The quotient map is given by restricting an automorphism of $\mathbf{E}$ to $\mathbf{B}$. We have already seen that $\mathrm{Gal}(\mathbf{B}/\mathbf{Q}) = \{\mathrm{id}, \varphi\}$. But note that $\sigma \mid \mathbf{B} = \varphi$ as $\sigma\big((1 + \sqrt{5})/2\big) = (1 - \sqrt{5}/2)$, so $\sigma(\sqrt{5}) = -\sqrt{5} = \varphi(\sqrt{5})$.

In this example, we are back to the case in which $\mathrm{Gal}(\mathbf{E}/\mathbf{F})$ is abelian, so by the *FTGT*, any intermediate field $\mathbf{B}$ will also be a Galois extension of $\mathbf{F}$, i.e., will have $|\,\mathrm{Gal}(\mathbf{B}/\mathbf{F})| = \dim_{\mathbf{F}} B$. Since $\mathrm{Gal}(\mathbf{E}/\mathbf{F})$ is cyclic of order 4, it has a unique subgroup of order 2 corresponding to a unique intermediate field **B** with $\dim_{\mathbf{B}} \mathbf{E} = 2$, and we have explicitly identified this field **B** as $\mathbf{Q}(\sqrt{5})$. Also, we have shown how to obtain **E** in two stages: First, adjoin θ (or $\sqrt{5}$) to **Q**, creating an extension of degree 2, and the adjoin ζ_5 to **B**, creating an extension **E** of **B** degree 2, and hence an extension **E** of **Q** of degree 4. (The degree of an extension is its dimension as a vector space over the base field.)

Finally, this example is interesting as it is an example of the relationship between a quadratic field **B** (i.e., a field obtained by adjoining a square root to **Q**) and a cyclotomic field **E** (i.e., a field obtained by adjoining a root of 1 to **Q**). ◇

2

Field Theory and Galois Theory

2.1 Generalities on Fields

We begin by defining the objects we will be studying.

Definition 2.1.1. $\mathbf{F}$ is a *field* if $\mathbf{F}$ is an abelian group under addition, $\mathbf{F} - \{0\}$ is an abelian group under multiplication, and multiplication distributes over addition.

In other words, $\mathbf{F}$ is a field if :

1. For any $a, b \in \mathbf{F}$, $a + b \in \mathbf{F}$.
2. For any $a, b \in \mathbf{F}$, $a + b = b + a$.
3. For any $a, b, c \in \mathbf{F}$, $(a + b) + c = a + (b + c)$.
4. There is a $0 \in \mathbf{F}$ such that $a + 0 = 0 + a = a$ for every $a \in \mathbf{F}$.
5. For every $a \in \mathbf{F}$ there is an element $-a \in \mathbf{F}$ with $a+(-a)=(-a)+a = 0$.
6. For any $a, b \in \mathbf{F}$, $ab \in \mathbf{F}$, and if $a \neq 0$ and $b \neq 0$, then $ab \neq 0$.
7. For any $a, b \in \mathbf{F}$, $ab = ba$.
8. For any $a, b, c \in \mathbf{F}$, $(ab)c = a(bc)$.
9. There is a $1 \in \mathbf{F}$ such that $a1 = 1a = a$ for every $a \in \mathbf{F}$.
10. For every $a \neq 0 \in \mathbf{F}$ there is an element $a^{-1} \in \mathbf{F}$ with $aa^{-1} = a^{-1}a = 1$.
11. For every $a \in \mathbf{F}$, $a0 = 0a = 0$.
12. For every $a, b, c \in \mathbf{F}$, $(a + b)c = ac + bc$.
13. For every $a, b, c \in \mathbf{F}$, $c(a + b) = ca + ba$.
14. $0 \neq 1$.

◇

Example 2.1.2. Here are some examples of fields. The first three are very familiar.

S.H. Weintraub, *Galois Theory*, DOI 10.1007/978-0-387-87575-0_2,

(1) The rational numbers $\mathbf{Q}$.
(2) The real numbers $\mathbf{R}$.
(3) The complex numbers $\mathbf{C}$.
(4) The field $\mathbf{F} = \mathbf{Q}(\sqrt{D})$ where D is not a perfect square.

By definition, $\mathbf{F} = \{a + b\sqrt{D} \mid a, b \in \mathbf{Q}\}$ with addition and multiplication defined "as usual". That is, $(a+b\sqrt{D}) + (c+d\sqrt{D}) = (a+c) + (b+d)\sqrt{D}$ and $(a+b\sqrt{D})(c+d\sqrt{D}) = (ac+bdD) + (ad+bc)\sqrt{D}$. We need to see that inverses exist, and this comes from the familiar method of "rationalizing the denominator":

$$\begin{aligned}\frac{1}{a+b\sqrt{D}} &= \frac{1}{a+b\sqrt{D}} \cdot \frac{a-b\sqrt{D}}{a-b\sqrt{D}} = \frac{a-b\sqrt{D}}{a^2-b^2D} \\ &= \frac{a}{a^2-b^2D} - \frac{b}{a^2-b^2D}\sqrt{D}.\end{aligned}$$

(5) The fields $\mathbf{F} = \mathbf{F}_p$ of integers modulo p, where p is a prime. $\mathbf{F}_p = \{0, 1, \ldots, p-1\}$ with addition and multiplication defined mod p, i.e., $i+j = k$ in $\mathbf{F}_p$ if $i + j \equiv k (\mathrm{mod}\ p)$ and $ij = k$ in $\mathbf{F}_p$ if $ij \equiv k (\mathrm{mod}\ p)$. ◇

Example 2.1.2 (4) illustrates a construction of particular importance to us. It is, as we shall see below, an "algebraic extension" of $\mathbf{Q}$, obtained by "adjoining" the element $\sqrt{D}$, an element that is a root of the irreducible polynomial $X^2 - D \in \mathbf{Q}[X]$. (In a similar way, $\mathbf{C} = \mathbf{R}(i)$ where i is a root of the irreducible polynomial $X^2 + 1 \in \mathbf{R}[X]$). In fact, this chapter will be entirely devoted to studying algebraic extensions of fields and their properties.

In this book, we shall not only be studying individual fields, but also, and especially, maps between them. We thus make the following important observation.

Lemma 2.1.3. *Let $\mathbf{F}$ and $\mathbf{F}'$ be fields and let $\varphi\colon \mathbf{F} \to \mathbf{F}'$ be a map with $\varphi(a+b) = \varphi(a) + \varphi(b)$ and $\varphi(ab) = \varphi(a)\varphi(b)$ for every $a, b \in \mathbf{F}$. Then one of the following two alternatives holds:*
(1) $\varphi(0) = 0$, $\varphi(1) = 0$, and φ is the zero map.
(2) $\varphi(0) = 0$, $\varphi(1) = 1$, and φ is an injection.

Proof. We have that $\varphi(0) = \varphi(0+0) = \varphi(0) + \varphi(0)$ so $0 = \varphi(0)$.

We also have that $\varphi(1) = \varphi(1 \cdot 1) = \varphi(1)\varphi(1)$ so $0 = \varphi(1)\varphi(1) - \varphi(1) = \varphi(1)(1-\varphi(1))$ and hence, by the contrapositive of property (6) of a field, either $\varphi(1) = 0$ or $\varphi(1) = 1$.

First, suppose $\varphi(1) = 0$. Then for any $a \in \mathbf{F}$, we have that $\varphi(a) = \varphi(a \cdot 1) = \varphi(a)\varphi(1) = \varphi(a) \cdot 0 = 0$, and φ is the zero map.

Next suppose $\varphi(1) = 1$. Then for any $c \in \mathbf{F}$, $c \neq 0$, we have that $1 = \varphi(1) = \varphi(cc^{-1}) = \varphi(c)\varphi(c^{-1})$ so $\varphi(c) \neq 0$. Now let $a, b \in \mathbf{F}$ with $a \neq b$, and set $c = b-a \neq 0$. Then $\varphi(b) = \varphi(a+(b-a)) = \varphi(a+c) = \varphi(a)+\varphi(c) \neq \varphi(a)$, and φ is an injection. □

Definition 2.1.4. The *characteristic* char(**F**) of the field **F** is the smallest positive integer n such that $n \cdot 1 = 0 \in \mathbf{F}$, or 0 if no such n exists. ◇

Examining Example 2.1.2, we see that the fields in parts (1)–(4) have characteristic 0, while the fields in part (5) have characteristic p.

Lemma 2.1.5. *Let* **F** *be a field. Then* char(**F**) *is either* 0 *or a prime.*

Proof. Suppose $\operatorname{char}(\mathbf{F}) = n$ is neither 0 nor a prime, and let $n = ab$ with $1 < a, b < n$. Then $a \cdot 1 \neq 0$ and $b \cdot 1 \neq 0$ but $(a \cdot 1)(b \cdot 1) = (ab) \cdot 1 = n \cdot 1 = 0$. But this is impossible as the product of any two nonzero elements in a field is nonzero. □

Remark 2.1.6. If **F** is any field, there is a canonical map φ from the integers $\mathbb{Z}$ to **F** determined by $\varphi(1) = 1$. If $\operatorname{char}(\mathbf{F}) = 0$, then φ is an injection; using φ we regard $\mathbb{Z}$ as a subring of **F**, and then we also regard $\mathbf{F}_0 = \mathbf{Q}$ as a subfield of **F**. If $\operatorname{char}(\mathbf{F}) = p$, then φ gives an injection from $\mathbf{F}_p$ into **F** and we regard $\mathbf{F}_0 = \mathbf{F}_p$ as a subfield of **F**. In either case, $\mathbf{F}_0$ is called the *prime field.* ◇

Definition 2.1.7. A field **E** is an *extension* of the field **F** if **F** is a subfield of **E**. ◇

We will often be in the situation of considering a field **F** and an extension field **E**. To help the reader keep track of what is going on, we will adopt the notational convention that, in this situation, we will denote elements of **F** by Roman letters $(a, b, \ldots)$ and elements of **E** by Greek letters $(\alpha, \beta, \ldots)$.

It easy to check that if **E** is an extension of **F**, then $\operatorname{char}(\mathbf{E}) = \operatorname{char}(\mathbf{F})$ and furthermore **E** is an **F**-vector space. This leads us to the following definition.

Definition 2.1.8. (1) The *degree* $(\mathbf{E}/\mathbf{F})$ of **E** over **F** is $\dim_{\mathbf{F}} \mathbf{E}$, the dimension of **E** as an **F**-vector space.

(2) **E** is a *finite* extension of **F** (or **E**/**F** is *finite*) if $(\mathbf{E}/\mathbf{F})$ is finite. ◇

(Although we will not use it here, another common notation for $(\mathbf{E}/\mathbf{F})$ is $[\mathbf{E} : \mathbf{F}]$.)

Examining Example 2.1.2 (4), we see that $(\mathbf{Q}(\sqrt{D})/\mathbf{Q}) = 2$ as $\mathbf{Q}(\sqrt{D})$ has basis $\{1, \sqrt{D}\}$ as a **Q**-vector space (and similarly $(\mathbf{C}/\mathbf{R}) = 2$ as **C** has basis $\{1, i\}$ as an **R**-vector space). Note that 2 is also the degree of the polynomial

$X^2 - D$ (and of the polynomial X^2+1). As we shall see, this is no coincidence. On the other hand, $(\mathbf{C}/\mathbf{Q}) = \infty$. As we shall also see, with considerably more work, for every prime p and every positive integer n there is a field $\mathbf{F}_{p^n}$ with p^n elements, and $\mathbf{F}_{p^n}$ is unique up to isomorphism. Then $(\mathbf{F}_{p^n}/\mathbf{F}_p) = n$.

Here is an important observation about arithmetic in fields of characteristic p.

Lemma 2.1.9. *Let* $\mathbf{F}$ *be a field of characteristic* p. *Then* $(a+b)^p = a^p + b^p$ *for any elements* a, b *of* $\mathbf{F}$.

Proof. By the Binomial Theorem, $(a+b)^p = \Sigma_{i=0}^{p}\binom{p}{i}a^i b^{p-i} = a^p + b^p$ as $\binom{p}{i} = p!/(i!(p-i)!)$ is divisible by p for all $1 \leq i \leq p-1$. □

Definition 2.1.10. Let $\mathbf{F}$ be a field of characteristic p. The *Frobenius* map $\Phi\colon \mathbf{F} \to \mathbf{F}$ is the map defined by $\Phi(X) = X^p$. ◇

Corollary 2.1.11. *Let* $\mathbf{F}$ *be a field of characteristic* p. *Then* $\Phi\colon \mathbf{F} \to \mathbf{F}$ *is an endomorphism. If* $\mathbf{F}$ *is finite,* Φ *is an automorphism.*

Proof. Clearly $\Phi(1) = 1$ and $\Phi(ab) = \Phi(a)\Phi(b)$. But also $\Phi(a+b) = \Phi(a) + \Phi(b)$ by Lemma 2.1.9. Now, by Lemma 2.1.3, Φ is injective, so $|\operatorname{Im}(\Phi)| = |\mathbf{F}|$. Thus, if $|\mathbf{F}|$ is finite, $\operatorname{Im}(\Phi) = \mathbf{F}$. □

Lemma 2.1.12. *Let* $\mathbf{F} = \mathbf{F}_p$. *Then* $\Phi\colon \mathbf{F} \to \mathbf{F}$ *is the identity automorphism.*

Proof. By Fermat's Little Theorem, $x^p \equiv x \pmod{p}$ for $x = 0, 1, \ldots, p-1$. □

Thus, for the field $\mathbf{F}_p$, the Frobenius automorphism Φ is uninteresting. But, as we shall see, for other fields of characteristic p it plays an important role.

We shall find it convenient to introduce the following notion.

Definition 2.1.13. An *integral domain* is a ring R that satisfies properties (1)–(9) and (11)–(14) of Definition 2.1.1. ◇

In particular, in an integral domain R, property (6) holds, so the product of any two nonzero elements of R is nonzero. However, property (11) may not hold, so a nonzero element of R may not have an inverse. Those elements of R that do have inverses are called the *units* of R. As an example of an integral domain, we have the integers $\mathbb{Z}$, from which the name integral domain arises.

2.2 Polynomials

In this section we consider polynomial rings. We shall assume the reader has already encountered them, and so will not prove the more basic facts. As a reference for them, we recommend [AW, Sections 2.4 and 2.5].

For any field $\mathbf{F}$, we let $\mathbf{F}[X]$ be the ring of polynomials in the variable X with coefficients in $\mathbf{F}$. We observe that the product $f(X)g(X)$ of any two nonzero polynomials $f(X)$, $g(X) \in \mathbf{F}[X]$ is nonzero, and so $\mathbf{F}[X]$ is an integral domain. We regard $\mathbf{F} \subset \mathbf{F}[X]$ by identifying the element $a \in \mathbf{F}$ with the constant polynomial $a \in \mathbf{F}[X]$, and observe that, under this identification, the units of $\mathbf{F}[X]$ (i.e., the invertible elements of $\mathbf{F}[X]$) are precisely the nonzero elements of $\mathbf{F}$.

The most basic fact about $\mathbf{F}[X]$ is that we have the division algorithm: For any polynomials $f(X)$ and $g(X) \in \mathbf{F}[X]$ with $g(X) \neq 0$, there exist unique polynomials $q(X)$ and $r(X)$ with

$$f(X) = g(X)q(X) + r(X),$$

where either $r(X) = 0$ or $r(X)$ is a nonzero polynomial with $\deg(r(X)) < \deg(g(X))$. (Here deg is the degree of a polynomial.) We call $q(X)$ the quotient and $r(X)$ the remainder. Note that the quotient and remainder are independent of the field: If $\mathbf{E}$ is an extension of $\mathbf{F}$ and $f(X) = g(X)q(X) + r(X)$ is an equation in $\mathbf{F}[X]$, this equation holds in $\mathbf{E}[X]$ as well, so by the uniqueness of the quotient and remainder, $q(X)$ and $r(X)$ are the quotient and remainder in $\mathbf{E}[X]$ as well.

From the existence of the division algorithm we conclude that $\mathbf{F}[X]$ is a Euclidean domain, where the norm of a nonzero polynomial is equal to its degree. A Euclidean domain is so named because it is an integral domain in which Euclid's algorithm holds. One of the consequences of Euclid's algorithm is that any two nonzero elements of $\mathbf{F}[X]$ have a greatest common divisor (gcd). In general, the gcd of two elements of a Euclidean ring is only well defined up to multiplication by a unit, but in $\mathbf{F}[X]$ we specify the gcd by requiring it to be monic. Then the greatest common divisor (gcd) of two polynomials $f(X)$ and $g(X)$ is defined to be the unique monic polynomial of highest degree that divides both $f(X)$ and $g(X)$, and two polynomials are relatively prime if their gcd is 1.

In addition, since $\mathbf{F}[X]$ is a Euclidean domain, the gcd $d(X)$ of the polynomials $f(X)$ and $g(X)$ may be expressed as a linear combination of $f(X)$ and $g(X)$, i.e., there are polynomials $a(X)$ and $b(X)$ with $d(X) = a(X)f(X) + b(X)g(X)$.

Furthermore, every Euclidean domain is a principal ideal domain (PID). A principal ideal domain R is so named because it is an integral domain in which every ideal I is principal, i.e., consists of the multiples of a single element i, in which case we write $I = \langle i \rangle$. Applying that here, we see that every ideal of $\mathbf{F}[X]$ is of the form $\langle f(X) \rangle$ for some polynomial $f(X)$.

In an integral domain R, we have the notion of irreducible and prime elements: A nonunit $a \in R$ is irreducible if $a = bc$ implies that b or c is a unit, and a nonunit $a \in R$ is prime if a divides bc implies that a divides b or a divides c. In any integral domain R, every prime is irreducible, and in a PID R, every irreducible is prime, so in a PID R the notions of prime and irreducible are the same.

Finally, in any PID unique factorization holds. In particular, it holds in $\mathbf{F}[X]$. Concretely, in $\mathbf{F}[X]$ this means: Every nonconstant polynomial $f(X) \in \mathbf{F}[X]$ can be factored into irreducible polynomials in an essentially unique way. Furthermore, we may write $f(X)$ uniquely (up to the order of the terms) as

$$f(X) = u f_1(X) \cdots f_k(X) \quad \text{with } u \in \mathbf{F},\ f_i \in \mathbf{F}[X] \text{ monic irreducible}$$

(and in particular, if $f(X)$ is also monic then $u = 1$). We call this the prime factorization of $f(X)$.

Given this background, we now get to work.

Lemma 2.2.1. *Let $f(X) \in \mathbf{F}[X]$ be a nonzero polynomial.*

(1) For $a \in \mathbf{F}$, a is a root of $f(X)$, i.e., $f(a) = 0$, if and only if $X - a$ divides $f(X)$.

(2) The number of roots of $f(X)$ is less than or equal to the degree of $f(X)$.

Proof. (1) By the division algorithm, we may write $f(X) = (X - a)q(X) + b$, $b \in \mathbf{F}$. Setting $X = a$, we see $b = f(a)$.

(2) We prove this part of the lemma by induction on $d = \deg(f(X))$, beginning with $d = 0$. If $d = 0$, then $f(X)$ is a nonzero constant polynomial, so has 0 roots, and this part of the lemma holds. Now assume it holds for every polynomial of degree d and let $f(X)$ have degree $d + 1$. If $f(X)$ has no roots, we are done. Otherwise, let $f(\alpha) = 0$. Then $f(X) = (X - \alpha)g(X)$, so by induction $g(X)$ has $e \leq d$ roots. Then $f(X)$ has e' roots, with $e' = e$ or $e + 1$ according as α is or is not a root of $g(X)$, but in any case $e' \leq d + 1$ and we are done by induction. □

Lemma 2.2.1 has the following important consequence.

Corollary 2.2.2. *Let G be any finite subgroup of the multiplicative group of a field $\mathbf{F}$. Then G is cyclic.*

Proof. Let G have order n. Then any element g of G has order d for some d dividing n. Now an element g of G is a nonzero element of $\mathbf{F}$, and so the equation $g^d = 1$ in G is equivalent to the equation $g^d - 1 = 0$ in $\mathbf{F}$. Thus we see that if g has order d, then g is a root of the polynomial $X^d - 1$. By Lemma 2.2.1, this polynomial has at most d roots. Hence G is an abelian group of order n with the property that G has at most d elements of order d, for any d dividing n. But it is easy to check that an abelian group with this property must be cyclic (and furthermore that in this case G has exactly d elements of order d, for any d dividing n). □

Lemma 2.2.3. *Let* $\mathbf{F}$ *be a field and* R *an integral domain that is a finite-dimensional* $\mathbf{F}$*-vector space. Then* R *is a field.*

Proof. We need to show that any nonzero $r \in R$ has an inverse. Consider $\{1, r, r^2, \ldots, \}$. This is an infinite set of elements of R, and by hypothesis R is finite dimensional as an $\mathbf{F}$-vector space, so this set is linearly dependent. Hence $\Sigma_{i=0}^{n} c_i r^i = 0$ for some n and some $c_i \in \mathbf{F}$ not all zero. In other words, $f(r) = 0$ where $f(X) = \Sigma_{i=0}^{n} c_i X^i \in \mathbf{F}[X]$.

First, suppose $c_0 \neq 0$. Then

$$\begin{aligned}
c_n r^n + \cdots + c_1 r + c_0 &= 0,\\
c_0^{-1}(c_n r^n + \cdots + c_1 r + c_0) &= 0,\\
c_0^{-1} c_n r^n + \cdots + c_0^{-1} c_1 r + 1 &= 0,\\
-c_0^{-1} c_n r^n - \cdots - c_0^{-1} c_1 r &= 1,\\
r(-c_0^{-1} c_n r^{n-1} \cdots - c_0^{-1} c_1) &= 1,
\end{aligned}$$

so r has inverse $-c_0^{-1} c_n r^{n-1} \cdots - c_0^{-1} c_1 \in R$.

If $c_0 = 0$, write $f(X) = X^k q(X)$ where the constant term of $q(X)$ is nonzero. Then $0 = f(r) = r^k q(r)$ and R is an integral domain, so $q(r) = 0$ and we may apply the above argument to the polynomial $q(X)$. □

Proposition 2.2.4. *Let* $f(X) \in \mathbf{F}[X]$ *be an irreducible polynomial of degree* d*. Then* $\mathbf{F}[X]/\langle f(X)\rangle$ *is a field and is a* d*-dimensional* $\mathbf{F}$*-vector space.*

Proof. Let $\pi : \mathbf{F}[X] \to \mathbf{F}[X]/\langle f(X)\rangle$ be the canonical projection and set $\pi(f(X)) = \bar{f}(X)$.

By hypothesis, $f(X)$ has degree d. We claim that $S = \{\bar{1}, \bar{X}, \ldots, \bar{X}^{d-1}\}$ is a basis for $\mathbf{F}[X]/\langle f(X)\rangle$.

S spans $\mathbf{F}[X]/\langle f(X)\rangle$: Let $\bar{g}(X) \in \mathbf{F}[X]/\langle f(X)\rangle$. Then, by the division algorithm, $g(X) = f(X)q(X) + r(X)$ where either $r(X) = 0$ or $\deg r(X) < d$. Then $\bar{g}(X) = \bar{r}(X) = \Sigma_{i=0}^{d-1} c_i \bar{X}^i$ is a linear combination of elements of S.

S is linearly independent: Suppose $\Sigma_{i=0}^{d-1} c_i \bar{X}^i = \bar{g}(X) = 0$. Then $g(X)$ is divisible by $f(X)$. But $\deg g(X) < \deg f(X)$, so this is impossible unless $g(X) = 0$, in which case $c_0 = c_1 = \ldots = c_{d-1} = 0$.

Thus $\mathbf{F}[X]/\langle f(X)\rangle$ is a d-dimensional vector space over $\mathbf{F}$.

Next observe that $\mathbf{F}[X]/\langle f(X)\rangle$ is an integral domain: Suppose $\bar{g}(X)\bar{h}(X) = 0$. Then $g(X)h(X)$ is divisible by $f(X)$. Since $f(X) \in \mathbf{F}[X]$ is irreducible, and hence prime, we have that $f(X)$ divides $g(X)$, in which case $\bar{g}(X) = 0$, or $f(X)$ divides $h(X)$, in which case $\bar{h}(X) = 0$.

Then, by Lemma 2.2.3, we conclude that $\mathbf{F}[X]/\langle f(X)\rangle$ is a field. □

Remark 2.2.5. It is worthwhile to give an explicit description of the field $\mathbf{E} = \mathbf{F}[X]/\langle f(X)\rangle$. Let $f(X) \in \mathbf{F}[X]$ be an irreducible polynomial of degree n, $f(X) = a_n X^n + \cdots + a_0$. By taking representatives of equivalence classes, we may regard

$$\mathbf{E} = \{g(X) \in \mathbf{F}[X] \mid \deg g(X) < n\}$$

as a set. Addition in $\mathbf{E}$ is the usual addition of polynomials, while multiplication in $\mathbf{E}$ is defined as follows: Let $g_1(X), g_2(X) \in \mathbf{F}[X]$ and let $h(X) = g_1(X)g_2(X) \in \mathbf{F}[X]$. We may write uniquely $h(X) = f(X)q(X) + r(X)$ where either $r(X) = 0$ or $r(X)$ is a nonzero polynomial with $\deg r(X) < \deg f(X) = n$. Then, for $g_1(X), g_2(X) \in \mathbf{E}$,

$$g_1(X)g_2(X) = r(X) \in \mathbf{E},$$

where $r(X)$ is defined as above.

In particular, multiplication of polynomials $g_1(X), g_2(X) \in \mathbf{E}$ is the usual multiplication of polynomials if $\deg g_1(X) + \deg g_2(X) < n$, while

$$X^{n-1} \cdot X = (-1/a_n)(a_{n-1}X^{n-1} + \cdots + a_0) \in \mathbf{E}. \qquad \diamond$$

As we have remarked, we regard $\mathbf{F} \subset \mathbf{F}[X]$ by identifying the element a of $\mathbf{F}$ with the constant polynomial $f(X) = a$. If $\pi : \mathbf{F}[X] \to \mathbf{F}[X]/\langle f(X)\rangle$ is the canonical projection, then $\pi \mid \mathbf{F}$ is an injection, and using this we also regard $\mathbf{F} \subseteq \mathbf{F}[X]/\langle f(X)\rangle$.

In the proof of Proposition 2.2.4, we were careful to distinguish between $f(X) \in \mathbf{F}[X]$ and $\bar{f}(X) = \pi(f(X)) \in \mathbf{F}[X]/\langle f(X)\rangle$. But henceforth we shall follow the common practice of writing $f(X)$ for $\pi(f(X))$ and relying on the context to distinguish the two.

We now put these observations together to arrive at an important result.

Theorem 2.2.6 (Kronecker). *Let $f(X) \in \mathbf{F}[X]$ be any nonconstant polynomial. Then there is an extension field $\mathbf{E} \supseteq \mathbf{F}$ in which $f(X)$ has a root.*

Proof. Without loss of generality, we may assume that $f(X)$ is irreducible. Set $\mathbf{E} = \mathbf{F}[X]/\langle f(X)\rangle$. Then $\mathbf{E}$ is a field, $\mathbf{E} \supseteq \mathbf{F}$, and $f(X) = 0$ in $\mathbf{E}$, so X is a root of $f(X)$ in $\mathbf{E}$. □

Our last result in this section is a technical lemma that turns out to play an important role in our development of Galois theory.

Lemma 2.2.7 (Invariance of GCD Under Field Extensions). *Let* $\mathbf{E}$ *be an extension of* $\mathbf{F}$ *and let* $f(X)$ *and* $g(X)$ *be nonzero polynomials in* $\mathbf{F}[X]$. *Then the gcd of* $f(X)$ *and* $g(X)$ *as polynomials in* $\mathbf{E}[X]$ *is equal to the gcd of* $f(X)$ *and* $g(X)$ *as polynomials in* $\mathbf{F}[X]$. *In particular:*

(1) $f(X)$ *divides* $g(X)$ *in* $\mathbf{F}[X]$ *if and only if* $f(X)$ *divides* $g(X)$ *in* $\mathbf{E}[X]$.

(2) $f(X)$ *and* $g(X)$ *are relatively prime in* $\mathbf{F}[X]$ *if and only if* $f(X)$ *and* $g(X)$ *are relatively prime in* $\mathbf{E}[X]$.

Proof. Let $d(X)$ be the gcd of $f(X)$ and $g(X)$ in $\mathbf{F}[X]$ and let $\tilde{d}(X)$ be the gcd of $F(X)$ and $g(X)$ in $\mathbf{E}[X]$. Since $d(X)$ is a polynomial in $\mathbf{E}[X]$ dividing both $f(X)$ and $g(X)$, $\tilde{d}(X)$ divides $d(X)$. On the other hand, since $\tilde{d}(X)$ is the gcd of $f(X)$ and $g(X)$ in $\mathbf{E}[X]$, there are polynomials $\tilde{a}(X)$ and $\tilde{b}(X)$ in $\mathbf{E}[X]$ with

$$\tilde{d}(X) = f(X)\tilde{a}(X) + g(X)\tilde{b}(X).$$

Since $d(X)$ divides both $f(X)$ and $g(X)$, $d(X)$ divides $\tilde{d}(X)$ as well.

Hence $d(X)$ and $\tilde{d}(X)$ are two monic polynomials, each of which divides the other, and so they are equal. □

(Actually, Lemma 2.2.7 (1) follows directly from our observation that the quotient and remainder are independent of the field, and then Lemma 2.2.7 (1) can be used to prove the general case of Lemma 2.2.7 by considering Euclid's algorithm.)

Remark 2.2.8. We have chosen to present a concrete proof of Proposition 2.2.4 earlier in this section. But Proposition 2.2.4 is in fact a special case of a more general result: If R is a PID and $r \in R$ is an irreducible element, then $R/\langle r\rangle$ is a field. This general result is proved in Appendix B. ◇

2.3 Extension Fields

In this section we study basic properties of field extensions.

Let $\mathbf{F}$, $\mathbf{B}$, and $\mathbf{E}$ be fields with $\mathbf{F} \subseteq \mathbf{B} \subseteq \mathbf{E}$. Then, as $\mathbf{F}$-vector spaces, $\mathbf{B}$ is a subspace of $\mathbf{E}$, so $(\mathbf{B}/\mathbf{F}) \leq (\mathbf{E}/\mathbf{F})$. But in this situation we can say more.

Lemma 2.3.1. *Let* **F**, **B**, *and* **E** *be fields with* $\mathbf{F} \subseteq \mathbf{B} \subseteq \mathbf{E}$. *Then*

$$(\mathbf{E}/\mathbf{F}) = (\mathbf{E}/\mathbf{B})(\mathbf{B}/\mathbf{F}).$$

Proof. Let $\{\epsilon_1, \ldots, \epsilon_m\}$ be a basis for **E** as a **B**-vector space and $\{\beta_1, \ldots, \beta_n\}$ be a basis for **B** as an **F**-vector space. We claim that $\{\epsilon_i \beta_j \mid i = 1, \ldots, m,\ j = 1, \ldots, n\}$ is a basis for **E** as an **F**-vector space, yielding the lemma.

First, we show that this set spans **E**. Let $\alpha \in \mathbf{E}$. Then $\alpha = \Sigma_{i=1}^{m} x_i \epsilon_i$ with $x_i \in \mathbf{B}$. But for each i, $x_i = \Sigma_{j=1}^{n} y_{ij} \beta_j$ and so $\alpha = \Sigma_{i=1}^{m} (\Sigma_{j=1}^{n} y_{ij} \beta_j) \epsilon_i = \Sigma y_{ij}(\epsilon_i \beta_j)$.

Next we show this set is linearly independent. Suppose $\Sigma y_{ij}(\epsilon_i \beta_j) = 0$. Then $0 = \Sigma y_{ij}(\epsilon_i \beta_j) = \Sigma_{i=1}^{m} (\Sigma_{j=1}^{n} y_{ij} \beta_j) \epsilon_i$. Since $\{\epsilon_i\}$ are linearly independent, we have that, for each fixed i, $\Sigma_{i=1}^{n} y_{ij} \beta_j = 0$. Since $\{\beta_j\}$ are linearly independent, we have that $y_{ij} = 0$ for each j, completing the proof. □

Let **E** be an extension of **F**, and let **B** and **D** be subfields of **E**, both of which are extensions of **F**. Then $\mathbf{B} \cap \mathbf{D}$ is a subfield of **E** which is also an extension of **F**. Going in the opposite direction, we have the following definition.

Definition 2.3.2. Let **E** be an extension of **F** and let **B** and **D** be subfields of **E**, both of which are extensions of **F**. Then **BD**, the *composite* of **B** and **D**, is the smallest subfield of **E** that contains both **B** and **D**, and is an extension of **F**. ◇

Remark 2.3.3. (1) There *is* a smallest subfield as **BD** is the intersection of *all* subfields of **E** that contain both **B** and **D**.
(2) Clearly **BD** consists precisely of those elements of **E** of the form

$$(\Sigma_{i=1}^{m} b_i d_i)(\Sigma_{j=1}^{n} b'_j d'_j)^{-1}$$

with $b_i, b'_j \in \mathbf{B}$ and $d_i, d'_j \in \mathbf{D}$. ◇

There is an important case for which we do not need the denominators in Remark 2.3.3 (2).

Lemma 2.3.4. *Let* **E** *be an extension of* **F** *and let* **B** *and* **D** *be subfields of* **E**, *both of which are extensions of* **F**. *Suppose* $(\mathbf{D}/\mathbf{F})$ *is finite. Then* $(\mathbf{BD}/\mathbf{B}) \leq (\mathbf{D}/\mathbf{F})$ *and* $\mathbf{BD} = \{\Sigma_{i=1}^{m} b_i d_i \mid b_i \in \mathbf{B}, d_i \in \mathbf{D}\}$. *Furthermore,* $(\mathbf{BD}/\mathbf{F}) \leq (\mathbf{B}/\mathbf{F})(\mathbf{D}/\mathbf{F})$.

Proof. Let $(\mathbf{BD})_0 = \{\Sigma_{i=1}^{m} b_i d_i\}$. If $\{\delta_i\}$ is a basis for **D** as an **F**-vector space, then $\{\delta_i\}$ spans $(\mathbf{BD})_0$ as a **B**-vector space, so $\dim_{\mathbf{B}} (\mathbf{BD})_0 \leq \dim_{\mathbf{F}} \mathbf{D}$. In particular, if $\dim_{\mathbf{F}} \mathbf{D}$ is finite, then $\dim_{\mathbf{B}} (\mathbf{BD})_0$ is finite, so $(\mathbf{BD})_0$ is a field by Lemma 2.2.3, and hence $\mathbf{BD} = (\mathbf{BD})_0$.

Then, by Lemma 2.3.1,

$$(\mathbf{BD}/\mathbf{F}) = (\mathbf{BD}/\mathbf{B})(\mathbf{B}/\mathbf{F}) \leq (\mathbf{D}/\mathbf{F})(\mathbf{B}/\mathbf{F}),$$

as claimed. □

One of the most common, and important, ways of obtaining field extensions is by "adjoining" elements.

Definition 2.3.5. Let $\mathbf{E}$ be an extension of $\mathbf{F}$, and let $\{\alpha_i\}$ be a set of elements of $\mathbf{E}$. Then $\mathbf{F}(\{\alpha_i\})$ is the smallest subfield of $\mathbf{E}$ containing $\mathbf{F}$ and $\{\alpha_i\}$, and is the field obtained by *adjoining* (or by the *adjunction* of) $\{\alpha_i\}$. ◇

Remark 2.3.6. (1) Clearly $\mathbf{F}(\{\alpha_i\})$ is the set of elements of $\mathbf{E}$ that can be expressed as rational functions of the elements of $\{\alpha_i\}$ with coefficients in $\mathbf{F}$.

(2) Suppose that $\{\alpha_i\} = \{\alpha_1, \ldots, \alpha_n\}$ is finite. Then

$$\mathbf{F}(\alpha_1, \ldots, \alpha_n) = \mathbf{F}(\alpha_1)\mathbf{F}(\alpha_2)\cdots\mathbf{F}(\alpha_n).$$

(3) Suppose that $\{\alpha_i\} = \{\alpha_1, \ldots, \alpha_n\}$ is finite. Then

$$\mathbf{F}(\alpha_1, \ldots, \alpha_n) = \mathbf{F}(\alpha_1, \ldots, \alpha_{n-1})(\alpha_n).$$ ◇

Note that Remark 2.3.6 (3) says that in case $\{\alpha_i\}$ is finite, we may obtain $\mathbf{F}(\alpha_1, \ldots, \alpha_n)$ by adjoining the elements of $\{\alpha_i\}$ one at a time, and Remark 2.3.6 (2) says that the order in which we adjoin them does not matter.

Note also that $\mathbf{F}(\alpha) = \mathbf{F}$ if (and only if) $\alpha \in \mathbf{F}$. Thus we think of obtaining $\mathbf{F}(\alpha)$ by "adding in" α to $\mathbf{F}$, and if $\alpha \notin \mathbf{F}$, then this indeed gives us something new, but if $\alpha \in \mathbf{F}$ this does nothing.

Now that we know how to adjoin elements, let us give a concrete example of Lemma 2.3.4.

Example 2.3.7. (1) Let $\mathbf{F} = \mathbf{Q}$, $\mathbf{B} = \mathbf{Q}(\sqrt{2})$, $\mathbf{D} = \mathbf{Q}(\sqrt{3})$, and $\mathbf{E} = \mathbf{BD} = \mathbf{Q}(\sqrt{2}, \sqrt{3})$. Then $(\mathbf{BD}/\mathbf{B}) = 2 = (\mathbf{D}/\mathbf{F})$ and $(\mathbf{BD}/\mathbf{F}) = 4 = (\mathbf{B}/\mathbf{F})(\mathbf{D}/\mathbf{F})$.

(2) Let $\mathbf{F} = \mathbf{Q}$, $\mathbf{B} = \mathbf{Q}(\sqrt[3]{2})$, $\mathbf{D} = \mathbf{Q}(\omega\sqrt[3]{2})$, and $\mathbf{E} = \mathbf{BD} = \mathbf{Q}(\omega, \sqrt[3]{2})$. Here $\omega = (-1 + i\sqrt{3})/2$ is a primitive cube root of 1. Note that ω is a root of the quadratic polynomial $X^2 + X + 1$. Observe that $\mathbf{BD} = \mathbf{B}(\omega)$. Then $(\mathbf{BD}/\mathbf{B}) = 2 < (\mathbf{D}/\mathbf{F}) = 3$ and $(\mathbf{BD}/\mathbf{F}) = 6 < (\mathbf{B}/\mathbf{F})(\mathbf{D}/\mathbf{F}) = 9$.

Note that in both parts of this example $\mathbf{B} \cap \mathbf{D} = \mathbf{F}$. ◇

Remark 2.3.8. From Lemma 2.3.4 we have that, since $\mathbf{B}$ and $\mathbf{D}$ are both extensions of $\mathbf{B} \cap \mathbf{D}$, $(\mathbf{BD}/\mathbf{B} \cap \mathbf{D}) \leq (\mathbf{D}/\mathbf{B} \cap \mathbf{D})$. Hence if $\mathbf{B} \cap \mathbf{D} \supset \mathbf{F}$, we certainly have that $(\mathbf{BD}/\mathbf{B} \cap \mathbf{D}) < (\mathbf{D}/\mathbf{F})$, and also that $(\mathbf{BD}/\mathbf{F}) < (\mathbf{B}/\mathbf{F})(\mathbf{D}/\mathbf{F})$.

However, if $\mathbf{B} \cap \mathbf{D} = \mathbf{F}$, we may or may not have equality in Lemma 2.3.4, as Example 2.3.7 shows. There are many important cases in which we indeed do have equality. We will investigate this question further in Section 3.4 below. ◇

2.4 Algebraic Elements and Algebraic Extensions

Definition 2.4.1. (1) Let **E** be an extension of **F**. Then $\alpha \in \mathbf{E}$ is *algebraic* over **F** if α is a root of some polynomial $f(X) \in \mathbf{F}[X]$.

(2) **E** is an *algebraic extension* of **F** (or $\mathbf{E}/\mathbf{F}$ is *algebraic*) if every $\alpha \in \mathbf{E}$ is algebraic over **F**. ◇

A complex number that is algebraic over **Q** is an *algebraic number*.

Lemma 2.4.2. *The following are equivalent:*

(1) α *is algebraic over* **F**.
(2) $\mathbf{F}(\alpha)/\mathbf{F}$ *is finite.*
(3) $\mathbf{F}(\alpha) = \{$*polynomials in* α *with coefficients in* $\mathbf{F}\}$.
(4) $\alpha \in \mathbf{B}$ *for some finite extension* **B** *of* **F**.

Proof. Note that in general $\mathbf{F}(\alpha) = \{$rational functions in α with coefficients in $\mathbf{F}\}$. If we let $R = \{$polynomials in α with coefficients in $\mathbf{F}\}$, then $R \subseteq \mathbf{F}(\alpha)$.

First, suppose (2) is true. Then R is a finite-dimensional vector space over **F**, so by Lemma 2.2.3, R is a field and $R = \mathbf{F}(\alpha)$, proving (3). If (3) is true, then $\alpha^{-1} = f(\alpha)$ is a polynomial in α, so $\alpha f(\alpha) = 1$. Then $g(\alpha) = 0$ where $g(X) = Xf(X) - 1$, and α is algebraic, proving (1). If (1) is true, then $f(\alpha) = 0$ for some polynomial $f(X)$, so if $f(X)$ has degree d, then $\{1, \alpha, \ldots, \alpha^{d-1}\}$ spans R (as for any polynomial $g(X)$, $g(X) = f(X)q(X) + r(X)$ with $\deg r(X) < d$, and hence $g(\alpha) = f(\alpha)q(\alpha) + r(\alpha) = r(\alpha) = \Sigma_{i=0}^{d-1} c_i\alpha^i$); so again R is a field, $\mathbf{F}(\alpha) = R$ and $(\mathbf{F}(\alpha)/\mathbf{F}) \leq d$ is finite, proving (2). Thus we see that (1), (2), and (3) are equivalent. Also, (2) implies (4) since we may choose $\mathbf{B} = \mathbf{F}(\alpha)$, and (4) implies (2) since if $\alpha \in \mathbf{B}$, then $\mathbf{F}(\alpha) \subseteq \mathbf{B}$, so $\mathbf{B}/\mathbf{F}$ finite implies $\mathbf{F}(\alpha)/\mathbf{F}$ finite. □

Lemma 2.4.3. *Let* $\alpha \in \mathbf{E}$ *be algebraic. Let*

$f_1(X)$ *be the monic generator of the ideal* $\{f(X) \in \mathbf{F}[X] \mid f(\alpha) = 0\}$
$f_2(X)$ *be the monic polynomial of lowest degree with* $f_2(\alpha) = 0$.
$f_3(X)$ *be the monic irreducible polynomial with* $f_3(\alpha) = 0$.

Then $f_1(X)$, $f_2(X)$, *and* $f_3(X)$ *are well defined and* $f_1(X) = f_2(X) = f_3(X)$. □

The polynomial satisfying the equivalent conditions of this lemma plays a vital role, and we give it a name.

Definition 2.4.4. Let $\alpha \in \mathbf{E}$ be algebraic. The *minimum polynomial* $m_\alpha(X)$ is the polynomial satisfying the equivalent conditions of Lemma 2.4.3. ◇

Remark 2.4.5. (1) Note that $m_\alpha(X)$ depends on the field $\mathbf{F}$, but in order to keep the notation simple, we suppress that dependence.
(2) Note that $m_\alpha(X)$ is linear, in which case $m_\alpha(X) = X - \alpha$, if and only if $\alpha \in \mathbf{F}$. ◇

We have the following very important properties of $m_\alpha(X)$:

Proposition 2.4.6. *Let α be algebraic over $\mathbf{F}$. Then*

(1) $\{1, \alpha, \alpha^2, \ldots, \alpha^{d-1}\}$ is a basis for $\mathbf{F}(\alpha)$ as a vector space over $\mathbf{F}$, where $d = \deg m_\alpha(X)$.

(2) $(\mathbf{F}(\alpha)/\mathbf{F}) = \deg m_\alpha(X)$.

(3) Let $T : \mathbf{F}(\alpha) \to \mathbf{F}(\alpha)$ be multiplication by α, i.e., $T(\beta) = \alpha\beta$. Then T is a linear transformation of $\mathbf{F}$-vector spaces, and its minimum polynomial $m_T(X) = m_\alpha(X)$.

Proof. Let $d = \deg m_\alpha(X)$. Since $m_\alpha(\alpha) = 0$, the set $\{1, \alpha, \ldots, \alpha^{d-1}\}$ spans $\mathbf{F}(\alpha)$, as in the proof of Lemma 2.4.2. We claim this set is linearly independent. Suppose not. Then $\Sigma_{i=0}^{d-1} c_i\alpha^i = 0$ with not all $c_i = 0$, i.e., $f(\alpha) = 0$ where $0 \neq f(X) = \Sigma_{i=0}^{d-1} c_i X^i \in \mathbf{F}[X]$. But $\deg f(X) < \deg m_\alpha(X)$, a contradiction by Lemma 2.4.3. Hence $\{1, \alpha, \ldots, \alpha^{d-1}\}$ is a basis for $\mathbf{F}(\alpha)$, yielding (1), and this basis has d elements, so $(\mathbf{F}(\alpha)/\mathbf{F}) = d$, yielding (2).

(3) Let $m_\alpha(X) = \Sigma_{i=0}^{d} c_i X^i$. By definition, $m_\alpha(X)$ is monic, so $c_d = 1$. Then $0 = \Sigma_{i=0}^{d} c_i\alpha^i$, so $\alpha^d = \Sigma_{i=0}^{d-1}(-c_i)\alpha^i$.

Now consider the matrix of T in the basis $\{1, \alpha, \ldots, \alpha^{d-1}\}$ of $\mathbf{F}(\alpha)$. Since $T(\alpha^i) = \alpha^{i+1}$, we see T has matrix

$$\begin{bmatrix} 0 & 0 & \cdots & 0 & -c_0 \\ 1 & 0 & \cdots & 0 & -c_1 \\ 0 & 1 & \cdots & 0 & -c_2 \\ \vdots & \vdots & & \vdots & \vdots \\ 0 & 0 & \cdots & 1 & -c_{d-1} \end{bmatrix}$$

which we recognize as the companion matrix $C(m_\alpha(X))$ of the polynomial $m_\alpha(X)$. But for any polynomial $f(X)$, $C(f(X))$ has minimum polynomial (and characteristic polynomial) $f(X)$ (see [AW, Theorem 4.4.14]). □

Lemma 2.4.7. *Let $\mathbf{F} \subseteq \mathbf{B} \subseteq \mathbf{E}$ and let $\alpha \in \mathbf{E}$. Then $(\mathbf{B}(\alpha)/\mathbf{B}) \leq (\mathbf{F}(\alpha)/\mathbf{F})$.*

Proof. If $(\mathbf{F}(\alpha)/\mathbf{F})$ is infinite there is nothing to prove, so suppose $(\mathbf{F}(\alpha)/\mathbf{F})$ is finite. Then this follows immediately from Lemma 2.3.4. □

Corollary 2.4.8. *Let $\alpha_1, \ldots, \alpha_n \in \mathbf{E}$ with $\mathbf{F}(\alpha_i)/\mathbf{F}$ finite for each i. Then $\mathbf{F}(\alpha_1, \ldots, \alpha_n)/\mathbf{F}$ is finite.*

Proof. Trivial for $n = 1$. For $n = 2$,

$$(\mathbf{F}(\alpha_1, \alpha_2)/\mathbf{F}(\alpha_1)) = ((\mathbf{F}(\alpha_1)(\alpha_2)/\mathbf{F}(\alpha_1)) \leq (\mathbf{F}(\alpha_2)/\mathbf{F})$$

by Lemma 2.3.4. Then proceed by induction. □

Corollary 2.4.9. *(1) If* $\mathbf{E}/\mathbf{F}$ *is finite, then* $\mathbf{E}/\mathbf{F}$ *is algebraic.*

(2) If $\mathbf{E} = \mathbf{F}(\{\alpha_i\})$ *with each* α_i *algebraic over* $\mathbf{F}$*, then* $\mathbf{E}/\mathbf{F}$ *is algebraic.*

Proof. (1) For any $\alpha \in \mathbf{E}$, $\mathbf{F}(\alpha) \subseteq \mathbf{E}$, so $(\mathbf{F}(\alpha)/\mathbf{F}) \leq (\mathbf{E}/\mathbf{F})$ is finite, and so, by Lemma 2.4.2, $\mathbf{E}/\mathbf{F}$ is algebraic.

(2) Let $\alpha \in \mathbf{E}$. Then α is a rational function of $\{\alpha_i\}$, so in particular $\alpha \in \mathbf{F}(\alpha_1, \ldots, \alpha_n)$ for some n.

But, by Corollary 2.4.8, $\mathbf{F}(\alpha_1, \ldots, \alpha_n)/\mathbf{F}$ is finite, so, by part (1) and Lemma 2.4.2, α is algebraic over $\mathbf{F}$. Since α is arbitrary, $\mathbf{E}/\mathbf{F}$ is algebraic. □

Remark 2.4.10. (1) It follows from Lemma 2.4.2 and Corollary 2.4.8 that if $\mathbf{E}$ is an algebraic extension of $\mathbf{F}$ obtained by adjoining finitely many algebraic elements to $\mathbf{F}$, then $\mathbf{E}$ is a finite extension of $\mathbf{F}$. But *not every* algebraic extension $\mathbf{E}$ of $\mathbf{F}$ is a finite extension of $\mathbf{F}$. For example, if $\mathbf{E} = \mathbf{Q}(\sqrt{2}, \sqrt{3}, \sqrt{5}, \sqrt{7}, \ldots)$, then $\mathbf{E}/\mathbf{Q}$ is infinite. As a second example, if $\mathbf{E} = \mathbf{Q}(\sqrt{2}, \sqrt[3]{2}, \sqrt[4]{2}, \sqrt[5]{2}, \ldots)$, then $\mathbf{E}/\mathbf{Q}$ is infinite. (While these two claims are quite plausible, they of course require proof, but we shall not prove them now.)

(2) We also remark that a complex number that is algebraic over $\mathbf{Q}$ is known as an *algebraic number*, and that a subfield $\mathbf{E}$ of the field of complex numbers $\mathbf{C}$ that is obtained by adjoining finitely many algebraic numbers to $\mathbf{Q}$ is known as an *algebraic number field*. We then immediately see that $\mathbf{E} \subseteq \mathbf{C}$ is an algebraic number field if and only if $\mathbf{E}$ is a finite extension of $\mathbf{Q}$. ◇

The next two results illustrate a theme which we will often see repeated. Let us consider fields $\mathbf{F}$, $\mathbf{B}$, and $\mathbf{E}$ with $\mathbf{F} \subseteq \mathbf{B} \subseteq \mathbf{E}$. Thus $\mathbf{B}$ is an extension of $\mathbf{F}$ and $\mathbf{E}$ is an extension of $\mathbf{B}$. Also, $\mathbf{E}$ is an extension of $\mathbf{F}$ and $\mathbf{B}$ is a subextension. We have some "nice" property of extensions (in this case, being algebraic). We ask if a "nice" extension of a "nice" extension is necessarily a "nice" extension, and also if a subextension of a "nice" extension is necessarily "nice". We will see that for some properties (e.g., being algebraic) the answers to both questions are yes, but for others (e.g., being Galois) they are not.

Proposition 2.4.11. *Let* $\mathbf{E}$ *be an algebraic extension of* $\mathbf{F}$ *and let* $\mathbf{B}$ *be any field intermediate between* $\mathbf{F}$ *and* $\mathbf{E}$*. Then* $\mathbf{B}$ *is an algebraic extension of* $\mathbf{F}$ *and* $\mathbf{E}$ *is an algebraic extension of* $\mathbf{B}$*.*

Proof. Since $\mathbf{E}$ is an algebraic extension of $\mathbf{F}$, every element of $\mathbf{E}$ is algebraic over $\mathbf{F}$, and $\mathbf{B} \subseteq \mathbf{E}$, so certainly every element of $\mathbf{B}$ is algebraic over $\mathbf{F}$, and so $\mathbf{B}$ is an algebraic extension of $\mathbf{F}$.

Now let $\alpha \in \mathbf{E}$. Then α is a root of some polynomial $f(X) \in \mathbf{F}[X]$, and $\mathbf{F}[X] \subseteq \mathbf{B}[X]$, so α is certainly a root of some polynomial in $\mathbf{B}[X]$, and so α is algebraic over $\mathbf{B}$. Since α is arbitrary, $\mathbf{E}$ is algebraic over $\mathbf{B}$. □

Theorem 2.4.12. *Let* $\mathbf{B}$ *be an algebraic extension of* $\mathbf{F}$ *and let* $\mathbf{E}$ *be an algebraic extension of* $\mathbf{B}$. *Then* $\mathbf{E}$ *is an algebraic extension of* $\mathbf{F}$.

Proof. Let $\alpha \in \mathbf{E}$. Then α is a root of some polynomial $f(X) \in \mathbf{B}[X]$. Let $\mathbf{B}_0 \subseteq \mathbf{B}$ be the field obtained from $\mathbf{F}$ by adjoining the finitely many coefficients of $f(X)$, each of which is algebraic over $\mathbf{F}$. Then $(\mathbf{B}_0/\mathbf{F})$ is finite and $(\mathbf{B}_0(\alpha)/\mathbf{B}_0)$ is finite, so $(\mathbf{B}_0(\alpha)/\mathbf{F}) = (\mathbf{B}_0(\alpha)/\mathbf{B}_0)(\mathbf{B}_0/\mathbf{F})$ is finite as well. But $\alpha \in \mathbf{B}_0(\alpha)$, so α is an element of a finite extension of $\mathbf{F}$ and hence, by Lemma 2.4.2, α is algebraic over $\mathbf{F}$. Since α is arbitrary, $\mathbf{E}$ is algebraic over $\mathbf{F}$. □

For any fixed element $\alpha \in \mathbf{E}$, we may define a ring homomorphism $\varphi\colon \mathbf{F}[X] \to \mathbf{E}$ by $\varphi \mid \mathbf{F} = \text{id}$ and $\varphi(X) = \alpha$ (so that $\varphi(f(X)) = f(\alpha)$ for every $f(X) \in \mathbf{F}[X]$).

Lemma 2.4.13. *Let* $\alpha \in \mathbf{E}$ *be algebraic over* $\mathbf{F}$. *Then* φ *induces*

$$\bar{\varphi} : \mathbf{F}[X]/\langle m_\alpha(X)\rangle \to \mathbf{F}(\alpha)$$

an isomorphism of fields and of $\mathbf{F}$*-vector spaces.*

Proof. Let $\pi : \mathbf{F}[X] \to \mathbf{F}[X]/\langle m_\alpha(X)\rangle$ be the canonical projection and set $\pi(f(X)) = \bar{f}(X)$.

First, we show that $\bar{\varphi}$ is well defined. Let $\bar{f}(X) \in \mathbf{F}[X]/\langle m_\alpha(X)\rangle$ and suppose that $\pi(f(X)) = \bar{f}(X) = \pi(g(X))$. Then $g(X) - f(X)$ is divisible by $m_\alpha(X)$, so $g(X) = f(X) + m_\alpha(X)q(X)$ for some polynomial $q(X) \in \mathbf{F}[X]$, and then $\varphi(g(X)) = g(\alpha) = f(\alpha) + m_\alpha(\alpha)q(\alpha) = f(\alpha) = \varphi(f(X))$, as required.

Now it is easy to check that $\bar{\varphi}$ is a homomorphism of rings and a linear map of $\mathbf{F}$-vector spaces, so to complete the proof of the lemma we need only show that it is both one-to-one and onto.

Since $\varphi(1) = 1$, we see, by Lemma 2.1.3, that φ is injective.

Now, by Proposition 2.2.4 and Proposition 2.4.6 (2), $\mathbf{F}[X]/\langle m_\alpha(X)\rangle$ and $\mathbf{F}(\alpha)$ are both $\mathbf{F}$-vector spaces of the same finite dimension, and hence, since φ is injective, it is surjective as well. □

Remark 2.4.14. We have chosen to present a direct proof of Lemma 2.4.13, but it is a special case of a more general algebraic result: Let $\varphi : R \to S$ be a map of rings. Then φ induces an isomorphism $\bar{\varphi} : R/\operatorname{Ker}(\varphi) \to \operatorname{Im}(\varphi)$. ◇

Remark 2.4.15. Let $f(X) \in \mathbf{F}[X]$ be a monic irreducible polynomial, and let $\mathbf{E}$ be an extension field in which $f(X)$ has a root α. Then $f(X) = m_\alpha(X)$ is the minimum polynomial of α. Of course, $\mathbf{F}(\alpha) \subseteq \mathbf{E}$, and $\mathbf{F}(\alpha) = \mathbf{E}$ if and only if $(\mathbf{E}/\mathbf{F}) = \deg m_\alpha(X)$, by Proposition 2.4.6 (2). ◇

2.5 Splitting Fields

Definition 2.5.1. (1) Let $f(X) \in \mathbf{F}[X]$ be a nonconstant polynomial. If $\mathbf{E} \supseteq \mathbf{F}$ is such that $f(X)$ factors into (possibly repeated) linear factors in $\mathbf{E}[X]$, then $f(X)$ *splits* in $\mathbf{E}[X]$.

(2) If $f(X)$ splits in $\mathbf{E}[X]$ but not in $\mathbf{B}[X]$ for any proper subfield $\mathbf{B}$ of $\mathbf{E}$, then $\mathbf{E}$ is a *splitting field* of $f(X)$. ◇

Note that $X - \alpha$ is a factor of $f(X)$ if and only if $f(\alpha) = 0$, i.e., if and only if α is a root of $f(X)$. Thus, we regard $f(X)$ as splitting in $\mathbf{E}$ if "all" the roots of $f(X)$ are in $\mathbf{E}$, and we regard $\mathbf{E}$ as a splitting field if "all" the roots of $f(X)$ lie in $\mathbf{E}$ but not in any proper subfield of $\mathbf{E}$.

Lemma 2.5.2. *Let $f(X) \in \mathbf{F}[X]$ be a nonconstant polynomial. Then $f(X)$ has a splitting field.*

Proof. By induction on $\deg f(X)$, and all fields simultaneously. Without loss of generality we may assume that $f(X)$ is monic. If $\deg f(X) = 1$, then $f(X) = X - a \in \mathbf{F}[X]$ and $\mathbf{F}$ is a splitting field for $f(X)$.

Assume the lemma holds for all polynomials of degree less than d, and all fields, and let $f(X) \in \mathbf{F}[X]$ have degree d.

By Theorem 2.2.6, there is an extension field $\mathbf{B}$ of $\mathbf{F}$ in which $f(X)$ has a root α_1, and moreover $\mathbf{B} = \mathbf{F}(\alpha_1)$. Then $f(X) = (X - \alpha_1)g(X) \in \mathbf{B}[X]$. By induction, $g(X)$ has a splitting field $\mathbf{E}$ over $\mathbf{B}$. Hence $g(X) = (X-\alpha_2)\cdots(X-\alpha_d) \in \mathbf{E}[X]$ and $f(X) = (X - \alpha_1)\cdots(X - \alpha_d) \in \mathbf{E}[X]$ splits.

Clearly $\mathbf{E} \supseteq \mathbf{B}(\alpha_2, \ldots, \alpha_n)$, and then $\mathbf{E} = \mathbf{B}(\alpha_2, \ldots, \alpha_n)$, since certainly $g(X)$ does not split over any field not containing each of $\alpha_2, \ldots, \alpha_n$. Then $\mathbf{E} = \mathbf{F}(\alpha_1)(\alpha_2, \ldots, \alpha_n) = \mathbf{F}(\alpha_1, \ldots, \alpha_n)$, and $f(X)$ does not split in any proper subfield of $\mathbf{E}$, by the same logic, so $\mathbf{E}$ is a splitting field for $f(X)$. □

Note that in general it may well be the case that an extension $\mathbf{B}$ of $\mathbf{F}$ contains a root of an irreducible polynomial $f(X) \in \mathbf{F}[X]$ but that polynomial does not split in $\mathbf{B}[X]$, i.e., that "some" but not "all" of the roots of $f(X)$ are in $\mathbf{B}$. As an example of this, we may take $\mathbf{B} = \mathbf{Q}(\sqrt[4]{2})$. Then the polynomial $X^4 - 2$ is irreducible in $\mathbf{Q}[X]$ with roots $\sqrt[4]{2}, i\sqrt[4]{2}, -\sqrt[4]{2}$, and $-i\sqrt[4]{2}$, and $\mathbf{B}$ contains just two of these four roots ($\sqrt[4]{2}$ and $-\sqrt[4]{2}$).

We shall soon see that any two splitting fields of $f(X) \in \mathbf{F}[X]$ are isomorphic. Granting that, for the moment, we have the following result.

Lemma 2.5.3. *Let $f(X) \in \mathbf{F}[X]$ be a nonconstant polynomial, with* $\deg f(X) = d$. *Let $\mathbf{E}$ be a splitting field of $f(X)$. If $f(X)$ is irreducible, $(\mathbf{E}/\mathbf{F})$ is divisible by d. In any case, $(\mathbf{E}/\mathbf{F}) \leq d!$.*

Proof. Without loss of generality we may assume that $f(X)$ is monic.

Suppose that $f(X)$ is irreducible, and let $\alpha \in \mathbf{E}$ be a root of $f(X)$. Then $f(X) = m_\alpha(X)$, by Lemma 2.4.3, and $d = \deg m_\alpha(X) = (\mathbf{F}(\alpha)/\mathbf{F})$, by Proposition 2.4.6 (2). But $\mathbf{F}(\alpha) \subseteq \mathbf{E}$, and so $(\mathbf{F}(\alpha)/\mathbf{F})$ divides $(\mathbf{E}/\mathbf{F})$.

For the upper bound, we proceed by induction on d, and all fields simultaneously. If $d = 1$, $\mathbf{E} = \mathbf{F}$ and $(\mathbf{E}/\mathbf{F}) = 1$. Suppose the lemma holds for all polynomials of degree less than d, and all fields, and let $f(X) \in \mathbf{F}[X]$ have degree d. Let α be a root of $f(X)$, set $\mathbf{B} = \mathbf{F}(\alpha)$, and write $f(X) = (X - \alpha)g(X)$ with $g(X) \in \mathbf{B}[X]$. Since $f(\alpha) = 0$, $m_\alpha(X)$ divides $f(X)$, so $(\mathbf{B}/\mathbf{F}) = \deg m_\alpha(X) \leq \deg f(X) = d$.

Let $\mathbf{E}$ be a splitting field of $g(X)$, a polynomial of degree $d - 1$, over $\mathbf{B}$. Then, by induction, $(\mathbf{E}/\mathbf{B}) \leq (d-1)!$, and then $(\mathbf{E}/\mathbf{F}) = (\mathbf{E}/\mathbf{B})(\mathbf{B}/\mathbf{F}) \leq d!$. □

Example 2.5.4. (1) Let $\mathbf{F} = \mathbf{Q}$ and let $\mathbf{E}$ be the splitting field of the irreducible polynomial $X^3 - 2 \in \mathbf{F}[X]$. Then $\mathbf{E} = \mathbf{F}(\omega, \sqrt[3]{2})$ where $\omega^2 + \omega + 1 = 0$, and $(\mathbf{E}/\mathbf{F}) = 6 = 3!$.

(2) Let $\mathbf{F} = \mathbf{Q}(\omega)$ and let $\mathbf{E}$ be the splitting field of the irreducible polynomial $X^3 - 2 \in \mathbf{F}[X]$. Then $\mathbf{E} = \mathbf{F}(\sqrt[3]{2})$, and $(\mathbf{E}/\mathbf{F}) = 3 < 3!$. ◇

Corollary 2.5.5. *Let $\mathbf{E}$ be a splitting field of a nonconstant polynomial $f(X) \in \mathbf{F}[X]$. Then $\mathbf{E}/\mathbf{F}$ is algebraic.*

Proof. $\mathbf{E}/\mathbf{F}$ is finite. □

Remark 2.5.6. Suppose that $f(X) \in \mathbf{F}[X]$ splits in some extension $\mathbf{A}$ of $\mathbf{F}$, and let $\alpha_1, \ldots, \alpha_k$ be the roots of $f(X)$ in $\mathbf{A}$ (so that $f(X) = (X - \alpha_1)^{d_1} \cdots (X - \alpha_k)^{d_k} \in \mathbf{A}[X]$). Then $\mathbf{E} = \mathbf{F}(\alpha_1, \ldots, \alpha_k)$ is a splitting field for $f(X)$, as $f(X)$ visibly splits over $\mathbf{E}$, and any field over which $f(X)$ splits must contain $\alpha_1, \ldots, \alpha_k$, and hence must contain $\mathbf{E}$. ◇

2.6 Extending Isomorphisms

We now prove two results about extensions of field isomorphisms that will be key technical tools.

Lemma 2.6.1. *Let $\sigma_0 : \mathbf{F}_1 \to \mathbf{F}_2$ be an isomorphism of fields. Let $f_1(X) \in \mathbf{F}_1[X]$ be irreducible and let $\mathbf{E}_1 = \mathbf{F}_1(\beta_1)$ where $f_1(\beta_1) = 0$. Let $f_2(X) = \sigma_0(f_1(X))$ and let $\mathbf{E}_2 = \mathbf{F}_2(\beta_2)$ where $f_2(\beta_2) = 0$. Then σ_0 extends to a unique isomorphism $\sigma : \mathbf{E}_1 \to \mathbf{E}_2$ with $\sigma(\beta_1) = \beta_2$.*

Proof. Clearly σ_0 gives an isomorphism $\sigma_0 : \mathbf{F}_1[X] \to \mathbf{F}_2[X]$ and $f_2(X)$ is irreducible if and only if $f_1(X)$ is. Now, by Lemma 2.4.13, we have isomorphisms

$$\bar{\varphi}_i : \mathbf{F}_i[X]/\langle f_i(X)\rangle \to \mathbf{F}_i(\beta_i) = \mathbf{E}_i, \quad i = 1, 2,$$

and we let

$$\sigma = \bar{\varphi}_2 \sigma_0 \bar{\varphi}_1^{-1}. \qquad \square$$

Corollary 2.6.2. *Let $f(X) \in \mathbf{F}[X]$ be irreducible. Suppose β_1 and β_2 are both roots of $f(X)$ and let $\mathbf{E}_i = \mathbf{F}(\beta)$, $i = 1, 2$. Then there is a unique isomorphism $\sigma : \mathbf{E}_1 \to \mathbf{E}_2$ with $\sigma \mid \mathbf{F} = \mathrm{id}$ and $\sigma(\beta_1) = \beta_2$. In particular, if $\mathbf{E} = \mathbf{F}(\beta_1) = \mathbf{F}(\beta_2)$, there is a unique automorphism σ of $\mathbf{E}$ with $\sigma \mid \mathbf{F} = \mathrm{id}$ and $\tilde{\sigma}(\beta_1) = \beta_2$.* $\square$

This corollary says that, algebraically, there is no distinction between roots of the same irreducible polynomial — choosing any two roots gives us isomorphic extensions.

Lemma 2.6.3. *Let $\sigma_0 : \mathbf{F}_1 \to \mathbf{F}_2$ be an isomorphism of fields. Let $f_1(X) \in \mathbf{F}_1[X]$ and let $f_2(X) = \sigma_0(f_1(X)) \in \mathbf{F}_2[X]$. Let $\mathbf{E}_1$ be a splitting field of $f_1(X)$ and let $\mathbf{E}_2$ be a splitting field of $f_2(X)$. Then σ_0 extends to an isomorphism $\sigma : \mathbf{E}_1 \to \mathbf{E}_2$.*

Proof. Factor $f_1(X)$ into irreducibles in $\mathbf{F}[X]$, and let there be k factors. Set $d_{\mathbf{F}}(f_1) = \deg(f_1) - k$. We prove the theorem by induction on $d_{\mathbf{F}}(f_1)$, and all fields.

If $d_{\mathbf{F}}(f_1) = 0$, then $f_1(X)$ is a product of linear factors, and $\mathbf{E}_1 = \mathbf{F}_1$, $\mathbf{E}_2 = \mathbf{F}_2$, so $\sigma = \sigma_0$.

Suppose $d_{\mathbf{F}}(f_1) > 0$. Then $f_1(X)$ has an irreducible factor $g_1(X)$ of degree greater than 1. Let α_1 be a root of $g_1(X)$ in $\mathbf{E}_1$ and let α_2 be a root of $\sigma_0(g_1(X))$ in $\mathbf{E}_2$. Then $\mathbf{E}_1 \supseteq \mathbf{F}(\alpha_1)$ and $\mathbf{E}_2 \supseteq \mathbf{F}(\alpha_2)$. By Lemma 2.6.1, there is an isomorphism $\sigma_1 : \mathbf{F}(\sigma_1) \to \mathbf{F}(\sigma_2)$ with $\sigma_1 \mid \mathbf{F} = \sigma_0$ and $\sigma_1(\alpha_1) = \alpha_2$.

Set $\mathbf{B} = \mathbf{F}(\alpha_1)$, and consider $f_1(X) \in \mathbf{B}[X]$. Now $g_1(X) \in \mathbf{B}[X]$ has the factor $X - \alpha_1$, so $f_1(X)$ has at least $k+1$ irreducible factors in $\mathbf{B}[X]$ and hence $d_{\mathbf{B}}(f_1) < d_{\mathbf{F}}(f_1)$. Since $\mathbf{E}_1$ was a splitting field of $f_1(X)$, regarded as polynomial in $\mathbf{F}_1[X]$, and $\mathbf{B} \subseteq \mathbf{E}_1$, then, by Remark 2.5.6, $\mathbf{E}_1$ is still a splitting field of $f_1(X)$ regarded as a polynomial in $\mathbf{B}[X]$, and similarly for $\mathbf{E}_2$, so by induction σ_1 extends to σ, as claimed. □

Corollary 2.6.4. *Let $f(X) \in \mathbf{F}[X]$. Then any two splitting fields of $f(X)$ are isomorphic.* □

Note that we are certainly not claiming that the isomorphism $\tilde{\sigma}$ of Lemma 2.6.3 is unique! In fact, studying automorphisms $\mathbf{E}$ that fix $\mathbf{F}$ is what Galois theory is all about. To give a simple example of this, let $\mathbf{F} = \mathbf{Q}$ and $\mathbf{E} = \mathbf{Q}(\sqrt{D})$. Then $\mathbf{E}$ has an automorphism σ given by $\sigma(a + b\sqrt{D}) = a - b\sqrt{D}$.

2.7 Normal, Separable, and Galois Extensions

We begin by defining two (independent) properties of field extensions.

Definition 2.7.1. An algebraic extension $\mathbf{E}$ of $\mathbf{F}$ is a *normal* extension if every irreducible polynomial $f(X) \in \mathbf{F}[X]$ with $f(\alpha) = 0$ for some $\alpha \in \mathbf{E}$ splits in $\mathbf{E}[X]$. ◇

Remark 2.7.2. Observe that Definition 2.7.1 is equivalent to: An algebraic extension $\mathbf{E}$ of $\mathbf{F}$ is normal if $m_\alpha(X)$ splits in $\mathbf{E}[X]$ for every $\alpha \in \mathbf{E}$. ◇

Definition 2.7.3. (1) If the irreducible factors of $f(X) \in \mathbf{F}[X]$ split into a product of distinct linear factors in some (hence in any) splitting field for $f(X)$, then $f(X)$ is a *separable* polynomial. Otherwise, it is an *inseparable* polynomial.

(2) Let $\mathbf{E}$ be an algebraic extension of $\mathbf{F}$. Then $\alpha \in \mathbf{E}$ is a *separable* element if $m_\alpha(X)$ is a separable polynomial. Otherwise, α is an *inseparable* element.

(3) Let $\mathbf{E}$ be an algebraic extension of $\mathbf{F}$. Then $\mathbf{E}$ is a *separable* extension of $\mathbf{F}$ if every $\alpha \in \mathbf{E}$ is a separable element. Otherwise, it is an *inseparable* extension. ◇

Remark 2.7.4. We are most interested in extensions that are both normal and separable, as these have the closest connection with Galois theory, so we will defer our consideration of extensions that satisfy only one of these two properties until later. However, some discussion of both of these properties is in order now.

First, we might observe that the definitions of normal and separable are rather discouraging on the face of it, as they each require us to verify a condition on $m_\alpha(X)$ for *each* $\alpha \in \mathbf{E}$. But, as we will soon see (in Theorem 2.7.14), we have an easy criterion for extensions to be both normal and separable.

Even if we are interested in extensions that are normal or separable, the situation is not so bad. We note some results that we shall prove later on.

Theorem 5.2.1. $\mathbf{E}$ *is a finite normal extension of* $\mathbf{F}$ *if and only if* $\mathbf{E}$ *is the splitting field of a polynomial* $f(X) \in \mathbf{F}[X]$.

Theorem 5.1.9. $\mathbf{E}$ *is a finite separable extension of* $\mathbf{F}$ *if and only if* $\mathbf{E}$ *is obtained from* $\mathbf{F}$ *by adjoining root(s) of a separable polynomial* $f(X) \in \mathbf{F}[X]$.

Actually, sometimes separability is automatic. There are some fields $\mathbf{F}$ *all* of whose algebraic extensions are separable. Such a field $\mathbf{F}$ is called *perfect*, and we will show below that the following is true.

Theorem 3.2.6. $\mathbf{F}$ *is a perfect field if and only if*
(1) $\mathrm{char}(\mathbf{F}) = 0$, *or*
(2) $\mathrm{char}(\mathbf{F}) = p$ *and* $\mathbf{F} = \mathbf{F}^p$.

Corollary 3.2.7. *Every finite field is perfect.* ◇

We now define one of the most important objects in mathematics.

Definition 2.7.5. Let $\mathbf{E}$ be an algebraic extension of $\mathbf{F}$. The *Galois group* $\mathrm{Gal}(\mathbf{E}/\mathbf{F})$ is the group of all automorphisms of $\mathbf{E}$ that restrict to the identity on $\mathbf{F}$,

$$\mathrm{Gal}(\mathbf{E}/\mathbf{F}) = \{\sigma : \mathbf{E} \to \mathbf{E} \text{ automorphism } \mid \sigma \mid \mathbf{F} = \mathrm{id}\}. \qquad \diamond$$

(Strictly speaking, we should have defined $\mathrm{Gal}(\mathbf{E}/\mathbf{F})$ to be the set of these automorphisms, but it is easy to verify that this set is indeed a group under composition.)

Lemma 2.7.6. *Let* $\mathbf{E}$ *be an algebraic extension of* $\mathbf{F}$ *and let* $\mathbf{B}$ *be a field with* $\mathbf{F} \subseteq \mathbf{B} \subseteq \mathbf{E}$. *Then* $\mathrm{Gal}(\mathbf{E}/\mathbf{B})$ *is a subgroup of* $\mathrm{Gal}(\mathbf{E}/\mathbf{F})$.

Proof. Any automorphism $\sigma : \mathbf{E} \to \mathbf{E}$ with $\sigma \mid \mathbf{B} = \mathrm{id}$ certainly satisfies $\sigma \mid \mathbf{F} = \mathrm{id}$. □

Definition 2.7.7. Let G be a group of automorphisms of a field $\mathbf{E}$. Then the *fixed field* $\mathrm{Fix}(G) = \{\alpha \in \mathbf{E} \mid \sigma(\alpha) = \alpha \text{ for all } \sigma \in G\}$. ◇

Lemma 2.7.8. *Let G be a group of automorphisms of a field* **E**. *Then* $\mathrm{Fix}(G)$ *is a subfield of* **E**. □

Note that, by definition, $\mathbf{F} \subseteq \mathrm{Fix}(\mathrm{Gal}(\mathbf{E}/\mathbf{F}))$.

Definition 2.7.9. Let **E** be an algebraic extension of **F**. Then **E** is a *Galois extension* of **F** if $\mathrm{Fix}(\mathrm{Gal}(\mathbf{E}/\mathbf{F})) = \mathbf{F}$. ◇

Our first main goal is the Fundamental Theorem of Galois Theory (*FTGT*), which, for a finite Galois extension **E** of **F**, establishes a one-to-one correspondence between intermediate fields **B** (i.e., fields **B** with $\mathbf{F} \subseteq \mathbf{B} \subseteq \mathbf{E}$) and subgroups H of $G = \mathrm{Gal}(\mathbf{E}/\mathbf{F})$. On the way there, we need to obtain a criterion for **E** to be a Galois extension of **F**, an important result in its own right.

Definition 2.7.10. Let **E** be an algebraic extension of **F** and let $\alpha \in \mathbf{E}$. Then $\{\sigma(\alpha) \mid \sigma \in \mathrm{Gal}(\mathbf{E}/\mathbf{F})\}$ is the set of *(Galois) conjugates* of α in **E**. ◇

Lemma 2.7.11. *Let $H = \{\sigma \in \mathrm{Gal}(\mathbf{E}/\mathbf{F}) \mid \sigma(\alpha) = \alpha\}$. Then H is a subgroup of $G = \mathrm{Gal}(\mathbf{E}/\mathbf{F})$ and the number of conjugates of α in* **E** *is equal to the index $[G : H]$.* □

The following lemma is particularly useful.

Lemma 2.7.12. *Let* **E** *be a Galois extension of* **F** *and let $\alpha \in \mathbf{E}$. Then α has finitely many conjugates in* **E**. *If $\{\alpha_i\}_{i=1,\ldots,r}$ is the set of conjugates of $\alpha = \alpha_1$, then*

$$m_\alpha(X) = \prod_{i=1}^{r}(X - \alpha_i).$$

Proof. Since $m_\alpha(X) \in \mathbf{F}[X]$, for each $\sigma \in G = \mathrm{Gal}(\mathbf{E}/\mathbf{F})$, $\sigma(\alpha)$ is a root of $\sigma(m_\alpha(X)) = m_\alpha(X)$. Since $m_\alpha(X)$ has only finitely many roots, $\{\sigma(\alpha)\}$ must be a finite set. Now let $n_\alpha(X) = \prod_{i=1}^{r}(X - \alpha_i)$. Since $\sigma \in G$ permutes $\{\alpha_i\}$, $\sigma(n_\alpha(X)) = n_\alpha(X)$. In other words, every $\sigma \in G$ fixes each coefficient of $n_\alpha(X)$. Since **E** is a Galois extension of **F**, this implies that each of these coefficients is in **F**, i.e., that $n_\alpha(X) \in \mathbf{F}[X]$. As we have observed, $m_\alpha(\alpha_i) = 0$ for each i, so $X - \alpha_i$ divides $m_\alpha(X)$ for each i, and hence $n_\alpha(X)$ divides $m_\alpha(X)$ in $\mathbf{E}[X]$, and hence in $\mathbf{F}[X]$, by Lemma 2.2.7. But $m_\alpha(X)$ is irreducible in $\mathbf{F}[X]$, so $m_\alpha(X) = n_\alpha(X)$. □

Corollary 2.7.13. *Let* **E** *be a Galois extension of* **F** *and let $\alpha \in \mathbf{E}$. Then $(\mathbf{F}(\alpha)/\mathbf{F})$ is equal to the number of conjugates of α in* **E**.

Proof. $(\mathbf{F}(\alpha)/\mathbf{F}) = \deg m_\alpha(X)$ by Proposition 2.4.6 (2) and $\deg m_\alpha(X)$ is equal to the number of conjugates of α in **E** by Lemma 2.7.12. □

The following theorem is a key result.

Theorem 2.7.14. *Let* **E** *be a finite extension of* **F**. *The following are equivalent:*

(1) **E** *is a Galois extension of* **F**.

(2) **E** *is a normal and separable extension of* **F**.

(3) **E** *is the splitting field of a separable polynomial* $f(X) \in \mathbf{F}[X]$.

Proof. (1) $\Rightarrow$ (2): Let **E** be a Galois extension of **F** and let $\alpha \in \mathbf{E}$. Let $\{\alpha_i\}$ be the set of conjugates of α in **E**. By Lemma 2.7.12, $m_\alpha(X) = \prod_{i=1}^{r}(X - \alpha_i)$, and this is a separable polynomial (as its roots are distinct) that splits in **E**.

(2) $\Rightarrow$ (3): Let $\{\epsilon_i\}$ be a basis for **E** over **F**, and let $f(X)$ be the product of the distinct elements of $\{m_{\epsilon_i}(X)\}$. Then $f(X)$ is separable, as each $m_{\epsilon_i}(X)$ has distinct roots, and $f(X)$ certainly splits in **E**. Moreover, any field in which $f(X)$ splits must contain each ϵ_i, so **E** is the splitting field of $f(X)$.

(3) $\Rightarrow$ (1): We prove this by induction on $(\mathbf{E}/\mathbf{F})$ and all fields simultaneously. Let **E** be the splitting field of the separable polynomial $f(X)$.

If $(\mathbf{E}/\mathbf{F}) = 1$, then $\mathbf{E} = \mathbf{F}$, $\mathrm{Gal}(\mathbf{E}/\mathbf{F}) = \{\mathrm{id}\}$, and we are done.

Suppose now that the theorem is true for all extensions of degree less than n and let $(\mathbf{E}/\mathbf{F}) = n > 1$.

Factor $f(X)$ into irreducibles $f(X) = f_1(X) \cdots f_k(X)$ in $\mathbf{F}[X]$. Some $f_i(X)$ has degree greater than one, as otherwise $f(X)$ would split in $\mathbf{F}[X]$ and so $\mathbf{E} = \mathbf{F}$. Renumbering if necessary, we may suppose that this factor is $f_1(X)$. Let it have degree r. Let $\alpha = \alpha_1$ be a root of $f_1(X)$, $\alpha \in \mathbf{E}$. Then $f_1(X) = m_\alpha(X)$, since $f_1(X)$ is irreducible. Of course, $\alpha \notin \mathbf{F}$.

By assumption, $m_\alpha(X)$ is separable, so has distinct roots $\alpha_1, \ldots, \alpha_r$. Now α_i is a root of the irreducible polynomial $m_\alpha(X)$ by assumption, and α_i is a root of the irreducible polynomial $m_{\alpha_i}(X)$ by definition, so, by Lemma 2.4.3, $m_{\alpha_i}(X) = m_\alpha(X)$, for each i. Furthermore, by Corollary 2.6.2, there are isomorphisms $\sigma_i : \mathbf{F}(\alpha_1) \to \mathbf{F}(\alpha_i)$ with $\sigma_i \mid \mathbf{F} = \mathrm{id}$ and $\sigma_i(\alpha_1) = \alpha_i$, $i = 1, \ldots, r$.

Consider $\mathbf{E} \supseteq \mathbf{F}(\alpha_i) \supset \mathbf{F}$. Then **E** is the splitting field of $m_\alpha(X) \in \mathbf{F}(\alpha_i)[X]$, so, by Lemma 2.6.3, there is an automorphism of **E** extending the isomorphism σ_i. Denote this extension by σ_i as well. Then $\sigma_i \mid \mathbf{F} = \mathrm{id}$, so $\sigma_i \in \mathrm{Gal}(\mathbf{E}/\mathbf{F})$. Note that $\sigma_i(\mathbf{F}(\alpha_j)) = \mathbf{F}(\alpha_k)$ for some k, as the automorphism σ_i of **E** must permute the roots of the irreducible polynomial $m_\alpha(X)$.

Now consider **E** as an extension of $\mathbf{F}(\alpha)$. Then $(\mathbf{E}/\mathbf{F}(\alpha)) < (\mathbf{E}/\mathbf{F})$, and **E** is the splitting field of the separable polynomial $m_\alpha(X) \in \mathbf{F}(\alpha)[X]$, so by induction **E** is a Galois extension of $\mathbf{F}(\alpha)$, i.e., $\mathbf{F}(\alpha)$ is the fixed field of $\mathrm{Gal}(\mathbf{E}/\mathbf{F}(\alpha))$. By Lemma 2.7.6, $\mathrm{Gal}(\mathbf{E}/\mathbf{F}(\alpha))$ is a subgroup of $\mathrm{Gal}(\mathbf{E}/\mathbf{F})$, so this implies that $\mathbf{B} = \mathrm{Fix}(\mathrm{Gal}(\mathbf{E}/\mathbf{F})) \subseteq \mathbf{F}(\alpha)$. We wish to show that $\mathbf{B} = \mathbf{F}$.

Let $\tilde{m}_\alpha(X) \in \mathbf{B}[X]$ be the minimum polynomial of α regarded as an element of $\mathbf{B}$. Since $\mathbf{B} \subseteq \mathbf{F}(\alpha)$, we see that $\mathbf{B}(\alpha) = \mathbf{F}(\alpha)$, and then

$$\deg \tilde{m}_\alpha(X) = (\mathbf{B}(\alpha)/\mathbf{B}) = (\mathbf{F}(\alpha)/\mathbf{B}) \leq (\mathbf{F}(\alpha)/\mathbf{F}) = \deg m_\alpha(X)$$

with equality if and only if $\mathbf{B} = \mathbf{F}$. Thus it remains to show that $\deg \tilde{m}_\alpha(X) = \deg m_\alpha(X)$. In fact we show that $\tilde{m}_\alpha(X) = m_\alpha(X)$.

Now $\tilde{m}_\alpha(X)$ divides $m_\alpha(X)$ and $m_\alpha(X) = \prod_{i=1}^{r}(X - \alpha_i)$ by Lemma 2.7.12.

Consider α_i. We have constructed an element $\sigma_i \in \mathrm{Gal}(\mathbf{E}/\mathbf{F})$ with $\sigma_i(\alpha) = \alpha_i$, i.e., there is an automorphism σ_i of $\mathbf{E}$ with $\sigma_i(\alpha) = \alpha_i$ and $\sigma_i \mid \mathbf{F} = \mathrm{id}$. But $\mathbf{B} = \mathrm{Fix}(\mathrm{Gal}(\mathbf{E}/\mathbf{F}))$ so $\sigma_i \mid \mathbf{B} = \mathrm{id}$, and hence $\sigma_i \in \mathrm{Gal}(\mathbf{E}/\mathbf{B})$. Then $\sigma_i(\tilde{m}_\alpha(X)) = \tilde{m}_\alpha(X)$, as $\sigma_i(\tilde{m}_\alpha(X)) \in \mathbf{B}[X]$; thus

$$\tilde{m}_\alpha(\alpha_i) = \tilde{m}_\alpha(\sigma_i(\alpha)) = \sigma_i(\tilde{m}_\alpha(\alpha)) = \sigma_i(0) = 0,$$

so α_i is a root of $\tilde{m}_\alpha(X)$ for each i, i.e., $X - \alpha_i$ is a factor of $\tilde{m}_\alpha(X)$ for each i, and so $\tilde{m}_\alpha(X) = m_\alpha(X)$, as required. □

2.8 The Fundamental Theorem of Galois Theory

In this section we reach our first main goal. Before we get there, we have some work to do.

We let $\mathbf{F}^*$ denote the set of nonzero elements of $\mathbf{F}$, and recall that $\mathbf{F}^*$ is a group under multiplication.

Definition 2.8.1. A *character* of a group G in a field $\mathbf{F}$ is a homomorphism $\sigma: G \to \mathbf{F}^*$. ◇

Example 2.8.2. (1) Let $G = \mathbf{F}^*$. Then $\mathrm{id}: G \to \mathbf{F}^*$ is a character.

(2) Let $G = \mathbf{F}^*$. For any automorphism σ of $\mathbf{F}$, $\sigma: G \to \mathbf{F}^*$ is a character.

(3) Let $G = \mathbf{F}^*$. For any map of fields $\sigma: \mathbf{F} \to \mathbf{E}$, $\sigma: G \to \mathbf{E}^*$ is a character. ◇

Definition 2.8.3. A set of characters $\{\sigma_i\}_{i=1,\dots,n}$ is *dependent* if there is a set of elements $\{a_i\}_{i=1,\dots,n}$ of $\mathbf{F}$, not all zero, with $a_1\sigma_1 + \cdots + a_n\sigma_n = 0$ (i.e., with $(a_1\sigma_1 + \cdots + a_n\sigma_n)(x) = 0$ for every $x \in G$). Otherwise it is *independent*. ◇

Theorem 2.8.4 (Dirichlet). *Let $\{\sigma_i\}_{i=1,\dots,n}$ be a set of mutually distinct characters of G in $\mathbf{F}$. Then $\{\sigma_i\}$ is independent.*

Proof (Artin). By induction on n.

Suppose $n = 1$. Then $a_1\sigma_1 = 0$ implies $a_1 = 0$ (as $\sigma(1) = 1$).

Suppose now that the theorem is true for any set of fewer than n characters and consider $\{\sigma_i\}_{i=1,\dots,n}$. Suppose $a_1\sigma_1 + \dots + a_n\sigma_n = 0$. Some a_i is nonzero. Let that be a_1. Since σ_1 and σ_n are distinct, there is an $x \in G$ with $\sigma_1(x) \neq \sigma_n(x)$. Now by hypothesis, for every $y \in G$,

$$a_1\sigma_1(y) + \dots + a_n\sigma_n(y) = 0,$$

so, multiplying by $\sigma_n(x)$,

$$(*) \quad a_1\sigma_n(x)\sigma_1(y) + \dots + a_n\sigma_n(x)\sigma_n(y) = 0.$$

On the other hand, for every $y \in G$,

$$a_1\sigma_1(xy) + \dots + a_n\sigma_n(xy) = 0;$$

using the fact that $\sigma_i(xy) = \sigma_i(x)\sigma_i(y)$,

$$(**) \quad a_1\sigma_1(x)\sigma_1(y) + \dots + a_n\sigma_n(x)\sigma_n(y) = 0,$$

and then, subtracting $(*)$ from $(**)$,

$$a_1(\sigma_1(x) - \sigma_n(x))\sigma_1(y) + \dots + a_{n-1}(\sigma_{n-1}(x) - \sigma_n(x))\sigma_{n-1}(y) = 0$$

for every $y \in G$. By induction, each coefficient is zero, so in particular $\sigma_1(x) = \sigma_n(x)$, a contradiction. □

Theorem 2.8.5. *Let* $\Sigma = \{\sigma_i,\ i = 1, \dots, n\}$ *be a group of mutually distinct automorphisms of a field* $\mathbf{E}$*, and let* $\mathbf{F}$ *be the fixed field of* Σ*. Then* $(\mathbf{E}/\mathbf{F}) = n$.

Proof. (*Artin*) First we show $n \leq (\mathbf{E}/\mathbf{F})$ (which only uses the fact that Σ is a set) and then we show $n \geq (\mathbf{E}/\mathbf{F})$ (which uses the fact that Σ is a group). We prove both inequalities by contradiction. Let $r = (\mathbf{E}/\mathbf{F})$.

For the first step, suppose $r < n$. Let $\{\epsilon_1, \dots, \epsilon_r\}$ be a basis for $\mathbf{E}$ over $\mathbf{F}$. Consider the system of equations in $\mathbf{E}$:

$$\begin{bmatrix} \sigma_1(\epsilon_1) & \cdots & \sigma_n(\epsilon_1) \\ \vdots & & \\ \sigma_1(\epsilon_r) & \cdots & \sigma_n(\epsilon_r) \end{bmatrix} \begin{bmatrix} \alpha_1 \\ \vdots \\ \alpha_n \end{bmatrix} = 0.$$

This system has more unknowns than equations, so it has a nontrivial solution which we still denote by $\alpha_1, \dots, \alpha_n$. Then

$$\sum_{i=1}^{n} \alpha_i \sigma_i(\epsilon_k) = 0 \text{ for each } k = 1, \ldots, r.$$

Let $\epsilon \in \mathbf{E}$ be arbitrary. Since $\{\epsilon_k\}$ is a basis, we have that $\epsilon = \sum_{k=1}^{r} f_k \epsilon_k$ with $f_k \in \mathbf{F}$, and since $\mathbf{F}$ is the fixed field of Σ, $\sigma_i(f_k) = f_k$ for each i, k. Then

$$\sum_{i=1}^{n} \alpha_i \sigma_i(\epsilon) = \sum_{i=1}^{n} \alpha_i \sigma_i(\sum_{k=1}^{r} f_k \epsilon_k) = \sum_{i,k} \alpha_i \sigma_i(f_k \epsilon_k) = \sum_{i,k} \alpha_i f_k \sigma_i(\epsilon_k)$$
$$= \sum_{k=1}^{r} f_k (\sum_{i=1}^{n} \alpha_i \sigma_i(\epsilon_k)) = \sum_{k=1}^{r} f_k(0) = 0,$$

so $\sum_{i=1}^{n} \alpha_i \sigma_i = 0$, contradicting Theorem 2.8.4.

For the second step, suppose $r > n$. Since Σ is a group, some σ_i is the identity. Let it be σ_1. Let $\{\epsilon_1, \ldots, \epsilon_{n+1}\}$ be a set of $\mathbf{F}$-linearly independent elements of $\mathbf{E}$. Consider the system of equations in $\mathbf{E}$:

$$\begin{bmatrix} \sigma_1(\epsilon_1) & \cdots & \sigma_1(\epsilon_{n+1}) \\ \vdots & & \\ \sigma_n(\epsilon_1) & \cdots & \sigma_n(\epsilon_{n+1}) \end{bmatrix} \begin{bmatrix} \beta_1 \\ \vdots \\ \beta_{n+1} \end{bmatrix} = 0.$$

This system has more unknowns than equations, so it has a nontrivial solution which we still denote by $\beta_1, \ldots, \beta_{n+1}$. Among all such solutions choose one with the smallest number s of nonzero values. By permuting the subscripts, we may assume this solution is $\beta_1, \ldots, \beta_s, 0, \ldots, 0$. Note that $s > 1$, since if $s = 1$, $0 = \beta_1 \sigma_1(\epsilon_1) = \beta_1 \epsilon_1$ implies $\beta_1 = 0$ as $\epsilon_1 \neq 0$. Then we may further assume (multiplying by β_s^{-1} if necessary) that $\beta_s = 1$. Finally, not all β_i are in $\mathbf{F}$, since if they were, we would have $0 = \beta_1 \sigma_1(\epsilon_1) + \cdots + \beta_s \sigma_1(\epsilon_s) = \beta_1 \epsilon_1 + \cdots + \beta_s \epsilon_s$, an $\mathbf{F}$-linear dependence of $\{\epsilon_1, \ldots, \epsilon_n\}$. Thus we may finally assume that $\beta_1 \notin \mathbf{F}$. Then we have

$$(*_i) \quad \beta_1 \sigma_i(\epsilon_1) + \cdots + \beta_{s-1} \sigma_i(\epsilon_{s-1}) + \sigma_i(\epsilon_s) = 0, \ i = 1, \ldots, n.$$

Since $\beta_1 \notin \mathbf{F} = \mathrm{Fix}(\Sigma)$, there is a $\sigma_k \in \Sigma$ with $\sigma_k(\beta_1) \neq \beta_1$. Since Σ is a group, for any $\sigma_i \in \Sigma$ there is a $\sigma_j \in \Sigma$ with $\sigma_i = \sigma_k \sigma_j$. Then applying σ_k to $(*_j)$, we have

$$(**_i) \quad \sigma_k(\beta_1) \sigma_i(\epsilon_1) + \cdots + \sigma_k(\beta_{s-1}) \sigma_i(\epsilon_{s-1}) + \sigma_i(\epsilon_s) = 0,$$

for $i = 1, \ldots, n$. Subtracting $(**_i)$ from $(*_i)$ gives

$$(***_i) \quad (\beta_1 - \sigma_k(\beta_1)) \sigma_i(\epsilon_1) + \cdots + (\beta_{s-1} - \sigma_k(\epsilon_{s-1})) \sigma_i(\epsilon_{s-1}) = 0,$$

for $i = 1, \ldots, n$. Now $\beta_1 - \sigma_k(\beta_1) \neq 0$ and this equation has fewer than s nonzero coefficients, a contradiction. □

Corollary 2.8.6. *(1) Let G be a finite group of automorphisms of* $\mathbf{E}$ *and let* $\mathbf{F} = \mathrm{Fix}(G)$. *Then any automorphism of* $\mathbf{E}$ *fixing* $\mathbf{F}$ *must belong to* G.

(2) Let G_1 *and* G_2 *be two distinct finite groups of automorphisms of* $\mathbf{E}$, *and let* $\mathbf{F}_1 = \mathrm{Fix}(G_1)$, $\mathbf{F}_2 = \mathrm{Fix}(G_2)$. *Then* $\mathbf{F}_1$ *and* $\mathbf{F}_2$ *are distinct.*

Proof. (1) $(\mathbf{E}/\mathbf{F}) = |G| = n$, say, by Theorem 2.8.5, so if there were some automorphism σ of $\mathbf{E}$ fixing $\mathbf{F}$ that was not in G, then $\mathbf{F}$ would be fixed by at least $n+1 > (\mathbf{E}/\mathbf{F})$ automorphisms of $\mathbf{E}$, a contradiction.

(2) We prove the contrapositive: If $\mathbf{F}_1 = \mathbf{F}_2$, then $\mathbf{F}_1$ is fixed by G_2, so $G_2 \subseteq G_1$ by part (1), and vice versa, so $G_1 = G_2$. □

Lemma 2.8.7. *Let G be a group of automorphisms of* $\mathbf{E}$ *fixing* $\mathbf{F}$ *and let H be a subgroup of G. Let* $\mathbf{B} = \mathrm{Fix}(H)$. *Then for any* $\sigma \in G$, $\sigma(\mathbf{B}) = \mathrm{Fix}(\sigma H \sigma^{-1})$.

Proof. Let $\beta \in \mathbf{B}$ and $\tau \in H$. Then $\sigma\tau\sigma^{-1}(\sigma(\beta)) = \sigma\tau(\beta) = \sigma(\beta)$, so $\sigma(\mathbf{B}) \subseteq \mathrm{Fix}(\sigma H \sigma^{-1})$. On the other hand, if $\sigma\tau\sigma^{-1}(\beta') = \beta'$ for some $\beta' \in \mathbf{E}$ and all $\tau \in H$, then $\tau\sigma^{-1}(\beta') = \sigma^{-1}(\beta')$; so $\sigma^{-1}(\beta') = \beta \in \mathrm{Fix}(H) = \mathbf{B}$ and $\beta' = \sigma(\beta)$, so $\mathrm{Fix}(\sigma H \sigma^{-1}) \subseteq \sigma(\mathbf{B})$. □

Here is the most important theorem in this book, and one of the fundamental theorems in all of mathematics.

Theorem 2.8.8 (Fundamental Theorem of Galois Theory = FTGT). *Let* $\mathbf{E}$ *be a finite Galois extension of* $\mathbf{F}$ *and let* $G = \mathrm{Gal}(\mathbf{E}/\mathbf{F})$.

(1) There is a one-to-one correspondence between intermediate fields $\mathbf{E} \supseteq \mathbf{B} \supseteq \mathbf{F}$ *and subgroups* $\{1\} \subseteq G_{\mathbf{B}} \subseteq G$ *given by*

$$\mathbf{B} = \mathrm{Fix}(G_{\mathbf{B}}).$$

(2) $\mathbf{B}$ *is a normal extension of* $\mathbf{F}$ *if and only if* $G_{\mathbf{B}}$ *is a normal subgroup of G. This is the case if and only if* $\mathbf{B}$ *is a Galois extension of* $\mathbf{F}$. *In this case*

$$\mathrm{Gal}(\mathbf{B}/\mathbf{F}) \cong G/G_{\mathbf{B}}.$$

(3) For each $\mathbf{E} \supseteq \mathbf{B} \supseteq \mathbf{F}$, $(\mathbf{B}/\mathbf{F}) = [G : G_{\mathbf{B}}]$ *and* $(\mathbf{E}/\mathbf{B}) = |G_{\mathbf{B}}|$.

We say that the intermediate field $\mathbf{B}$ and the subgroup $G_{\mathbf{B}}$ "belong" to each other.

Before proving this theorem, we make some observations.

Remark 2.8.9. (1) If $\mathbf{B}_1 \subseteq \mathbf{B}_2$, then $G_{\mathbf{B}_1} \supseteq G_{\mathbf{B}_2}$, so the correspondence is "inclusion-reversing".

(2) $\mathbf{E}$ is always a Galois extension of $\mathbf{B}$ and $G_{\mathbf{B}} = \mathrm{Gal}(\mathbf{E}/\mathbf{B})$, both claims being true by the definition of a Galois extension (Definition 2.7.9).

(3) Since $\mathbf{E}/\mathbf{F}$ is separable, $\mathbf{B}/\mathbf{F}$ is separable for any $\mathbf{B}$ intermediate between $\mathbf{F}$ and $\mathbf{E}$ (recall Definition 2.8.3 (3)), so, by Theorem 2.7.14, $\mathbf{B}/\mathbf{F}$ is normal if and only if $\mathbf{B}/\mathbf{F}$ is Galois. This justifies the second sentence of Theorem 2.8.8 (2).

(4) As we shall see from the proof, in case $\mathbf{B}$ is a Galois extension of $\mathbf{F}$, the quotient map $\text{Gal}(\mathbf{E}/\mathbf{F}) \to \text{Gal}(\mathbf{B}/\mathbf{F})$ is simply given by restriction, $\sigma \mapsto \sigma \mid \mathbf{B}$. ◇

Proof. (1) For each subgroup H and G, let

$$\mathbf{B}_H = \text{Fix}(H).$$

This gives a map

$$\Gamma\colon \{\text{subgroups of } G\} \to \{\text{fields intermediate between } \mathbf{F} \text{ and } \mathbf{E}\}.$$

We show Γ is a one-to-one correspondence.

Γ is one-to-one: If $H_1 \neq H_2$, then $\text{Fix}(H_1) \neq \text{Fix}(H_2)$ by Corollary 2.8.6 (2).

Γ is onto: Let $\mathbf{F} \subseteq \mathbf{B} \subseteq \mathbf{E}$ and let

$$H = \{\sigma \in G \mid \sigma \mid \mathbf{B} = \text{id}\} = \text{Gal}(\mathbf{E}/\mathbf{B}) \subseteq \text{Gal}(\mathbf{E}/\mathbf{F}).$$

Since $\mathbf{E}/\mathbf{F}$ is Galois, it is the splitting field of a separable polynomial $f(X) \in \mathbf{F}[X]$ by Theorem 2.7.14. But $\mathbf{F} \subseteq \mathbf{B}$ so $f(X) \in \mathbf{B}[X]$. Then $\mathbf{E}$ is the splitting field of the separable polynomial $f(X) \in \mathbf{B}[X]$, so $\mathbf{E}/\mathbf{B}$ is Galois, again by Theorem 2.7.14, and hence $\mathbf{B} = \text{Fix}(\text{Gal}(\mathbf{E}/\mathbf{B})) = \text{Fix}(H)$.

Thus Γ is one-to-one and onto, and if $\Gamma^{-1}(\mathbf{B}) = G_{\mathbf{B}}$, then $\mathbf{B} = \text{Fix}(G_{\mathbf{B}})$.

(3) Since $\mathbf{E}/\mathbf{B}$ is Galois, we know that $(\mathbf{E}/\mathbf{B}) = |\text{Gal}(\mathbf{E}/\mathbf{B})| = |G_{\mathbf{B}}|$, by Theorem 2.8.5, and $(\mathbf{E}/\mathbf{F}) = (\mathbf{E}/\mathbf{B})(\mathbf{B}/\mathbf{F})$, while $|G| = |G_{\mathbf{B}}|[G : G_{\mathbf{B}}]$, so $(\mathbf{B}/\mathbf{F}) = [G : G_{\mathbf{B}}]$.

(2) Suppose that $G_{\mathbf{B}}$ is a normal subgroup of G. For any $\sigma \in \text{Gal}(\mathbf{E}/\mathbf{F})$, $\sigma(\mathbf{B}) = \text{Fix}(\sigma G_{\mathbf{B}} \sigma^{-1}) = \text{Fix}(G_{\mathbf{B}}) = \mathbf{B}$, by Lemma 2.8.7. Hence we have the restriction map $R : \text{Gal}(\mathbf{E}/\mathbf{F}) \to \text{Gal}(\mathbf{B}/\mathbf{F})$ given by $\sigma \mapsto \sigma \mid \mathbf{B}$, and $\text{Ker}(R) = \{\sigma \in G \mid \sigma \mid \mathbf{B} = \text{id}\} = G_{\mathbf{B}}$ by definition. Thus $\text{Im}(R) \cong \text{Gal}(\mathbf{E}/\mathbf{F})/\text{Ker}(R) = G/G_{\mathbf{B}}$. But let $\sigma_0 \in \text{Gal}(\mathbf{B}/\mathbf{F})$. Then $\mathbf{E}/\mathbf{B}$ is Galois, so $\mathbf{E}$ is the splitting field of a (separable) polynomial $f(X)$. Then $\sigma_0\colon \mathbf{B} \to \mathbf{B}$ extends to $\sigma\colon \mathbf{E} \to \mathbf{E}$ by Lemma 2.6.3, and $\sigma \mid \mathbf{F} = \text{id}$, so $\sigma \in \text{Gal}(\mathbf{E}/\mathbf{F})$. Then $R(\sigma) = \sigma_0$, so $R\colon \text{Gal}(\mathbf{E}/\mathbf{F}) \to \text{Gal}(\mathbf{B}/\mathbf{F})$ is onto and hence $\text{Gal}(\mathbf{B}/\mathbf{F}) \cong G/G_{\mathbf{B}}$.

Conversely, suppose that $\mathbf{B}/\mathbf{F}$ is Galois. Then $\mathbf{B}$ is the splitting field of a separable polynomial $f(X)$. Let $f(X)$ have roots $\beta_1, \ldots, \beta_r$ in $\mathbf{B}$, and note, by Remark 2.5.6, that $\mathbf{B} = \mathbf{F}(\beta_1, \ldots, \beta_r)$. Let $\sigma \in G = \mathrm{Gal}(\mathbf{E}/\mathbf{F})$.

Now $f(X) \in \mathbf{F}[X]$, so $\sigma(f(X)) = f(X)$ and hence σ permutes the roots of $f(X)$, i.e., for every i, $\sigma(\beta_i) = \beta_j$ for some j. In particular this implies that $\sigma(\mathbf{B}) = \mathbf{B}$. But now, by Lemma 2.8.7, $\mathrm{Fix}(G_{\mathbf{B}}) = \mathbf{B} = \sigma(\mathbf{B}) = \mathrm{Fix}(\sigma G_{\mathbf{B}} \sigma^{-1})$, so, by part (1), $G_{\mathbf{B}} = \sigma G_{\mathbf{B}} \sigma^{-1}$. Since σ is arbitrary, $G_{\mathbf{B}}$ is a normal subgroup of G. □

In the next section, we shall give a number of examples of the *FTGT*, but first we continue here with our theoretical development.

Definition 2.8.10. Two fields $\mathbf{B}_1$ and $\mathbf{B}_2$ intermediate between $\mathbf{E}$ and $\mathbf{B}$ are *conjugate* if there is a $\sigma \in \mathrm{Gal}(\mathbf{E}/\mathbf{F})$ with $\sigma(\mathbf{B}_1) = \mathbf{B}_2$. ◇

Corollary 2.8.11. *(1) Two intermediate fields $\mathbf{B}_1$ and $\mathbf{B}_2$ are conjugate if and only if $G_{\mathbf{B}_1}$ and $G_{\mathbf{B}_2}$ are conjugate subgroups of* $\mathrm{Gal}(\mathbf{E}/\mathbf{F})$.

(2) $\mathbf{B}$ is its only conjugate if and only if $G_{\mathbf{B}}$ is a normal subgroup of $\mathrm{Gal}(\mathbf{E}/\mathbf{F})$.

Proof. Immediate from Lemma 2.8.7 and Theorem 2.8.8. □

Returning to the situation of the Fundamental Theorem of Galois Theory, let $\mathbf{E}$ be a finite Galois extension of $\mathbf{F}$, and let $\mathbf{B}$ be an intermediate field between $\mathbf{E}$ and $\mathbf{F}$.

Definition 2.8.12. Let

$$Q_{\mathbf{B}} = \{\sigma_0 \colon \mathbf{B} \to \mathbf{B}' \text{ an isomorphism } \mid \mathbf{F} \subseteq \mathbf{B}' \subseteq \mathbf{E},\ \ \sigma_0 \mid \mathbf{F} = \mathrm{id}\}.$$

Let $\Lambda \colon Q_{\mathbf{B}} \to \{\text{left cosets of } G_{\mathbf{B}} \text{ in } G\}$ be defined as follows:

Let $\sigma_0 \in Q_{\mathbf{B}}$. Since $\mathbf{E}$ is a splitting field, σ_0 extends to $\sigma \in \mathrm{Gal}(\mathbf{E}/\mathbf{F})$ by Lemma 2.6.3. Then set

$$\Lambda(\sigma_0) = [\sigma] \in G/G_{\mathbf{B}}.$$ ◇

Proposition 2.8.13. *(1) Λ is well defined.*

(2) Λ is one-to-one and onto.

(3) $\mathrm{Gal}(\mathbf{B}/\mathbf{F}) \subseteq Q_{\mathbf{B}}$ *with equality if and only if $G_{\mathbf{B}}$ is normal, or equivalent if and only if $\mathbf{B}/\mathbf{F}$ is Galois. In this case, Λ is an isomorphism of groups, $\Lambda \colon Q_{\mathbf{B}} \to G/G_{\mathbf{B}}$.*

Proof. (1) We need to check that $\Lambda(\sigma_0)$ is independent of the choice of extension σ. Let σ' be another extension of σ_0. Then $\sigma' \mid \mathbf{B} = \sigma_0 = \sigma \mid \mathbf{B}$, i.e., $\sigma'(\beta) = \sigma(\beta)$ for all $\beta \in \mathbf{B}$, so $(\sigma^{-1}\sigma')(\beta) = \beta$ for all $\beta \in \mathbf{B}$, and hence $\sigma^{-1}\sigma' \in G_{\mathbf{B}}$, so $\sigma' \in \sigma G_{\mathbf{B}}$ and $[\sigma'] = [\sigma]$.

(2) Suppose $\Lambda(\sigma_0) = [\sigma]$ and $\Lambda(\tau_0) = [\tau]$ with $[\sigma] = [\tau]$. Then $\tau = \sigma\rho$ for some $\rho \in G_{\mathbf{B}}$, so $\tau \mid \mathbf{B} = \sigma\rho \mid \mathbf{B}$. But $\rho \in G_{\mathbf{B}}$ so $\rho \mid \mathbf{B} = \text{id}$; hence $\tau \mid \mathbf{B} = \sigma \mid \mathbf{B}$, i.e., $\tau_0 = \sigma_0$, and Λ is one-to-one. Also, let $\sigma G_{\mathbf{B}}$ be a left coset of $G_{\mathbf{B}}$ in G, and let $\sigma_0 = \sigma \mid \mathbf{B}, \sigma_0 \colon \mathbf{B} \to \mathbf{B}' = \sigma_0(\mathbf{B}) \subseteq \mathbf{E}$. Then $\Lambda(\sigma) = [\sigma_0]$, so Λ is onto.

(3) Clearly $\text{Gal}(\mathbf{B}/\mathbf{F}) \subseteq Q_{\mathbf{B}}$ as $\text{Gal}(\mathbf{B}/\mathbf{F}) = \{\sigma_0 \in Q_{\mathbf{B}} \mid \mathbf{B}' = \mathbf{B}\}$. The first sentence then follows directly from Corollary 2.8.11. As for the second, if $\Lambda(\sigma) = [\sigma_0]$ and $\Lambda(\tau) = \tau_0$, then $\sigma \mid \mathbf{B} = \sigma_0$ and $\tau \mid \mathbf{B} = \tau_0$, so $\sigma\tau \mid \mathbf{B} = \sigma_0\tau_0$ and hence $\Lambda(\sigma\tau) = [\sigma_0\tau_0]$. But if $G_{\mathbf{B}}$ is normal, $(\sigma_0 G_{\mathbf{B}})(\tau_0 G_{\mathbf{B}}) = \sigma_0\tau_0 G_{\mathbf{B}}$, i.e., $[\sigma_0\tau_0] = [\sigma_0][\sigma_0]$, so $\Lambda(\sigma\tau) = \Lambda(\sigma)\Lambda(\tau)$. Thus Λ is a group homomorphism that is one-to-one and onto as a map of sets, so Λ is an isomorphism. □

Proposition 2.8.14. *Let* $\mathbf{E}$ *be a finite Galois extension of* $\mathbf{F}$ *and let* $\mathbf{B}_1$ *and* $\mathbf{B}_2$ *be intermediate fields with* $\mathbf{B}_1 = \text{Fix}(H_1)$ *and* $\mathbf{B}_2 = \text{Fix}(H_2)$.

Then:

(1) $\mathbf{B}_1\mathbf{B}_2 = \text{Fix}(H_1 \cap H_2)$

(2) $\mathbf{B}_1 \cap \mathbf{B}_2 = \text{Fix}(H_3)$ *where* H_3 *is the subgroup generated by* H_1 *and* H_2.

Proof. (1) If $\sigma \in H_1 \cap H_2$, then $\sigma \in H_1$, so σ fixes $\mathbf{B}_1$, and $\sigma \in H_2$, so σ fixes $\mathbf{B}_2$; hence σ fixes $\mathbf{B}_1\mathbf{B}_2$. On the other hand, if σ fixes $\mathbf{B}_1\mathbf{B}_2$, then σ fixes $\mathbf{B}_1$, so $\sigma \in H_1$, and σ fixes $\mathbf{B}_2$, so $\sigma \in H_2$; hence $\sigma \in H_1 \cap H_2$.

(2) If $\sigma(\beta) = \beta$ for every $\sigma \in H_3$, then $\sigma(\beta) = \beta$ for every $\sigma \in H_1$, so $\beta \in \mathbf{B}_1$, and $\sigma(\beta) = \beta$ for every $\sigma \in H_2$, so $\beta \in \mathbf{B}_2$; hence $\beta \in \mathbf{B}_1 \cap \mathbf{B}_2$. On the other hand, if $\beta \in \mathbf{B}_1 \cap \mathbf{B}_2$, then $b \in \mathbf{B}_1$, so $\sigma(\beta) = \beta$ for every $\sigma \in H_1$, and $b \in \mathbf{B}_2$, so $\sigma(\beta) = \beta$ for every $\sigma \in H_2$; hence $\sigma(\beta) = \beta$ for every $\sigma \in H_3$. □

In the original development of group theory, and in particular in the work of Galois, groups were not regarded as abstract groups, but rather as groups of symmetries, and in particular as groups of permutations of the roots of a polynomial. This viewpoint, still a most useful one, is encapsulated in the following result. In order to state it, we first recall that an action of a group G on a set $S = \{s_i\}$ is *transitive* if for any two elements s_i and s_j of S, there is an element σ of G with $\sigma(s_i) = s_j$.

Proposition 2.8.15. *Let* $\mathbf{E}$ *be a finite Galois extension of* $\mathbf{F}$, *so* $\mathbf{E}$ *is the splitting field of a separable polynomial* $f(X) \in \mathbf{F}[X]$. *Then* $G = \text{Gal}(\mathbf{E}/\mathbf{F})$ *is isomorphic to a permutation group on the roots of* $f(X)$. *If* $f(X)$ *is irreducible, then* G *is isomorphic to a transitive permutation group on the roots of* $f(X)$.

Proof. Let $\alpha_1, \ldots, \alpha_d$ be the roots of $f(X)$ in $\mathbf{E}$. For any $\sigma \in \mathrm{Gal}(\mathbf{E}/\mathbf{F})$, $\sigma(\alpha_i)$ is a root of $\sigma(f(X)) = f(X)$, so $\mathrm{Gal}(\mathbf{E}/\mathbf{F})$ permutes $\{\alpha_i\}$. Furthermore, $\mathbf{E} = \mathbf{F}(\alpha_1, \ldots, \alpha_d)$ (Remark 2.5.6), so if $\sigma(\alpha_i) = \alpha_i$ for each i, then $\sigma = \mathrm{id}$.

Now if $f(X)$ is irreducible, and α_i and α_j are any two roots of $f(X)$, there is isomorphism $\sigma_0 \colon \mathbf{F}(\alpha_i) \to \mathbf{F}(\alpha_j)$ with $\sigma \mid \mathbf{F} = \mathrm{id}$ and $\sigma_0(\alpha_i) = \alpha_j$ (Lemma 2.6.1), and σ_0 extends to $\sigma : \mathbf{E} \to \mathbf{E}$ (Lemma 2.6.3). □

Remark 2.8.16. Note Proposition 2.8.15 implies Lemma 2.5.3 in case $f(X)$ is separable. ◇

Remark 2.8.17. Note that Proposition 2.8.15 is about how the Galois group is realized, not about its structure as an abstract group. For example, consider $\mathbf{E} = \mathbf{Q}(\sqrt{2}, \sqrt{3})$ and $\mathbf{F} = \mathbf{Q}$. Then $G = \mathrm{Gal}(\mathbf{E}/\mathbf{F}) \cong (\mathbb{Z}/2\mathbb{Z}) \oplus (\mathbb{Z}/2\mathbb{Z})$, and $\mathbf{E}$ is the splitting field of the polynomial $(X^2 - 2)(X^2 - 3)$, which is not irreducible, so G does not act transitively on $\{\sqrt{2}, -\sqrt{2}, \sqrt{3}, -\sqrt{3}\}$. On the other hand $\mathbf{E} = \mathbf{Q}(\sqrt{2} + \sqrt{3})$ and $\mathbf{E}$ is the splitting field of the polynomial $(X - (\sqrt{2}+\sqrt{3}))(X - (\sqrt{2}-\sqrt{3}))(X - (-\sqrt{2}+\sqrt{3}))(X - (-\sqrt{2}-\sqrt{3})) = X^4 - 10X^2 + 1$, which is irreducible, and G does indeed act transitively on $\{\sqrt{2}+\sqrt{3}, \sqrt{2}-\sqrt{3}, -\sqrt{2}+\sqrt{3}, -\sqrt{2}-\sqrt{3}\}$. ◇

Corollary 2.8.18. *Let $f(X) \in \mathbf{F}[X]$ be a separable polynomial and write $f(X) = f_1(X)^{e_1} \cdots f_k(X)^{e_k}$, a factorization into irreducibles in $\mathbf{F}[X]$. Let $f_i(X)$ have degree d_i. Let $\mathbf{E}$ be the splitting field of $f(X) \in \mathbf{F}[X]$. Then $\mathrm{Gal}(\mathbf{E}/\mathbf{F})$ is isomorphic to a subgroup of $S_{d_1} \times \cdots \times S_{d_k}$ (where S_d denotes the symmetric group on d elements). In particular, if $f(X) \in \mathbf{F}[X]$ is a separable polynomial of degree d, and $\mathbf{E}$ is the splitting field of $f(X) \in \mathbf{F}[X]$, then $\mathrm{Gal}(\mathbf{E}/\mathbf{F})$ is isomorphic to a subgroup of S_d.* □

Theorem 2.8.19. *Let $\mathbf{E}$ be a finite Galois extension of $\mathbf{F}$ of degree d. Then $G = \mathrm{Gal}(\mathbf{E}/\mathbf{F})$ is isomorphic to a transitive subgroup of S_d, the symmetric group on d elements. Also, G has a subgroup H of index d.*

Proof. Since $\mathbf{E}$ is a Galois extension of $\mathbf{F}$, it is certainly a separable extension of $\mathbf{F}$ (Proposition 2.8.14).We now apply a result we shall prove later (Corollary 3.5.3) to conclude that $\mathbf{E} = \mathbf{F}(\alpha)$ for some $\alpha \in \mathbf{E}$. Then $d = (\mathbf{E}/\mathbf{F}) = (\mathbf{F}(\alpha)/\mathbf{F}) = \deg m_\alpha(X)$ (Proposition 2.4.6 (2)), so the theorem follows from Proposition 2.8.15 and the following general fact: If G acts transitively on $S = \{s_i\}$, a set of cardinality d, then $H = \{\sigma \in G \mid \sigma(s_1) = s_1\}$ is a subgroup of index d. □

2.9 Examples

In this section we give a number of examples of the Fundamental Theorem of Galois Theory, and compute a number of Galois groups.

Example 2.9.1. See Example 1.1.1. ◇

Example 2.9.2. See Example 1.1.2. ◇

Example 2.9.3. See Example 1.1.3. ◇

Example 2.9.4. See Example 1.1.4. ◇

Example 2.9.5. Let $\mathbf{E}_a$ be the splitting field of $f(X) = X^3 - a$ over $\mathbf{Q}$, where for simplicity we take a to be an integer. There are several cases.

$a = 0$: Then $f(X) = (X)(X)(X)$ is a product of linear factors, so $\mathbf{E}_a = \mathbf{Q}$. ($\mathbf{E}_a$ is obtained from $\mathbf{Q}$ by adjoining 0 three times, but adjoining 0 does nothing.) Note $(\mathbf{E}_a/\mathbf{Q}) = 1$.

$a \neq 0$ a perfect cube: Let $a = b^3$. Then $f(X) = X^3 - b^3 = b^3((X/b)^3 - 1)$ so $\mathbf{E}_a = \mathbf{E}_1$. Thus we need only find the splitting field of $f(X) = X^3 - 1$. This we have done already. $\mathbf{E}_1 = \mathbf{Q}(\omega)$ where $\omega = (-1 + i\sqrt{3})/2$ is a primitive cube root of 1. Note $(\mathbf{E}_a/\mathbf{Q}) = 2$ and $\mathrm{Gal}(\mathbf{E}_a/\mathbf{Q}) \cong \mathbb{Z}/2\mathbb{Z}$.

$a \neq 0$ not a perfect cube: This situation is entirely analogous to Example 1.1.3. $\mathbf{E}_a = \mathbf{Q}(\omega, \sqrt[3]{a})$ and $(\mathbf{E}_a/\mathbf{Q}) = 6$. Also, $G = \mathrm{Gal}(\mathbf{E}_a/\mathbf{Q}) \cong D_6$, the dihedral group of order 6. Note that we can determine G without doing any computation. We know that $|G| = (\mathbf{E}_a/\mathbf{Q}) = 6$, and there are only two groups of order 6, $\mathbb{Z}/6\mathbb{Z}$, which is abelian, and D_6, which is not. Now $\mathbf{E}_a \supset \mathbf{Q}(\sqrt[3]{a})$ and $\mathbf{Q}(\sqrt[3]{a})$ is not a normal extension of $\mathbf{Q}$, so G has a subgroup that is not normal and hence G is nonabelian. ◇

Example 2.9.6. $\mathbf{E}$ is the splitting field of $f(X) = X^6 - 3$ over $\mathbf{Q}$. Then $\mathbf{E} = \mathbf{Q}(\zeta, \sqrt[6]{3})$ where ζ is a primitive sixth root of 1 and $(\mathbf{E}/\mathbf{Q}) = 12$. We may choose $\zeta = (1 + i\sqrt{3})/2$. Then $\omega = \zeta^2$ and $-1 = \zeta^3$. But then notice also that $\zeta = 1 + \omega$ so in particular $\mathbf{Q}(\zeta) = \mathbf{Q}(\omega)$. An element σ of $\mathrm{Gal}(\mathbf{E}/\mathbf{Q})$ is determined by its effect on ζ and on $\sqrt[6]{3}$. One can check that $G = \mathrm{Gal}(\mathbf{E}/\mathbf{Q})$ has elements σ and τ with:

$$\begin{aligned} \sigma(\zeta) &= \zeta^{-1} & \tau(\zeta) &= \zeta \\ \sigma(\sqrt[6]{3}) &= \sqrt[6]{3} & \tau(\sqrt[6]{3}) &= \zeta\sqrt[6]{3}. \end{aligned}$$

Clearly $\sigma^2 = 1$ and $\tau^6 = 1$. Also $\sigma\tau(\zeta) = \zeta^5 = \tau^{-1}\sigma(\zeta)$ and $\sigma\tau(\sqrt[6]{3}) = \zeta^{-1}\sqrt[6]{3} = \tau^{-1}\sigma(\sqrt[6]{3})$. Since any element of G must take ζ to ζ or ζ^{-1}, and must take $\sqrt[3]{6}$ to $\zeta^i\sqrt[3]{6}$ for some i, we see that σ and τ generate G. Thus we conclude that

$$G = \langle \sigma, \tau \mid \sigma^2 = 1,\ \tau^6 = 1, i\ \sigma\tau = \tau^{-1}\sigma \rangle$$

is isomorphic to the dihedral group D_{12}. It is then routine but lengthy to check that G has 15 subgroups, which we list below, where subgroups denoted by the some letter are mutually conjugate (so in particular A_1, D_1, E_1, G_1, H_1 and I_1 are normal):

$A_1 = \{1\}$
$B_1 = \{1, \sigma\},\ B_2 = \{1, \tau^2\sigma\},\ B_3 = \{1, \tau^4\sigma\}$
$C_1 = \{1, \tau\sigma\},\ C_2 = \{1, \tau^3\sigma\},\ C_3 = \{1, \tau^5\sigma\}$
$D_1 = \{1, \tau^3\}$
$E_1 = \{1, \tau^2, \tau^4\}$
$F_1 = 1, \sigma, \tau^3, \tau^3\sigma\},\ F_2 = \{1, \tau\sigma, \tau^3, \tau^4\sigma\},\ F_3 = \{1, \tau^2\sigma, \tau^3, \tau^5\sigma\}$
$G_1 = \{1, \tau^2, \tau^4, \sigma, \tau^2\sigma, \tau^4\sigma\}$
$H_1 = \{1, \tau, \tau^2, \tau^3, \tau^4, \tau^5\}$
$I_1 = G$

These subgroups have the following fixed fields:

$\mathrm{Fix}(A_1) = \mathbf{E}$
$\mathrm{Fix}(B_1) = \mathbf{Q}(\sqrt[6]{3}),\ \mathrm{Fix}(B_2) = \mathbf{Q}(\zeta\sqrt[6]{3}),\ \mathrm{Fix}(B_3) = \mathbf{Q}(\omega\sqrt[6]{3})$
$\mathrm{Fix}(C_1) = \mathbf{Q}((1+\zeta)\sqrt[6]{3}),\ \mathrm{Fix}(C_2) = \mathbf{Q}(i\sqrt[6]{3}),$
$\mathrm{Fix}(C_3) = \mathbf{Q}((1+\zeta^{-1})\sqrt[6]{3})$
$\mathrm{Fix}(D_1) = \mathbf{Q}(\zeta, \sqrt[3]{3})$
$\mathrm{Fix}(E_1) = \mathbf{Q}(\zeta, \sqrt{3})$
$\mathrm{Fix}(F_1) = \mathbf{Q}(\sqrt[3]{3}),\ \mathrm{Fix}(F_2) = \mathbf{Q}(\zeta\sqrt[3]{3}),\ \mathrm{Fix}(F_3) = \mathbf{Q}(\omega\sqrt[3]{3})$
$\mathrm{Fix}(G_1) = \mathbf{Q}(\sqrt{3})$
$\mathrm{Fix}(H_1) = \mathbf{Q}(\zeta)$
$\mathrm{Fix}(I_1) = \mathbf{Q}$

The 6 subfields fixed by normal subgroups are Galois extensions of $\mathbf{Q}$, so are splitting fields of separable polynomials:

$\mathrm{Fix}(A_1) =$ splitting field of $X^6 - 3$
$\mathrm{Fix}(D_1) =$ splitting field of $X^3 - 3$
$\mathrm{Fix}(E_1) =$ splitting field of $(X^2 - X + 1)(X^2 - 3)$
$\mathrm{Fix}(G_1) =$ splitting field of $X^2 - 3$
$\mathrm{Fix}(H_1) =$ splitting field of $X^2 - X + 1$
$\mathrm{Fix}(I_1) =$ splitting field of 1

Note here there are also interesting relationships between the fields intermediate between $\mathbf{E}$ and $\mathbf{Q}$. For example, let $\mathbf{F} = \mathrm{Fix}(F_1) = \mathbf{Q}(\sqrt[3]{3})$. Then B_1,

C_2, and D_1 are all normal subgroups of F_1 (although B_1 and C_2 are not normal subgroups of G). Hence $\text{Fix}(B_1)/\mathbf{F}_1$, $\text{Fix}(C_2)/\mathbf{F}_1$, and $\text{Fix}(D_1)/\mathbf{F}_1$ are all Galois extensions, all of degree two, hence with Galois groups isomorphic to $\mathbb{Z}/2\mathbb{Z}$, and they are the splitting fields of the polynomials $X^2 - \sqrt[3]{3}$, $X^2 + \sqrt[3]{3}$, and $X^2 + X + 1 \in \mathbf{Q}(\sqrt[3]{3})[X]$, respectively. ◇

Example 2.9.7. $\mathbf{E}$ is the splitting field of $X^6 + 3$ over $\mathbf{Q}$. Again $\mathbf{E} = \mathbf{Q}(\zeta, \sqrt[6]{-3})$ where ζ is a primitive sixth root of 1. However, $(\sqrt[6]{-3})^3 = i\sqrt{3}$, so $\zeta \in \mathbf{Q}(\sqrt[6]{-3})$ and $\mathbf{E} = \mathbf{Q}(\sqrt[6]{-3})$. Thus in this case $(\mathbf{E}/\mathbf{Q}) = 6$, and here $\text{Gal}(\mathbf{E}/\mathbf{Q}) = D_6$ by an argument as in Example 2.9.5.

Let us examine this example more closely. Let $\alpha = \sqrt[6]{-3}$. Then $\mathbf{E} = \mathbf{Q}(\alpha)$, so $\lambda \in G = \text{Gal}(\mathbf{E}/\mathbf{Q})$ is determined by $\lambda(\alpha)$. Since $\lambda(\alpha)$ must be a root of $X^6 + 3$, we must have $\lambda(\alpha) = \zeta^k\alpha$, for some k, where $\zeta = (1 + i\sqrt{3})/2$ is a primitive sixth root of 1, and all possible values of k, $0 \leq k \leq 5$ occur. If $\lambda(\alpha) = \zeta^k\alpha$, with $0 \leq k \leq 5$, set $\lambda = \lambda_k$. Thus $G = \{\text{id}, \lambda_1, \ldots, \lambda_5\}$. This makes G "look" abelian, but we can see by direct calculation that it is not.

Note that $\zeta = (1 + \alpha^3)/2$, so

$$\begin{aligned}\lambda_j(\zeta) &= (1 + \lambda_j(\alpha^3))/2 = (1 + (\lambda_j(\alpha))^3)/2 = (1 + (\zeta^j\alpha)^3)/2\\ &= (1 + \zeta^{3j}\alpha^3)/2 = \zeta \text{ for } j \text{ even}, \ \zeta^{-1} \text{ for } j \text{ odd}.\end{aligned}$$

In particular we see that

$$\begin{aligned}\lambda_j\lambda_i(\alpha) &= \lambda_j(\lambda_i(\alpha)) = \lambda_j(\zeta^i\alpha) = \lambda_j(\zeta^i)\lambda_j(\alpha) = \lambda_j(\zeta)^i\zeta^j\alpha\\ &= \zeta^k\alpha \text{ where } k = (-1)^j i + j,\end{aligned}$$

directly exhibiting the noncommutativity of G.

In fact, we may obtain more precise information on the structure of G, again by direct computation. Note that $\zeta^2 = \omega = (-1 + i\sqrt{3})/2$, so, setting $\gamma = i\sqrt{3} = \sqrt{-3}$, we see that $\mathbf{Q}(\zeta) = \mathbf{Q}(\omega) = \mathbf{Q}(\gamma)$. Note also that $\alpha = \gamma/\beta$. Then $\mathbf{E} = \mathbf{Q}(\alpha) = (\mathbf{Q}(\beta))(\gamma) = (\mathbf{Q}(\gamma))(\beta)$. Hence there is an element φ of G with $\varphi(\gamma) = -\gamma$, $\varphi(\beta) = \beta$, and an element ψ of G with $\psi(\gamma) = \gamma$, $\psi(\beta) = \omega\beta$. Then

$$\begin{aligned}\varphi(\alpha) &= \varphi(\gamma/\beta) = \varphi(\gamma)/\varphi(\beta) = -\gamma/\beta = \zeta^3(\gamma/\beta) = \zeta^3\alpha \text{ so } \varphi = \lambda_3,\\ \psi(\alpha) &= \psi(\gamma/\beta) = \psi(\gamma)/\psi(\beta) = \gamma/(\omega\beta) = \zeta^4(\gamma/\beta) = \zeta^4\alpha \text{ so } \psi = \lambda_4.\end{aligned}$$

Observe that $\varphi^2 = \text{id}$, $\psi^2 = \lambda_2$, and $\psi^3 = \text{id}$. Furthermore

$$\begin{aligned}\varphi^{-1}\psi\varphi(\alpha) &= \varphi^{-1}\psi\varphi(\gamma/\beta) = \varphi^{-1}\psi(-\gamma/\beta)\\ &= \varphi^{-1}(-\gamma/(\omega\beta)) = \gamma/(\omega^2\beta) = \omega\alpha = \zeta^2\alpha,\end{aligned}$$

so we conclude $\lambda_3^{-1}\lambda_4\lambda_3 = \lambda_4^2$, giving an explicit presentation of G as the dihedral group D_6. ◇

Example 2.9.8. If **E** is the splitting field of an irreducible cubic polynomial over **Q**, then, by Theorem 2.8.19, Gal(**E**/**Q**) is a subgroup of S_3 whose order is divisible by 3, so it is either S_3 or $\mathbb{Z}/3\mathbb{Z}$. The construction in Example 2.9.5 always gave S_3. Here is a case where the Galois group is $\mathbb{Z}/3\mathbb{Z}$. Let ζ be a primitive ninth root of 1 and consider the polynomial $f(X) = (X-(\zeta+\zeta^{-1}))(X-(\zeta^2+\zeta^{-2}))(X-(\zeta^4+\zeta^{-4})) = X^3-3X+1 \in \mathbf{Q}[X]$. This polynomial is irreducible over **Q** (as otherwise it would have a linear factor, i.e., a root in **Q**, and it does not). Let λ_0, λ_1, and λ_2 be the roots of this polynomial, in the order written, and let $\mathbf{E} = \mathbf{F}(\lambda_0, \lambda_1, \lambda_2)$ be its splitting field.

Then $(\mathbf{E}/\mathbf{Q}) = \deg f(X) = 3$, and since $|\,\mathrm{Gal}(\mathbf{E}/\mathbf{Q})| = (\mathbf{E}/\mathbf{Q})$, the Galois group must be isomorphic to $\mathbb{Z}/3\mathbb{Z}$. We can see this concretely as follows: Note that $\lambda_0^2 = \lambda_1 + 2$ and $\lambda_1^2 = \lambda_2 + 2$ (and also $\lambda_2^2 = \lambda_0 + 2$), so in fact $\mathbf{E} = \mathbf{F}(\lambda_0)$, and $\sigma \in \mathrm{Gal}(\mathbf{E}/\mathbf{Q})$ is determined by $\sigma(\lambda_0)$. Since $f(X)$ is irreducible, there is an element $\sigma_i \in \mathrm{Gal}(\mathbf{E}/\mathbf{Q})$ with $\sigma_i(\lambda_0) = \lambda_i$, $i = 0, 1, 2$. Clearly $\sigma_0 = \mathrm{id}$. If $\sigma_1(\lambda_0) = \lambda_1$, then $\sigma_1(\lambda_1) = \sigma_1(\lambda_0^2 - 2) = \sigma_1(\lambda_0)^2 - \sigma_1(2) = \lambda_1^2 - 2 = \lambda_2$ and similarly $\sigma_1(\lambda_2) = \lambda_0$. Then $\sigma_1^2(\lambda_0) = \sigma_1(\sigma_1(\lambda_0)) = \sigma_1(\lambda_1) = \lambda_2$, so $\sigma_1^2 = \sigma_2$; similarly $\sigma_1^3(\lambda_0) = \lambda_0$, so $\sigma_1^3 = \sigma_0$, and we see explicitly that $\mathrm{Gal}(\mathbf{E}/\mathbf{Q}) = \{1, \sigma_1, \sigma_1^2\}$. ◇

2.10 Exercises

Exercise 2.10.1. In each case, find a polynomial in $\mathbf{Q}[X]$ that has α as a root. Then find all complex roots of this polynomial.

(a) $\alpha = 1 + \sqrt{2}$.
(b) $\alpha = \sqrt{2} + \sqrt{3}$.
(c) $\alpha = \sqrt{1 + \sqrt{2}}$.
(d) $\alpha = \sqrt{3 - 2\sqrt{2}}$.
(e) $\alpha = \sqrt[3]{1 + \sqrt{-3}}$.
(f) $\alpha = \sqrt[3]{1 + \sqrt{2}}$.
(g) $\alpha = \sqrt{1 + \sqrt[3]{2}}$.
(h) $\alpha = \sqrt{6 + \sqrt{3}} + \sqrt{6 - \sqrt{3}}$.
(i) $\alpha = \sqrt{4 + \sqrt{7}} + \sqrt{4 - \sqrt{7}}$.

Exercise 2.10.2. Let $f(X) = X^3 + 7X + 1 \in \mathbf{Q}[X]$. This polynomial is irreducible. Let $\mathbf{E} = \mathbf{Q}[\alpha]$ for some root α of $f(X)$. Consider the following elements of $\mathbf{E}$: $\beta_1 = \alpha + 3,\ \beta_2 = \alpha^2 + 1,\ \beta_3 = \alpha^2 - 2\alpha + 3$.
(a) Express each of the following in the form $g(\alpha)$ for some polynomial $g(X) \in \mathbf{Q}[X]$: $\beta_1^{-1},\ \beta_2^{-1},\ \beta_3^{-1},\ \beta_1\beta_2,\ \beta_1\beta_3,\ \beta_2\beta_3$.
(b) Find $m_{\beta_1}(X),\ m_{\beta_2}(X),\ m_{\beta_3}(X) \in \mathbf{Q}[X]$.

Exercise 2.10.3. Let $f(X) = X^3 + X^2 + 2 \in \mathbf{F}_5[X]$. This polynomial is irreducible. Let $\mathbf{E} = \mathbf{F}_5[\alpha]$ for some root α of $f(X)$. Consider the following elements of $\mathbf{E}$: $\beta_1 = 2\alpha + 1,\ \beta_2 = \alpha^2 - 1,\ \beta_3 = \alpha^2 - \alpha + 2$.
(a) Express each of the following in the form $g(\alpha)$ for some polynomial $g(X) \in \mathbf{F}_5[X]$: $\beta_1^{-1},\ \beta_2^{-1},\ \beta_3^{-1},\ \beta_1\beta_2,\ \beta_1\beta_3,\ \beta_2\beta_3$.
(b) Find $m_{\beta_1}(X),\ m_{\beta_2}(X),\ m_{\beta_3}(X) \in \mathbf{F}_5[X]$.

Exercise 2.10.4. As we shall see in Section 3.3, for every prime power $q = p^r$, there is a unique (up to isomorphism) field $\mathbf{F}_q$ with q elements.
(a) Write down an explicit addition and multiplication table for $\mathbf{F}_2$, $\mathbf{F}_3$, and $\mathbf{F}_5$.
(b) Write down an explicit addition and multiplication table for $\mathbf{F}_4$ and $\mathbf{F}_9$.
(c) For each element $\alpha \in \mathbf{F}_4$, find its minimum polynomial $m_\alpha(X) \in \mathbf{F}_2[X]$, and for each element $\alpha \in \mathbf{F}_9$, find its minimum polynomial $m_\alpha(X) \in \mathbf{F}_3[X]$.
(d) Explicitly compute the Frobenius endomorphism Φ on $\mathbf{F}_4$ and $\mathbf{F}_9$.
(e) Write down an explicit addition and multiplication table for $\mathbf{F}_8$.
(f) For each element $\alpha \in \mathbf{F}_8$, find its minimum polynomial $m_\alpha(X) \in \mathbf{F}_2[X]$.
(g) Explicitly compute the Frobenius endomorphism Φ on $\mathbf{F}_8$.

Exercise 2.10.5. (a) Find all irreducible quadratic, cubic, and quartic polynomials $f(X) \in \mathbf{F}_2[X]$.
(b) Find all irreducible quadratic, and cubic polynomials $f(X) \in \mathbf{F}_3[X]$.
(c) Find all irreducible quadratic polynomials $f(X) \in \mathbf{F}_5[X]$.
(d) Find all irreducible quadratic polynomials $f(X) \in \mathbf{F}_4[X]$.

Exercise 2.10.6. Let $f(X) = a_n X^n + \cdots + a_0 \in \mathbf{F}[X]$ split in $\mathbf{E}$. Assume that $\mathrm{char}(\mathbf{F}) \neq 2$. Show that (1) implies (2):
(1) For every $\alpha \in \mathbf{E}$, if α is a root of $f(X)$, then $-\alpha \in \mathbf{E}$ is also a root of $f(X)$ of the same multiplicity.
(2) $a_{n-k} = 0$ for all odd k.

Exercise 2.10.7. Let $f(X) = a_n X^n + \cdots + a_0 \in \mathbf{F}[X]$ split in $\mathbf{E}$. Assume that $a_0 \neq 0$. Show that (1) implies (2):
(1) For every $\alpha \neq 0 \in \mathbf{E}$, if α is a root of $f(X)$, then $1/\alpha \in \mathbf{E}$ is also a root of $f(X)$ of the same multiplicity.
(2) For $e = \pm 1$, $a_{n-k} = e a_k$ for all k.

Exercise 2.10.8. (a) Let $f(X) \in \mathbb{Z}[X]$ be a monic polynomial. Show that if its mod p reduction $\bar{f}(X) \in \mathbf{F}_p[X]$ is irreducible for some p, then $f(X)$ is irreducible.
(b) Use this to show that $f(X) = X^3 - 5X + 36$ is irreducible.

Exercise 2.10.9. A priori, there is a distinction between polynomial *expressions* and polynomial *functions*, i.e., between $f(X)$ regarded as an element of $\mathbf{F}[X]$ and $f(X)$ as defining a function on $\mathbf{F}$. Clearly, if $f(X) = g(X)$ as elements of $\mathbf{F}[X]$, then $f(X) = g(X)$ as functions on $\mathbf{F}[X]$ (i.e., $f(a) = g(a)$ for every $a \in \mathbf{F}$).
(a) Show that, if $\mathbf{F}$ is infinite, the converse is true: If $f(a) = g(a)$ for every $a \in \mathbf{F}$, then $f(X) = g(X) \in \mathbf{F}[X]$.
(b) Find a counterexample to the converse if $\mathbf{F}$ is finite. That is, if $\mathbf{F}$ is finite, find polynomials $f(X) \neq g(X) \in \mathbf{F}[X]$ but with $f(a) = g(a)$ for every $a \in \mathbf{F}$.
(c) Use the result of (b) to show that no finite field is algebraically closed, i.e., that if $\mathbf{F}$ is any finite field, there is a nonconstant polynomial $f(X) \in \mathbf{F}[X]$ that does not have a root in $\mathbf{F}$.

Exercise 2.10.10. (a) Let $f(X) \in \mathbf{F}[X]$ have roots $\alpha_1, \ldots, \alpha_n$ in some splitting field $\mathbf{E}$. Show that $f(X)$ is irreducible in $\mathbf{F}[X]$ if and only if

$$\prod_{\alpha_i \in T} (X - \alpha_i) \notin \mathbf{F}[X]$$

for any nonempty proper subset T of $S = \{\alpha_1, \ldots, \alpha_n\}$.
(b) Observe that $\mathbf{E} = \mathbf{Q}[\sqrt{3}, \sqrt{5}]$ is the splitting field of $f_1(X) = X^4 - 46X^2 + 289$, with roots $\sqrt{3} + 2\sqrt{5}$, $\sqrt{3} - 2\sqrt{5}$, $-\sqrt{3} + 2\sqrt{5}$, and $-\sqrt{3} - 2\sqrt{5}$, and also of $f_2(X) = X^4 - 6X^3 + 5X^2 + 10X + 2$, with roots $1 + \sqrt{3}$, $1 - \sqrt{3}$, $2 + \sqrt{5}$, and $2 - \sqrt{5}$. Use part (a) to show that $f_1(X)$ is irreducible in $\mathbf{Q}[X]$ but $f_2(X)$ is not.

Exercise 2.10.11. Let $f(X), g(X) \in \mathbf{F}[X]$ be monic irreducible polynomials. Show that if $f(X)$ and $g(X)$ have a common root in some extension $\mathbf{E}$ of $\mathbf{F}$, then $f(X) = g(X)$.

Exercise 2.10.12. Let $\mathbf{E} \supseteq \mathbf{F}$ and let $\alpha \neq 0 \in \mathbf{E}$. Show that α is algebraic over $\mathbf{F}$ if and only if $1/\alpha = f(\alpha)$ for some polynomial $f(X) \in \mathbf{F}[X]$.

Exercise 2.10.13. Let R be a subring of $\mathbf{F}$ and let $\mathbf{E}$ be an extension of $\mathbf{F}$. Let $\alpha \in \mathbf{E}$ be algebraic over $\mathbf{F}$. If $m_\alpha(X) \in R[X]$, then α is called *integral* over R. In case $R = \mathbb{Z}$ and $\mathbf{F} = \mathbf{Q}$, show that $\alpha \in \mathbf{E}$ is integral over R if and only if $f(\alpha) = 0$ for some monic polynomial $f(X) \in \mathbb{Z}[X]$. (Such a complex number is called an *algebraic integer*.)

Exercise 2.10.14. Suppose that $R = \mathbb{Z}$ and $\mathbf{F} = \mathbf{Q}$. Show that, for any $\alpha \in \mathbf{E}$, there is an integer $n \neq 0 \in \mathbb{Z}$ such that $n\alpha$ is an algebraic integer.

Exercise 2.10.15. Let $\mathbf{E}$ be a field and let R be a subring of $\mathbf{E}$. If every element of $\mathbf{E}$ is integral over R, show that R is a field.

Exercise 2.10.16. Let $\mathbf{E}$ be an extension of $\mathbf{F}$ and let $f(X) \in \mathbf{F}[X]$ be an irreducible polynomial. Let $f(X) = f_1(X)f_2(X)\cdots f_t(X) \in \mathbf{E}[X]$ be factored as a product of irreducible polynomials.
(a) If $\mathbf{E}$ is a Galois extension of $\mathbf{F}$, show that $\{f_1(X), f_2(X), \ldots, f_t(X)\}$ are mutually conjugate in $\mathbf{E}[X]$. (Two polynomials $f_1(X), f_2(X) \in \mathbf{E}[X]$ are mutually conjugate if there exists an element $\sigma \in \mathrm{Gal}(\mathbf{E}/\mathbf{F})$ with $\sigma(f_1(X)) = f_2(X)$.)
(b) Give a counterexample to this if $\mathbf{E}$ is not a Galois extension of $\mathbf{F}$.

Exercise 2.10.17. Let $\mathbf{E}$ be a splitting field of the separable irreducible polynomial $f(X) \in \mathbf{F}[X]$.
(a) For a root $\alpha \in \mathbf{E}$ of $f(X)$, let $r(\alpha)$ be the number of roots of $f(X)$ in $\mathbf{F}(\alpha)$. Show that $r(\alpha)$ is independent of the choice of α. Call this common value r.
(b) Let d be the number of distinct fields $\{\mathbf{F}(\alpha) \mid \alpha \in \mathbf{E} \text{ a root of } f(X)\}$. Show that $dr = \deg f(X)$.
(c) Give examples where $r = 1$, $1 < r < \deg f(X)$, and $r = \deg f(X)$.

Exercise 2.10.18. Let $\mathbf{E}$ be the splitting field of the separable irreducible polynomial $f(X) \in \mathbf{F}[X]$ and let $\mathbf{B}$ be a Galois extension of $\mathbf{F}$ with $\mathbf{F} \subseteq \mathbf{B} \subseteq \mathbf{E}$. Let $f(X) = f_1(X)\cdots f_k(X) \in \mathbf{B}[X]$ be a factorization into irreducible polynomials. Show that each $f_i(X)$ has the same degree d. Furthermore, if $\alpha \in \mathbf{E}$ is any root of $f(X)$, show that $d = (\mathbf{B}(\alpha)/\mathbf{B})$ and that $k = (\mathbf{B} \cap \mathbf{F}(\alpha)/\mathbf{F})$.

Exercise 2.10.19. Let $f(X) \in \mathbf{Q}[X]$ be an irreducible cubic with roots α, β, γ in some splitting field $\mathbf{E}$. Show that one of the following two alternatives holds:
(1) $\mathrm{Gal}(\mathbf{E}/\mathbf{Q}) \cong \mathbb{Z}/3\mathbb{Z}$, and there is no field $\mathbf{B}$ with $\mathbf{Q} \subset \mathbf{B} \subset \mathbf{E}$.
(2) $\mathrm{Gal}(\mathbf{E}/\mathbf{Q}) \cong S_3$, and there are exactly four fields $\mathbf{B}$ with $\mathbf{Q} \subset \mathbf{B} \subset \mathbf{E}$, these fields being $\mathbf{Q}(\alpha)$, $\mathbf{Q}(\beta)$, $\mathbf{Q}(\gamma)$, and $\mathbf{Q}(\sqrt{D})$ for some $D \in \mathbf{Q}$ that is not a perfect square.
(In fact, it is possible to decide between these two alternatives, and to find D, without finding the roots of $f(X)$. See Section 4.8.)

Exercise 2.10.20. Let $f(X) = X^n - p \in \mathbf{Q}[X]$. By Proposition 4.1.7 below, $f(X)$ is irreducible. Let $\mathbf{E}$ be a splitting field of $f(X)$. Show that for every $n \geq 3$, the Galois group $G = \mathrm{Gal}(\mathbf{E}/\mathbf{Q})$ is not abelian.

Exercise 2.10.21. Let **E** be a splitting field of the separable irreducible polynomial $f(X) \in \mathbf{F}[X]$. Let α be any root of $f(X)$ in **E**. Let p be a prime dividing $(\mathbf{E}/\mathbf{F})$. Show there is a field $\mathbf{F} \subseteq \mathbf{B} \subseteq \mathbf{E}$ with $\mathbf{E} = \mathbf{B}(\alpha)$ and with $(\mathbf{E}/\mathbf{B}) = p$.

Exercise 2.10.22. Let **F** be an arbitrary field. Let m and n be positive integers and let $d = \gcd(m, n)$. Let $f(X) = X^m - 1,\ g(X) = X^n - 1 \in \mathbf{F}[X]$. Show that $\gcd(f(X), g(X)) = h(X)$ where $h(X) = X^d - 1 \in \mathbf{F}[X]$.

Exercise 2.10.23. Let $f(X) \in \mathbf{Q}[X]$. If $f(X) \in \mathbb{Z}[X]$, then certainly $f(n) \in \mathbb{Z}$ for every $n \in \mathbb{Z}$. The converse of this is false, as the example $f(X) = X(X+1)/2$ shows. Let $c_0(X) = 1$ and let $c_k(X) = X(X-1)\cdots(X-(k-1))/k!$ for $k > 0$. Show that any $f(X) \in \mathbf{Q}[X]$ with the property that $f(n) \in \mathbb{Z}$ for every $n \in \mathbb{Z}$ can be written a linear combination of $\{c_k(X)\}$ with integer coefficients. (This remains true if **Q** is replaced by any field of characteristic 0.) Note this implies that if $f(n)$ is a polynomial of degree at most k with $f(n)$ an integer divisible by $k!$ for every integer n, then $f(X) \in \mathbb{Z}[X]$.

3

Development and Applications of Galois Theory

3.1 Symmetric Functions and the Symmetric Group

We now apply our general theory to the case of symmetric functions. We let **D** be an arbitrary field and set $\mathbf{E} = \mathbf{D}(X_1, \ldots, X_d)$, the field of rational functions in the variables $X_1, \ldots, X_d$. Then the symmetric group S_d acts on **E** by permuting $\{X_1, \ldots, X_d\}$.

Definition 3.1.1. The subfield **F** of **E** fixed under the action of S_d is the field of *symmetric functions* in $X_1, \ldots, X_d$. ($\mathbf{F} = \{f(X_1, \ldots, X_d) \in \mathbf{E} \mid f(X_{\sigma(1)}, \ldots, X_{\sigma(d)}) = f(X_1, \ldots, X_d)$ for every $\sigma \in S_d\}$.) A polynomial $f(X_1, \ldots, X_d) \in \mathbf{F}$ is a *symmetric polynomial.* ◇

Since $\mathbf{F} = \text{Fix}(S_d)$, by definition **E** is a Galois extension of **F** with Galois group S_d. We may use this observation to easily show that any finite group is the Galois group of some field extension.

Lemma 3.1.2. *(1) Let G be a finite group. Then there is a Galois extension* $\mathbf{E}/\mathbf{B}$ *with* $\text{Gal}(\mathbf{E}/\mathbf{B}) \cong G$.

(2)There is a Galois extension $\mathbf{E}/\mathbf{B}$ *with* $\text{Gal}(\mathbf{E}/\mathbf{B}) \cong G$ *and with* **E** *(and hence* **B***) a subfield of the complex numbers* **C**.

Proof. (1) Any finite group G is isomorphic to a subgroup of some symmetric group S_d. Identify G with its image under this isomorphism. Let $\mathbf{E} = \mathbf{D}(X_1, \ldots, X_d)$, as above, and set $\mathbf{B} = \text{Fix}(G)$. Then $\mathbf{E}/\mathbf{B}$ is Galois with Galois group isomorphic to G.

(2) Let $\mathbf{D} = \mathbf{Q}$. Then we need only show there is subfield of **C** isomorphic to $\mathbf{Q}(X_1, \ldots, X_d)$. Let $t_1, \ldots, t_d$ be algebraically independent elements of **C**, i.e., elements that do not satisfy any polynomial equation.

S.H. Weintraub, *Galois Theory*, DOI 10.1007/978-0-387-87575-0_3,

Let $\varphi\colon \mathbf{Q}(X_1, \dots, X_d) \to \mathbf{Q}(t_1, \dots, t_d)$ be defined by $\varphi(X_i) = t_i$. If $r(X_1, \dots, X_d) = f(X_1, \dots, X_d)/g(X_1, \dots, X_d) \in \mathbf{Q}(X_1, \dots, X_d)$, then $\varphi(r(X_1, \dots, X_d)) = f(t_1, \dots, t_d)/g(t_1, \dots, t_d) = r(t_1, \dots, t_d)$. Then φ is well defined as the denominator $g(t_1, \dots, t_d)$ is never zero. It is obviously onto and it is 1–1 as the numerator $f(t_1, \dots, t_d)$ is never zero, so φ is an isomorphism. Then set $\mathbf{E} = \mathbf{Q}(t_1, \dots, t_d)$ and let $\mathbf{B}$ be the fixed field of G acting by permuting $t_1, \dots, t_d$.

It remains only to show that we can indeed find algebraically independent elements $t_1, \dots, t_d$ of $\mathbf{Q}$. Let $\mathbf{E}_0 = \mathbf{Q}$ and consider the extension field $\mathbf{F}_1 = \{z \in \mathbf{C} \mid z \text{ is algebraic over } \mathbf{E}_0\}$. $\mathbf{E}_0$ is countable; so there are only countably many polynomials in $\mathbf{E}_0[X_1]$, each of which has only finitely many roots. Hence $\mathbf{F}_1$ is countable. But $\mathbf{C}$ is uncountable. Thus we may choose any $t_1 \notin \mathbf{F}_1$ and let $\mathbf{E}_1 = \mathbf{E}_0(t_1)$. Then $X_1 \mapsto t_1$ induces an isomorphism $\mathbf{E}_0(X_1) \to \mathbf{E}_1$. But again $\mathbf{E}_0(X_1)$ has only countably many elements, so $\mathbf{E}_1$ is countable and we may iterate this process. □

We now return to the study of symmetric functions.

Definition 3.1.3. The *elementary symmetric polynomials* $s_1, \dots, s_d$ are

$$s_k = \sum_{1 \le i_1 < \dots < i_k \le d} X_{i_1} X_{i_2} \cdots X_{i_k}, \quad k = 1, \dots, d. \qquad \diamond$$

For example, if $d = 4$:

$$\begin{aligned}
s_1 &= X_1 + X_2 + X_3 + X_4,\\
s_2 &= X_1X_2 + X_1X_3 + X_1X_4 + X_2X_3 + X_2X_4 + X_3X_4,\\
s_3 &= X_1X_2X_3 + X_1X_2X_4 + X_1X_3X_4 + X_2X_3X_4,\\
s_4 &= X_1X_2X_3X_4.
\end{aligned}$$

Lemma 3.1.4. *The field of symmetric functions* $\mathbf{F} = \mathbf{D}(s_1, \dots, s_d)$.

Proof. Let $\mathbf{B} = \mathbf{D}(s_1, \dots, s_d)$. Clearly $\mathbf{F} \supseteq \mathbf{B}$, as each s_i is fixed by S_d.

Now $(\mathbf{E}/\mathbf{F}) = |S_d| = d!$, so $(\mathbf{E}/\mathbf{B}) \ge d!$. We will show that $\mathbf{F}$ is the splitting field of a separable polynomial $f(X) \in \mathbf{B}[X]$ of degree d. Then, by Lemma 2.8.18, $(\mathbf{E}/\mathbf{B}) = |\mathrm{Gal}(\mathbf{E}/\mathbf{B})| \le d!$. Hence $(\mathbf{E}/\mathbf{B}) = d!$ and $\mathbf{F} = \mathbf{B}$.

Now $f(X)$ is simply the polynomial

$$f(X) = (X - X_1)(X - X_2) \cdots (X - X_d),$$

which obviously has splitting field $\mathbf{E}$, as its roots are $X_1, \dots, X_d$. □

Remark 3.1.5. (1) Note in particular that we have constructed an explicit polynomial $f(X)$ of degree d whose Galois group is S_d. Explicit computation shows that

$$f(X) = X^d + \sum_{i=1}^{d} (-1)^i s_i X^{d-i}.$$

(2) Let us write $f(X) = X^d + a_{d-1}X^{d-1} + \cdots + a_0$. Then we see from the above expression that the coefficients of $f(X)$ are, up to sign, the elementary symmetric functions of its roots,

$$a_{d-i} = (-1)^i s_i, \quad i = 1, \ldots, d,$$

and, in fact, this is true for any polynomial $f(X)$. ◇

Lemma 3.1.6. *Let Y be the monomial*

$$Y = X_2^1 X_3^2 \cdots X_d^{d-1} (= X_1^0 X_2^1 \cdots X_d^{d-1}).$$

Then $\mathbf{E} = \mathbf{F}(Y)$.

Proof. Obviously, the only element of S_d that fixes Y is the identity, so Y has $d!$ conjugates. Then, by Corollary 2.7.13, $(\mathbf{F}(Y)/\mathbf{F}) = d! = (\mathbf{E}/\mathbf{F})$, so $\mathbf{E} = \mathbf{F}(Y)$. □

Corollary 3.1.7. *Any element $f(X_1, \ldots, X_d)$ of $\mathbf{E}$ can be written uniquely as*

$$f(X_1, \ldots, X_d) = \sum_{i=0}^{d!-1} g_i(s_i, \ldots, s_d) Y^i$$

where $g_i(s_1, \ldots, s_d)$ is a rational function in $s_1, \ldots, s_d$.

Proof. By Lemma 3.1.6 and Proposition 2.4.6 (1), $\{1, Y, \ldots, Y^{d!-1}\}$ is a vector space basis for $\mathbf{E}$ over $\mathbf{F}$, and any element of $\mathbf{F}$ is a rational function of $s_1, \ldots, s_d$, by Lemma 3.1.4. □

Remark 3.1.8. In the language of Section 3.5, Lemma 3.1.6 shows that $\mathbf{E}$ is a simple extension of $\mathbf{F}$ and that Y is a primitive element of $\mathbf{E}$. ◇

Lemma 3.1.9. *Let Y be the monomial $Y = X_2^1 \cdots X_d^{d-1}$ (as in Lemma 3.1.6) and for $\sigma \in S_d$, let Y_σ be the monomial*

$$Y_\sigma = X_{\sigma(2)}^1 \cdots X_{\sigma(d)}^{d-1}.$$

Then $\{Y_\sigma \mid \sigma \in S_d\}$ is a (vector space) basis for $\mathbf{E}$ over $\mathbf{F}$.

Proof. We prove the claim by induction on d.

For $d = 1$ this is simply the claim that $Y = 1$ is a basis for $\mathbf{E}$ over $\mathbf{E}$, which is trivial.

Now assume the result is true for $d - 1$, and suppose that

$$\sum_{\sigma \in S_d} c_\sigma(X_1, \ldots, X_d) X_{\sigma(2)} \cdots X_{\sigma(d)}^{d-1} = 0,$$

with $c_\sigma(X_1, \ldots, X_d) \in \mathbf{F}$. We wish to show that $c_\sigma(X_1, \ldots, X_d) = 0$ for each σ. Assume not. By "clearing denominators" we may assume that $\{c_\sigma(X_1, \ldots, X_d)\}$ are polynomials in $\mathbf{F}$ with no common polynomial factor.

For convenience, write the above sum as $\sum$. Let $H_i = \{\sigma \in S_d \mid \sigma(i) = 1\}$, $i = 1, \ldots, d$. Then $\sum = \sum_1 + \cdots + \sum_d$, where

$$\sum_i = \sum_{\sigma \in H_i} c_\sigma(X_1, \ldots, X_d) X_{\sigma(2)} \cdots X_{\sigma(d)}^{d-1}.$$

We may further write $\sum_i = \sum_i' + \sum_i''$, $i = 1, \ldots, d$, where $\sum_i'$ is taken over those $\sigma \in H_i$ with $c_\sigma(X_1, \ldots, X_d)$ not divisible by X_i, and $\sum_i''$ is taken over those $\sigma \in H_i$ with $c_\sigma(X_1, \ldots, X_d)$ divisible by X_i. Call these two subsets H_i' and H_i'', respectively.

Let us consider the equation $\sum = 0$. Note that the exponent of X_1 is at least one for each term in $\sum_i$, $i \neq 1$ (as if $\sigma(i) \neq 1$, then $\sigma(j) = 1$ for some j with $2 \leq j \leq d$), and also for each term in $\sum_1''$. Thus, considering the terms in $\sum$ that are not divisible by X_1, we thus see that we must have $\sum_1' = 0$, i.e., that

$$\sum_{\sigma \in H_1'} c_\sigma(X_1, \ldots, X_d) X_{\sigma(2)} \cdots X_{\sigma(d)}^{d-1} = 0.$$

Observe that H_1 is isomorphic to S_{d-1}, permuting the subset $\{2, \ldots, d\}$ of $\{1, \ldots, d\}$ (and each H_i is a left coset of H_1).

Let us set $X_1 = 0$ in $\sum_1'$, thereby obtaining

$$\begin{aligned} 0 &= \sum_{\sigma \in H_1'} c_\sigma(0, X_2, \ldots, X_d) X_{\sigma(2)} \ldots X_{\sigma(d)}^{d-1} \\ &= \Big(\sum_{\sigma \in H_1'} c_\sigma(0, X_2, \ldots, X_d) X_{\sigma(3)} \cdots X_{\sigma(d)}^{d-2} \Big) X_2 \cdots X_d. \end{aligned}$$

Now we apply the inductive hypothesis. $\{X_{\sigma(3)} \cdots X_{\sigma(d)}^{d-2}\}$ is a basis for $\mathbf{D}(X_2, \ldots, X_d)$ over the subfield of symmetric functions in $X_2, \ldots, X_d$, so in particular they are linearly independent. Hence we have that each coefficient

$c_\sigma(0, X_2, \ldots, X_d) = 0$. In other words, each $c_\sigma(X_1, \ldots, X_d)$ is a polynomial that vanishes when we set $X_1 = 0$, so it must have X_1 as a factor, contradicting the definition of H_1'. Hence we see that $H_1' = \emptyset$, and so $H_1'' = H_1$. In other words, $c_\sigma(X_1, \ldots, X_d)$ is divisible by X_1 for every $\sigma \in H_1$. But $c_\sigma(X_1, \ldots, X_d) \in \mathbf{F}$ is a symmetric polynomial in $X_1, \ldots, X_d$, so this implies that $c_\sigma(X_1, \ldots, X_d)$ is also divisible by $X_2, \ldots, X_d$, and hence by the product $X_1 \cdots X_d$, for each $\sigma \in H_1$.

Similarly, we see, for each $i = 1, \ldots, d$, that $H_i' = \emptyset, H_i'' = H_i$, so $c_\sigma(X_1, \ldots, X_d)$ is divisible by X_i, and thus by $X_1 \cdots X_d$, for each $\sigma \in H_i$. Thus $\{c_\sigma(X_1, \ldots, X_d) \mid \sigma \in S_d\}$ have the common polynomial factor $X_1 \cdots X_d$, contradicting our choice of $\{c_\sigma(X_1, \ldots, X_d)\}$. Thus, $\{X_{\sigma(2)} \cdots X_{\sigma(d)}^{d-1}\}$ is linearly independent, and thus is a vector space basis for $\mathbf{E}$ over $\mathbf{F}$ (as it has $d! = (\mathbf{E}/\mathbf{F})$ elements), completing the inductive step. □

Corollary 3.1.10. *Any element $f(X_1, \ldots, X_d)$ of $\mathbf{E}$ can be written uniquely as*

$$f(X_1, \ldots, X_d) = \sum_{\sigma \in S_d} h_\sigma(s_1, \ldots, s_d) X_{\sigma(2)} \cdots X_{\sigma(d)}^{d-1}$$

where $h_\sigma(s_1, \ldots, s_d)$ is a rational function of $s_1, \ldots, s_d$.

Proof. Lemma 3.1.4 and Lemma 3.1.9. □

Remark 3.1.11. In the language of Section 3.6, Lemma 3.1.9 shows that $\{X_{\sigma(2)} \cdots X_{\sigma(d)}^{d-1} \mid \sigma \in S_d\}$ is a normal basis of $\mathbf{E}$ over $\mathbf{F}$. ◇

Our discussion here is not yet complete. Corollaries 3.1.7 and 3.1.10 show how to write rational functions uniquely in terms of symmetric functions. What we have not yet shown is that symmetric functions can be written uniquely in terms of the elementary symmetric functions. We do that now.

Lemma 3.1.12. *Let $f(X_1, \ldots, X_d) \in \mathbf{F}$ be a symmetric function. Then $f(X_1, \ldots, X_d)$ may be written uniquely as a rational function $e(s_1, \ldots, s_d)$ of the elementary symmetric functions $s_1, \ldots, s_d$.*

Proof. We prove this by induction on d. In case $d = 1$, $s_1 = X_1$ and $e(s_1) = f(X_1)$ is the unique expression. Now suppose the result is true for $d - 1$.

We first need to make an observation about the elementary symmetric functions. We really should have written s_k as $s_{k,d}$, as the expressions in Definition 3.1.4 depended on d, but for simplicity we did not. We do that now. We then observe that if we let $\bar{s}_{k,d}$ be the expression obtained from $s_{k,d}$ by setting $X_d = 0$, then

$$\bar{s}_{k,d} = s_{k,d-1} \text{ for } k \le d-1 \text{ and } \bar{s}_{d,d} = 0.$$

By equating different expressions for $f(X_1, \ldots, X_d)$ and "clearing denominators", it is easy to see that to prove the lemma it suffices to show that the only polynomial $e(s_{1,d}, \ldots, s_{d,d})$ that is identically zero is the zero polynomial. Thus suppose $e(s_{1,d}, \ldots, s_{d,d}) = 0$ but $e(s_{1,d}, \ldots, s_{d,d})$ is not the zero polynomial. Since $\mathbf{F}$ is a field and $s_{d,d}$ is a nonzero element of $\mathbf{F}$, $e(s_{1,d}, \ldots, s_{d,d}) = 0$ if and only if $(s_{d,d})^{-k} e(s_{1,d}, \ldots, s_{d,d}) = 0$ for any k. Hence, factoring out by a power of $s_{d,d}$ if necessary, we may assume, expanding $e(s_{1,d}, \ldots, s_{d,d})$ in powers of $s_{d,d}$, that the constant term is nonzero. In other words, we have that

$$0 = e(s_{1,d}, \ldots, s_{d,d}) = \sum_{i=0}^{n} e_i(s_{1,d}, \ldots, s_{d-1,d}) s_{d,d}^i$$

with $e_i(s_{1,d}, \ldots, s_{d-1,d})$ polynomials in $s_{1,d}, \ldots, s_{d-1,d}$, and furthermore $e_0(s_{1,d}, \ldots, s_{d-1,d})$ not the zero polynomial. Now set $X_d = 0$. Then the above expressions become

$$\begin{aligned} 0 &= e(\bar{s}_{1,d}, \ldots, \bar{s}_{d-1,d}, 0) \\ &= e_0(\bar{s}_{1,d}, \ldots, \bar{s}_{d-1,d}) \\ &= e_0(s_{1,d-1}, \ldots, s_{d-1,d-1}), \end{aligned}$$

contradicting the $d-1$ case. □

Remark 3.1.13. We have just shown that the only polynomial in $s_1, \ldots, s_d$ that vanishes is the zero polynomial. This is usually phrased as saying that $s_1, \ldots, s_d$ are "algebraically independent" over $\mathbf{D}$. ◇

We can also strengthen Lemma 3.1.12 as follows.

Lemma 3.1.14. *Let $f(X_1, \ldots, X_d) \in \mathbf{F}$ be a symmetric polynomial. Then $f(X_1, \ldots, X_d)$ may be written uniquely as a polynomial $e(s_1, \ldots, s_d)$ in the elementary symmetric functions $s_1, \ldots, s_d$.*

Proof. Uniqueness follows from Lemma 3.1.12 so we must show existence. We do so by induction on d. We include in our inductive hypothesis that every term in $e(s_1, \ldots, s_d)$, when regarded as a polynomial in $X_1, \ldots, X_d$, has degree at most the degree of $f(X_1, \ldots, X_d)$.

For $d = 1$, $s_1 = X_1$, and the result is trivial.

Assume the result is true for $d-1$. We proceed by induction on the degree k of the symmetric polynomial. If $k = 0$ the result is trivial. Assume it is true for all symmetric polynomials in $X_1, \ldots, X_d$ of degree less than k, and let $f(X_1, \ldots, X_d)$ have degree k.

Now $f(X_1, \ldots, X_{d-1}, 0)$ is a symmetric polynomial in $X_1, \ldots, X_{d-1}$ so by induction $f(X_1, \ldots, X_{d-1}) = g(s_{1,d-1}, \ldots, s_{d-1,d-1})$ for some polynomial g (where we use the notation of the proof of Lemma 3.1.12).

Then $f(X_1, \ldots, X_d) - g(s_1, \ldots, s_{d-1}) = f_1(X_1, \ldots, X_d)$ is a symmetric polynomial in $X_1, \ldots, X_d$ with $f_1(X_1, \ldots, X_{d-1}, 0) = 0$. Hence $f_1(X_1, \ldots, X_d)$ is divisible by X_d in $\mathbf{D}[X_1, \ldots, X_d]$, and hence, since this polynomial is symmetric, it is divisible by the product $X_1 \ldots X_d$ as well. Hence $f_1(X_1, \ldots, X_d) = (X_1 \cdots X_d) f_2(X_1, \ldots, X_d)$ with $f_2(X_1, \ldots, X_d)$ a polynomial of lower degree. By induction $f_2(X_1, \ldots, X_d) = h(s_1, \ldots, s_d)$ for some symmetric polynomial $h(s_1, \ldots, s_d)$ satisfying our inductive hypothesis, and of course $X_1 \cdots X_d = s_d$ is a symmetric polynomial satisfying our hypothesis as well.

Then

$$\begin{aligned} f(X_1, \ldots, X_d) &= g(s_1, \ldots, s_{d-1}) + s_d h(s_1, \ldots, s_d) \\ &= e(s_1, \ldots, s_d) \end{aligned}$$

is a polynomial in $s_1, \ldots, s_d$, as required. □

We have been considering polynomials with coefficients in a field. But, in fact, we also have a result valid for polynomials with integer coefficients.

Lemma 3.1.15. *Let $f(X_1, \ldots, X_d)$ be a symmetric polynomial with integer coefficients. Then $f(X_1, \ldots, X_d)$ may be written uniquely as $e(s_1, \ldots, s_d)$ where $e(s_1, \ldots, s_d)$ is a polynomial in the elementary symmetric functions $s_1, \ldots, s_d$ with integer coefficients.*

Proof. The proof of Lemma 3.1.14 goes through unchanged to prove this result. □

In addition to the elementary symmetric functions, there is another set of symmetric functions that readily (perhaps even more readily) appears, namely sums of powers. The relationship between these is classical.

Theorem 3.1.16 (Newton's Identities). *Let $\{s_i\}_{i=1,\ldots,d}$ be the elementary symmetric polynomials in $X_1, \ldots, X_d$. Let $t_i = X_1^i + \cdots + X_d^i$ for $i > 0$. Then:*

$$\begin{aligned} t_1 - s_1 &= 0 \\ t_2 - s_1t_1 + 2s_2 &= 0 \\ t_3 - s_1t_2 + s_2t_1 - 3s_3 &= 0 \\ &\vdots \\ t_d - s_1t_{d-1} + \cdots + (-1)^{d-1}s_{d-1}t_1 + (-1)^d ds_d &= 0 \end{aligned}$$

and

$$\begin{aligned} t_{d+1} - s_1t_d + \cdots + (-1)^d s_dt_1 &= 0 \\ t_{d+2} - s_1t_{d+1} + \cdots + (-1)^d s_dt_2 &= 0 \\ &\vdots \end{aligned}$$

Proof. Throughout this proof, we shall abbreviate $\sum_{j=1}^{d}$ by $\sum$.

Let us set $f(X) = (X - X_1)(X - X_2)\cdots(X - X_d) = X^d + a_{d-1}X^{d-1} + \cdots + a_0$. We know from Remark 3.1.5 that $a_{d-i} = (-1)^i s_i$, $i = 1, \ldots, d$. From the division algorithm for polynomials we know that for any $Y \in \mathbf{E}$, $f(X) = (X - Y)q(X) + f(Y)$ for a unique polynomial $q(X) \in \mathbf{E}$ of degree $d-1$. Let $q(X) = b_{d-1}X^{d-1} + b^{d-2}X^{d-2} + \cdots + b_0$. Then long division shows that

$$\begin{aligned} b_{d-1} &= 1 \\ b_{d-2} &= Y + a_{d-1} \\ b_{d-3} &= Y^2 + a_{d-1}Y + a_{d-2} \\ &\vdots \end{aligned}$$

(Of course, each b_k is a function of Y, but we write b_k instead of $b_k(Y)$ for simplicity. However, we need to keep this dependence in mind below.)

We now successively set $Y = X_1, \ldots, X_d$ and form the sum

$$(*) \quad \sum q(X_j) = \left(\sum b_{d-1}\right) X^{d-1} + \left(\sum b_{d-2}\right) X^{d-2} + \cdots .$$

On the one hand, we see from $f(X) = (X - X_1)(X - X_2)\cdots(X - X_d)$ that

$$\begin{aligned} \sum q(X_j) &= f'(X) \\ &= dX^{d-1} + (d-1)a_{d-1}X^{d-2} + (d-2)a_{d-2}X^{d-3} + \cdots . \end{aligned}$$

On the other hand, adding the expressions on the right-hand side of (*), we obtain

$$\sum q(X_j) = dX^{d-1} + \left(\sum (X_j + a_{d-1})\right) X^{d-2} + \left(\sum (X_j^2 + a_{d-1}X_j + a_{d-2})\right) X^{d-3} + \cdots .$$

Comparing coefficients, we see

$$\begin{aligned} t_1 + da_{d-1} &= (d-1)a_{d-1} \\ t_2 + a_{d-1}t_1 + da_{d-2} &= (d-2)a_{d-2} \\ t_3 + a_{d-1}t_2 + a_{d-2}t_1 + da_{d-3} &= (d-3)a_{d-3} \end{aligned}$$

which yields

$$\begin{aligned} t_1 - s_1 &= 0 \\ t_2 - s_1 t_1 + 2s_2 &= 0 \\ t_3 - s_1 t_2 + s_2 t_1 - 3s_3 &= 0 \\ &\vdots \end{aligned}$$

which are the first $d-1$ Newton's Identities. The remaining Newton's Identities comes from considering coefficients in

$$0 = \sum X^m f(X_j)$$

for $m = 0, 1, 2, \ldots$. □

Corollary 3.1.17. *Let $\{s_i\}$, $\{t_i\}$, and d be as in Theorem 3.1.16.*

(1) For every $k > 0$, t_k is a polynomial in $s_1, \ldots, s_j$ with integer coefficients, where $j = \min(k, d)$.

(2) Let $f(X_1, \ldots, X_d) \in \mathbf{F}$ be a symmetric polynomial of degree at most k, and let $j = \min(k, d)$. If $\mathrm{char}(\mathbf{D}) = 0$ or $\mathrm{char}(\mathbf{D}) > j$, then $f(X)$ may be written uniquely as a polynomial in $t_1, \ldots, t_k$.

(3) In particular, if $\mathrm{char}(\mathbf{D}) = 0$ or $\mathrm{char}(\mathbf{D}) > d$, then every symmetric polynomial $f(X_1, \ldots, X_d) \in \mathbf{F}$ may be written uniquely as a polynomial in $t_1, \ldots, t_d$.

Proof. Newton's identities always give a "triangular" system for $\{t_i\}_{i=1,\ldots,d}$ in terms of $\{s_i\}_{i=1,\ldots,d}$. They give a "triangular" system for $\{s_i\}_{i=1,\ldots,d}$ in terms of $\{t_i\}_{i=1,\ldots,d}$ as long as the coefficient of s_i is nonzero for $i = 1, \ldots, d$. □

3.2 Separable Extensions

In this section we develop criteria for polynomials and extensions to be separable. Recall that an irreducible polynomial $f(X) \in \mathbf{F}[X]$ is separable if $f(X)$ splits into a product of distinct linear factors over a splitting field $\mathbf{E}$. Recall that a root α of a polynomial $f(X)$ is called *simple* if $X - \alpha$ divides $f(X)$ but $(X - \alpha)^2$ does not, (otherwise α is called *multiple*) so an irreducible polynomial $f(X)$ is separable if and only if all of its roots in a splitting field $\mathbf{E}$ are simple.

For a polynomial $f(X) = \sum_{i=0}^{n} c_i X^i$, we let $f'(X)$ be its derivative $f'(X) = \sum_{i=1}^{n} i c_i X^{i-1}$.

Lemma 3.2.1. *Let $\mathbf{E}$ be an extension of $\mathbf{F}$, let α be an algebraic element of $\mathbf{E}$, and let $f(X) \in \mathbf{F}[X]$ a nonzero polynomial with $f(\alpha) = 0$. Then α is a simple root of $f(X)$ if and only if $f'(\alpha) \neq 0$.*

Proof. Write $f(X) = (X - \alpha)g(X)$, a factorization in $\mathbf{E}[X]$. Then, by the usual rules for derivatives, $f'(X) = (X - \alpha)g'(X) + g(X)$ so $f'(\alpha) = g(\alpha)$.

By the division algorithm, we may write $g(X) = (X - \alpha)h(X) + g(\alpha)$, so here $g(X) = (X - \alpha)h(X) + f'(\alpha)$ and then $f(X) = (X - \alpha)g(X) = (X - \alpha)^2 h(X) + (X - \alpha) f'(\alpha)$. Thus $(X - \alpha)^2$ divides $f(X)$ if and only if $f'(\alpha) = 0$. □

Proposition 3.2.2. *Let $f(X) \in \mathbf{F}[X]$ be irreducible. If $f'(X) \neq 0$, then $f(X)$ is separable. In particular:*

(1) If char($\mathbf{F}$) $= 0$, *then $f(X)$ is separable.*

(2) If char($\mathbf{F}$) $= p$, *then $f(X)$ is separable or $f(X)$ is of the form*

$$f(X) = \sum_{i=0}^{k} c_i X^{ip}.$$

Proof. Let α be any root of $f(X)$ in an extension field $\mathbf{E}$ of $\mathbf{F}$. Since $f(X)$ is irreducible, $f(X)$ is the unique monic polynomial in $\mathbf{F}[X]$ of lowest degree with $f(\alpha) = 0$ (and indeed, $f(X) = m_\alpha(X)$). Thus, if $f'(X) \neq 0$, then $f'(X)$ is a polynomial of lower degree, so $f'(\alpha) \neq 0$ and α is a simple root of $f(X)$ by Lemma 3.2.1. Hence every root of $f(X)$ in a splitting field is simple and $f(X)$ is separable.

Now if $f(X)$ is a nonconstant polynomial and char $\mathbf{F} = 0$, then $f'(X) \neq 0$. If char $\mathbf{F} = p$, then $f'(X) \neq 0$ unless the exponent of every power of X appearing in $f(X)$ is a multiple of p, proving the proposition. □

Recall that an algebraic extension $\mathbf{E}$ of $\mathbf{F}$ is separable if $m_\alpha(X) \in \mathbf{E}[X]$ is a separable polynomial for every $\alpha \in \mathbf{E}$. Thus we see:

Corollary 3.2.3. *If* char $\mathbf{F} = 0$, *every algebraic extension of* $\mathbf{F}$ *is separable.* □

Example 3.2.4. Here is an example of an inseparable polynomial and, correspondingly, an inseparable extension. Let $\mathbf{F} = \mathbf{F}_p(t)$, the field of rational functions in the variable t with coefficients in $\mathbf{F}_p$. Let $\mathbf{E}$ be the field obtained by adjoining a root of the polynomial $X^p - t$, $\mathbf{E} = \mathbf{F}(s)$ with $s^p = t$. Then, in $\mathbf{E}[X]$, $X^p - t = X^p - s^p = (X - s)^p$, so s is not a simple root of $X^p - t$. Note that $\mathbf{E}$ is in fact a splitting field for this polynomial. To complete this example we need to show that $X^p - t$ is an irreducible polynomial in $\mathbf{F}[X]$, and this follows immediately from Lemma 3.2.8 below. ◇

Definition 3.2.5. A field $\mathbf{F}$ is a *perfect* field if every algebraic extension of $\mathbf{F}$ is separable. ◇

We have a simple criterion for a field to be perfect. In order to state it, recall that for a field $\mathbf{F}$ of characteristic p, we have the Frobenius map $\Phi : \mathbf{F} \to \mathbf{F}$ defined by $\Phi(a) = a^p$. We let $\mathbf{F}^p = \mathrm{Im}(\Phi)$ and note that, since Φ is an endomorphism, $\mathbf{F}^p$ is a subfield of $\mathbf{F}$. (This is easy to verify directly: $1 = 1^p$, if $c = a^p$ and $d = b^p$, then $c + d = (a + b)^p$ and $cd = (ab)^p$, and if $c \neq 0$, $c^{-1} = (a^{-1})^p$.)

Theorem 3.2.6. $\mathbf{F}$ *is a perfect field if and only if:*
(1) char($\mathbf{F}$) $= 0$, *or*
(2) char($\mathbf{F}$) $= p$ *and* $\mathbf{F} = \mathbf{F}^p$.

Proof. By Corollary 3.2.3, if char($\mathbf{F}$) $= 0$, then $\mathbf{F}$ is perfect. Let char($\mathbf{F}$) $= p$.

First assume $\mathbf{F} = \mathbf{F}^p$. Let $\alpha \in \mathbf{E}$ be inseparable and consider $m_\alpha(X)$. By Proposition 3.2.2 (2), $m_\alpha(X)$ must be of the form $\sum_{i=0}^{k} c_i X^{ip}$.

Since $\mathbf{F} = \mathbf{F}^p$, for each c_i there is a $d_i \in \mathbf{F}$ with $d_i^p = c_i$. Then

$$m_\alpha(X) = \sum_{i=0}^{k} c_i X^{ip} = \sum_{i=0}^{k} d_i^p X^{ip} = \Big(\sum_{i=0}^{k} d_i X^i\Big)^p$$

contradicting the irreducibility of $m_\alpha(X)$. Hence α is separable and $\mathbf{F}$ is perfect.

Next assume $\mathbf{F}$ is perfect. (The argument here parallels Example 3.2.4.) Let $a \in \mathbf{F}$, $a \notin \mathbf{F}^p$ and consider $f(X) = X^p - a \in \mathbf{F}[X]$. By Lemma 3.2.8 below, $f(X)$ is irreducible. Let $\mathbf{E} = \mathbf{F}(\alpha)$ where α is a root of $f(X)$. Then, in $\mathbf{E}$, $\alpha^p = a$ so $X^p - a = X^p - \alpha^p = (X - \alpha)^p$ and so $\alpha \in \mathbf{E}$ is an inseparable element, contradicting the fact that $\mathbf{F}$ is perfect. Hence $\mathbf{F} = \mathbf{F}^p$. □

Corollary 3.2.7. *Every finite field is perfect.*

Proof. Immediate from Corollary 2.1.11 and Theorem 3.2.6. □

Lemma 3.2.8. *Let* char(**F**) $= p$, *and let* $a \in \mathbf{F}$, $a \notin \mathbf{F}^p$. *Then* $f(X) = X^{p^t} - a \in \mathbf{F}[X]$ *is irreducible for every* $t \geq 1$.

Proof. Factor $f(X)$ into monic irreducible polynomials in $\mathbf{F}[X]$, $f(X) = g_1(X) \cdots g_k(X)$. Let $\mathbf{E} = \mathbf{F}(\alpha)$ with $g_1(\alpha) = 0$. Then $g_1(X) = m_\alpha(X) \in \mathbf{F}[X]$ as $g_1(X)$ is irreducible.

Now $0 = g_1(\alpha)$ implies $0 = f(\alpha) = \alpha^{p^t} - a$ so $a = \alpha^{p^t}$ and

$$f(X) = X^{p^t} - a = X^{p^t} - \alpha^{p^t} = (X - \alpha)^{p^t} \in \mathbf{E}[X].$$

For any i, $g_i(X)$ divides $f(X)$ in $\mathbf{F}[X]$, and hence $g_i(X)$ divides $f(X)$ in $\mathbf{E}[X]$ by Lemma 2.2.7, i.e., $g_i(X)$ divides $(X - \alpha)^{p^t}$ in $\mathbf{E}[X]$, so $g_i(X)$ is a power of $X - \alpha$ and hence $g_i(\alpha) = 0$. Since $g_i(\alpha) = 0$, $g_i(X)$ is divisible by $m_\alpha(X) = g_1(X)$, for every i. Since the choice of $g_1(X)$ was arbitrary, we see that $g_1(X) = g_2(X) = \cdots = g_k(X)$ and further that $f(X) = g(X)^k$ where $g(X) = (X - \alpha)^j$ for some j.

Thus $X^{p^t} - a = ((X - \alpha)^j)^k \in \mathbf{E}[X]$ so $j = p^s$ and $k = p^{t-s}$ for some $s \leq t$. Suppose $s \leq t - 1$. Then $g(X) = (X - \alpha)^{p^s} \in \mathbf{F}[X]$, so $g(X)^{p^{t-s-1}} \in \mathbf{F}[X]$ as well. But $g(X)^{p^{t-s-1}} = (X - \alpha)^{p^{t-1}} = X^{p^{t-1}} - b$ with $b = \alpha^{p^{t-1}}$ and hence $b \in \mathbf{F}$. But then $b^p = (\alpha^{p^{t-1}})^p = \alpha^{p^t} = a$, so $a \in \mathbf{F}^p$, a contradiction. Hence $s = t$, $k = 1$, and $f(X) = g(X)$ is irreducible in $\mathbf{F}[X]$. □

3.3 Finite Fields

Let us use the Fundamental Theorem of Galois Theory to completely determine the structure of finite fields. Recall we defined the Frobenius map Φ in Definition 2.1.10: Let **F** be a field of characteristic p. Then $\Phi : \mathbf{F} \to \mathbf{F}$ is the map defined by $\Phi(a) = a^p$.

Theorem 3.3.1. *Let p be a prime and r a positive integer.*

(1) There is a field $\mathbf{F}_{p^r}$ of p^r elements, unique up to isomorphism.

(2) Let s divide r. Then $\mathbf{F}_{p^r}$ contains a unique subfield of p^s elements. If s does not divide r, then $\mathbf{F}_{p^r}$ does not contain such a field.

(3) If $\mathbf{E} = \mathbf{F}_{p^r}$ and $\mathbf{F} = \mathbf{F}_{p^s}$ for s dividing r, then Gal(**E**/**F**) *is cyclic of order r/s, generated by $\Psi = \Phi^s$, Φ the Frobenius map. (Then $\Psi(\alpha) = \alpha^{p^s}$.) In particular, if $\mathbf{E} = \mathbf{F}_{p^r}$ and $\mathbf{F} = \mathbf{F}_p$, then* Gal(**E**/**F**) *is cyclic of order r, generated by the Frobenius map Φ.*

Proof. (1) Set $q = p^r$. Let $f(X) = X^q - X \in \mathbf{F}_p[X]$, and let $\mathbf{F}$ be the splitting field of this polynomial. We know that a splitting field exists and is unique up to isomorphism (Lemma 2.5.2 and Corollary 2.6.4). Let $g(X) = X^{q-1} - 1$, so $f(X) = Xg(X)$. Now $f'(X) = -1$ so, by Lemma 3.2.1, the roots of $f(X)$ are all simple. Thus $f(X)$ has $\deg f(X) = q$ distinct roots in $\mathbf{F}$, so in particular $\mathbf{F}$ must have at least q elements. Now let $\mathbf{D} \subseteq \mathbf{F}$ be the subset of $\mathbf{F}$ consisting of the roots of $f(X)$ and note that $|\mathbf{D}| = \deg f(X) = q$. Now we have that $\mathbf{D} = \{a \in \mathbf{F} \mid a^q = a\}$ and we observe that $\mathbf{D}$ is a field: $1^q = 1$, if $a^q = a$ and $b^q = b$, then $(ab)^q = ab$ and $(a + b)^q = a^q + b^q = a + b$, and for $a \neq 0$, $(a^{-1})^q = a^{-1}$. Thus $f(X)$ splits in $\mathbf{D}$ and since $\mathbf{F}$ is a splitting field, $\mathbf{F} \subseteq \mathbf{D}$, and so $\mathbf{F} = \mathbf{D}$. Hence $\mathbf{F} = \mathbf{D} = \mathbf{F}_q$ is a field of q elements.

On the other hand, if $\mathbf{B}$ is *any* field of q elements, then the multiplicative group of $\mathbf{B}$ has order $q - 1$, so every nonzero element of $\mathbf{B}$ is a root of $g(X) = X^{q-1} - 1$; hence every element of $\mathbf{B}$ is a root of $f(X)$, so $\mathbf{B}$ is a splitting field of $f(X)$ (again as $f(X)$ cannot split in any proper subfield of $\mathbf{D}$) and hence $\mathbf{B}$ is isomorphic to $\mathbf{F}$.

We have observed that the roots of $f(X)$ are all simple, and hence $f(X)$ is separable. (Alternatively, this follows from from Proposition 3.2.2 (2).) Now $\mathbf{F}_q$ is the splitting field of the separable polynomial $f(X) \in \mathbf{F}_p[X]$, so $\mathbf{F}_q$ is a Galois extension of $\mathbf{F}_p$ (Theorem 2.7.14) and $|\operatorname{Gal}(\mathbf{F}_q/\mathbf{F}_p)| = (\mathbf{F}_q/\mathbf{F}_p) = \dim_{\mathbf{F}_p} \mathbf{F}_q = r$. Now $\Phi : \mathbf{F}_q \to \mathbf{F}_q$ and $\Phi \mid \mathbf{F}_p = \mathrm{id}$ as $a^p = a$ for every $a \in \mathbf{F}_p$ (by Fermat's Little Theorem, see Corollary 2.1.12). Hence $\Phi \in \operatorname{Gal}(\mathbf{F}_q/\mathbf{F}_p)$. Also, $\Phi^r(\alpha) = \alpha^{p^r} = \alpha$ for every $\alpha \in \mathbf{F}$, as every $\alpha \in \mathbf{F}_q$ is a root of $f(X)$, i.e., $\Phi^r = \mathrm{id} : \mathbf{F}_q \to \mathbf{F}_q$. On the other hand, $\Phi^s \neq \mathrm{id}$ for $s < r$ as $\Phi^s(\alpha) = \alpha^{p^s} = \alpha$ if and only if α is a root of the polynomial $X^{p^s} - X$, and this polynomial cannot have p^r roots by Lemma 2.2.1. Hence $\{1, \Phi, \ldots, \Phi^{r-1}\}$ is a subgroup of $\operatorname{Gal}(\mathbf{F}_q/\mathbf{F}_p)$. Since this subgroup has $r = |\operatorname{Gal}(\mathbf{F}_q/\mathbf{F}_p)|$ elements, it must be the entire Galois group.

Now we may prove (2) and (3) by a direct application of the Fundamental Theorem of Galois Theory. The subfields of $\mathbf{F}_q$ are in $1 - 1$ correspondence with the subgroups of the cyclic group $\{1, \Phi, \ldots, \Phi^{r-1}\}$ and each subgroup is a cyclic group, of order r/s, generated by $\Psi = \Phi^s$ for some divisor s of r. Thus these subfields are the fixed fields of the cyclic subgroups generated by Ψ, i.e., the fixed fields of Ψ. But $\Psi(\alpha) = \alpha$ is the equation $\alpha^{p^s} = \alpha$, whose solutions are the roots of $X^{p^s} - X$, and these are the fields $\mathbf{F}_{p^s}$ as in (1). □

Remark 3.3.2. Theorem 3.3.1 (2) has an alternate, more elementary proof. Consider the multiplicative group G of $\mathbf{F}_{p^r}$, a group of order $p^r - 1$. Suppose $\mathbf{F}_{p^r}$ contains a subfield isomorphic to $\mathbf{F}_{p^s}$, and let H be the multiplicative group of that field, of order $p^s - 1$. Now, by elementary number theory, $p^s - 1$ divides $p^r - 1$ if and only if s divides r. Hence if s does not divide r, $|H|$

does not divide $|G|$, so no subgroup H, and hence no such subfield $\mathbf{F}_{p^s}$ can exist. On the other hand, we also know, by Corollary 2.2.2, that G is cyclic, so if s divides r, and hence $p^s - 1$ divides $p^r - 1$, there is a unique subgroup H of order $p^s - 1$, and then $H \cup \{0\}$ is the unique subfield of $\mathbf{F}_{p^r}$ with p^s elements. ◇

Corollary 3.3.3. *Let* $\mathbf{F} = \mathbf{F}_{p^s}$ *and let* $f(X) \in \mathbf{F}[X]$ *be an irreducible polynomial of degree* n. *Let* $r = ns$. *Then* $\mathbf{E} = \mathbf{F}_{p^r}$ *is a splitting field for* $f(X)$. *Furthermore,* $\mathrm{Gal}(\mathbf{E}/\mathbf{F}) = \mathbb{Z}/n\mathbb{Z}$ *generated by* $\Psi = \Phi^s$.

Proof. Let $\mathbf{E}_0 = \mathbf{F}(\alpha)$ be the field obtained by adjoining a single root α of $f(X)$ to $\mathbf{F}$. Then, since $\deg f(X) = n$, $(\mathbf{E}_0/\mathbf{F}) = n$, and $\mathbf{E}_0 = \mathbf{F}_{p^r}$.

Let $\mathbf{E} \supseteq \mathbf{E}_0$ be a splitting field for $f(X)$. We claim that $\mathbf{E} = \mathbf{E}_0$.

On the one hand, this claim follows from Theorem 3.3.1. By Theorem 3.3.1, $\mathbf{E}_0$ is a Galois extension of $\mathbf{F}$, so by Theorem 2.7.14, it is in particular a normal extension of $\mathbf{F}$. But $f(X)$ has a root in $\mathbf{E}_0$; by the definition of a normal extension this implies that $f(X)$ splits in $\mathbf{E}_0$.

On the other hand, we can verify this claim directly here. Ψ is an automorphism of $\mathbf{E}_0 = \mathbf{F}(\alpha)$ of order n, and that implies that $\alpha, \Psi(\alpha), \ldots, \Psi^{n-1}(\alpha)$ are all distinct. But for any k, $0 = \Psi^k(0) = \Psi^k(f(\alpha)) = f(\Psi^k(\alpha))$, so $f(X)$ has the n distinct roots $\alpha, \Psi(\alpha), \ldots, \Psi^{n-1}(\alpha)$ in $\mathbf{E}_0$; therefore $f(X) = \prod_{k=0}^{n-1}(X - \Psi^k(\alpha)) \in \mathbf{E}_0[X]$ is a product of linear factors, i.e., $f(X)$ splits in $\mathbf{E}_0$.

The last sentence in the statement of the corollary is then simply a restatement of Theorem 3.3.1 (3). □

Corollary 3.3.4. *(1) For any* p, s, *and* n, *there is an irreducible monic polynomial* $f(X) \in \mathbf{F}_{p^s}[X]$ *of degree* n.

(2) Let $k(p^s, n)$ *be the cardinality of the set*

$$\mathcal{K} = \{1 \leq k \leq p^{ns} - 1 \mid (p^{ns} - 1)/\gcd(k, p^{ns} - 1) \text{ does not divide } p^{ms} - 1 \text{ for any proper divisor } m \text{ of } n\}.$$

Then the number of distinct irreducible monic polynomials in $\mathbf{F}_{p^s}[X]$ *of degree* n *is* $k(p^s, n)/n$.

Proof. (1) By Corollary 2.2.2, the multiplicative group G of $\mathbf{F}_{p^{ns}}$ is cyclic. Let α_0 be a generator of G, so $\mathbf{F}_{p^{ns}} = \mathbf{F}_{p^s}(\alpha_0)$. Then $f(X) = m_{\alpha_0}(X) \in \mathbf{F}_{p^s}[X]$ is an irreducible monic polynomial of degree n.

(2) First we claim that $k(p^s, n)$ is the cardinality of the set

$$\mathcal{K}' = \{\alpha \in \mathbf{F}_{p^{ns}} \mid \alpha \notin \text{ any proper subfield of } \mathbf{F}_{p^{ns}} \text{ containing } \mathbf{F}_{p^s}\}.$$

To see this, consider a nonzero element $\alpha \in \mathbf{F}_{p^{ns}}$. Then, as an element of G, α has order dividing $p^{ns} - 1$. By Theorem 3.3.1, the proper subfields of $\mathbf{F}_{p^{ns}}$ containing $\mathbf{F}_{p^s}$ are the fields $\mathbf{F}_{p^{ms}}$ with m a proper divisor of n, and α is in such a subfield if and only if its order, as an element of G, divides $p^{ms} - 1$. Now $\alpha = \alpha_0^k$ for some k, and then α has order $(p^{ns} - 1)/\gcd(k, p^{ns} - 1)$. Thus we see that $\mathcal{K}' = \{\alpha_0^k \mid k \in \mathcal{K}\}$.

Now observe that $\mathbf{F}_{p^s}(\alpha) = \mathbf{F}_{p^{ns}}$ if and only if $\alpha \in \mathcal{K}'$, and also that $\mathbf{F}_{p^s}(\alpha) = \mathbf{F}_{p^{ns}}$ if and only if $\deg m_\alpha(X) = n$. In this case $f(X) = m_\alpha(X)$ is an irreducible monic polynomial of degree n, and every irreducible monic polynomial of degree n arises in this way. Also, every such polynomial has distinct roots (as it divides $X^{p^{ns}} - X$, which has distinct roots), and no two distinct such polynomials have a root in common (since if $f_1(\alpha) = f_2(\alpha) = 0$, then $f_1(X) = m_\alpha(X) = f_2(X)$). Since each such polynomial has n distinct roots in $\mathcal{K}'$, and $\mathcal{K}'$ has $k(p^s, n)$ elements, there are $k(p^s, n)/n$ such polynomials. □

Now let us investigate $k(p^s, n)$ a bit further.

Lemma 3.3.5. *(1) $k(p^s, n) \geq \varphi(p^{ns} - 1)$, where φ denotes the Euler totient function.*

(2) If n is prime, $k(p^s, n) = p^{ns} - p^s$.

(3) For any positive integer n,

$$k(p^s, n) = \sum_{d \mid n} \mu(n/d) p^{sd},$$

where μ is the Möbius μ function, defined by $\mu(j) = 0$ if j is divisible by the square of a prime, and otherwise $\mu(j) = (-1)^i$ where i is the number of distinct prime factors of j.

Proof. (1) Observe that $k \in \mathcal{K}$ for any $1 \leq k \leq p^{ns} - 1$ with $\gcd(k, p^{ns} - 1) = 1$, and by definition $\varphi(p^{ns} - 1)$ is the number of such integers.

(2) If n is a prime, the only fields intermediate between $\mathbf{F}_{p^s}$ and $\mathbf{F}_{p^{ns}}$ are these two fields themselves, so in this case $\mathcal{K}' = \{\alpha \in \mathbf{F}_{p^{ns}},\ \alpha \notin \mathbf{F}_{p^s}\}$ and there are $p^{ns} - p^s$ such elements.

(3) Let $f(p^s, n) = p^{ns}$ denote the cardinality of the field $\mathbf{F}_{p^{ns}}$. Then

$$f(p^s, n) = \sum_{d \mid n} k(p^s, d)$$

as any element α of $\mathbf{F}_{p^{ns}}$ generates $\mathbf{F}_{p^{ds}}$ for exactly one value of d dividing n. (This is the value of d for which $\alpha \in \mathbf{F}_{p^{ds}}$ but $\alpha \notin \mathbf{F}_{p^{cs}}$ for any proper divisor c

of d.) But then the Möbius Inversion Formula from elementary number theory shows that

$$k(p^s, n) = \sum_{d|n} \mu(n/d) f(p^s, d),$$

as claimed. □

3.4 Disjoint Extensions

In this section, we assume that all fields are subfields of some field $\mathbf{A}$.

Definition 3.4.1. Let $\mathbf{B}_1$ and $\mathbf{B}_2$ be extensions of $\mathbf{F}$. Then $\mathbf{B}_1$ and $\mathbf{B}_2$ are *disjoint* extensions of $\mathbf{F}$ if $\mathbf{B}_1 \cap \mathbf{B}_2 = \mathbf{F}$. ◇

Of course, $\mathbf{B}_1$ and $\mathbf{B}_2$ are always disjoint extensions of $\mathbf{B}_1 \cap \mathbf{B}_2$.

We first note the following case, which is very special, but easy and useful. Note also that in this case we can strengthen Lemma 2.3.4.

Lemma 3.4.2. *Let $\mathbf{B}_1$ and $\mathbf{B}_2$ be finite extensions of $\mathbf{F}$, with $d_i = (\mathbf{B}_i/\mathbf{F})$, $i = 1, 2$. If d_1 and d_2 are relatively prime, then $\mathbf{B}_1$ and $\mathbf{B}_2$ are disjoint extensions of $\mathbf{F}$, and, furthermore,*

$$(\mathbf{B}_1\mathbf{B}_2/\mathbf{F}) = d_1 d_2, \quad (\mathbf{B}_1\mathbf{B}_2/\mathbf{B}_2) = d_1, \text{ and } (\mathbf{B}_1\mathbf{B}_2/\mathbf{B}_1) = d_2.$$

Proof. Let $\mathbf{B}_0 = \mathbf{B}_1 \cap \mathbf{B}_2$ and $d_0 = (\mathbf{B}_0/\mathbf{F})$. Then $(\mathbf{B}_i/\mathbf{F}) = (\mathbf{B}_i/\mathbf{B}_0)(\mathbf{B}_0/\mathbf{F})$, for $i = 1, 2$, so d_0 divides both d_1 and d_2, so $d_0 = 1$ and $\mathbf{B}_0 = \mathbf{F}$.

Now let $d = (\mathbf{B}_1\mathbf{B}_2/\mathbf{F})$. We observe that $d = (\mathbf{B}_1\mathbf{B}_2/\mathbf{B}_i)(\mathbf{B}_i/\mathbf{F})$ for $i = 1, 2$. From this we immediately see that d is divisible by both d_1 and d_2, and from Lemma 2.3.4 we see that $d \leq d_1 d_2$, so $d = d_1 d_2$ from which the remaining two equalities immediately follow. □

As we shall now see, we can say quite a bit about the composite of disjoint extensions when at least one of the extensions is finite Galois, and even more when both are.

Theorem 3.4.3. *Suppose that $\mathbf{B}_2$ is a finite Galois extension of $\mathbf{B}_1 \cap \mathbf{B}_2$. Then $\mathbf{B}_1\mathbf{B}_2$ is a finite Galois extension of $\mathbf{B}_1$, and*

$$\mathrm{Gal}(\mathbf{B}_1\mathbf{B}_2/\mathbf{B}_1) \cong \mathrm{Gal}(\mathbf{B}_2/\mathbf{B}_1 \cap \mathbf{B}_2),$$

with the isomorphism given by $\sigma \mapsto \sigma \mid \mathbf{B}_2$.

Proof. Let $\mathbf{B}_0 = \mathbf{B}_1 \cap \mathbf{B}_2$. Since $\mathbf{B}_2$ is a Galois extension of $\mathbf{B}_0$, it is the splitting field of a separable polynomial $f(X) \in \mathbf{B}_0[X]$. If $f(X)$ has roots $\alpha_1, \ldots, \alpha_k$, then $\mathbf{B}_2 = \mathbf{B}_0(\alpha_1, \ldots, \alpha_k)$ (Remark 2.5.6), and so $\mathbf{B}_1\mathbf{B}_2 = \mathbf{B}_1(\alpha_1, \ldots, \alpha_k)$. Then $\mathbf{B}_1\mathbf{B}_2$ is the splitting field of the separable polynomial $f(X) \in \mathbf{B}_1[X]$, so $\mathbf{B}_1\mathbf{B}_2$ is a finite Galois extension of $\mathbf{B}_1$.

Let $\sigma \in \mathrm{Gal}(\mathbf{B}_1\mathbf{B}_2/\mathbf{B}_1)$. Then $\sigma \mid \mathbf{B}_1 = \mathrm{id}$, so certainly $\sigma \mid \mathbf{B}_0 = \mathrm{id}$, and hence $\sigma(f(X)) = f(X)$. In particular, σ permutes the roots $\alpha_1, \ldots, \alpha_k$ of $f(X)$, so $\sigma(\mathbf{B}_2) = \mathbf{B}_0(\sigma(\alpha_1) \ldots, \sigma(\alpha_k)) = \mathbf{B}_0(\alpha_1', \ldots, \alpha_k') = \mathbf{B}_2$, where $\{\alpha_1', \ldots, \alpha_k'\}$ is a permutation of $\{\alpha_1, \ldots, \alpha_k\}$. Thus $\sigma \mid \mathbf{B}_2 : \mathbf{B}_2 \to \mathbf{B}_2$, and $\sigma \mid \mathbf{B}_0 = \mathrm{id}$, so $\sigma \mid \mathbf{B}_2 \in \mathrm{Gal}(\mathbf{B}_2/\mathbf{B}_0)$.

We claim that the map $\sigma \mapsto \sigma \mid \mathbf{B}_2$ is an isomorphism.

First, suppose that $\sigma \mid \mathbf{B}_2 = \mathrm{id}$. Then $\sigma(\alpha_i) = \alpha_i$ for each i. Also, since $\sigma \in \mathrm{Gal}(\mathbf{B}_1\mathbf{B}_2/\mathbf{B}_1)$, $\sigma \mid \mathbf{B}_1 = \mathrm{id}$ by definition. But we have observed that $\mathbf{B}_1\mathbf{B}_2 = \mathbf{B}_1(\alpha_1, \ldots, \alpha_k)$, so $\sigma = \mathrm{id}$.

Next, let $\tau \in \mathrm{Gal}(\mathbf{B}_2/\mathbf{B}_0)$. Then $\mathbf{B}_1\mathbf{B}_2 = \mathbf{B}_1(\alpha_1, \ldots, \alpha_k)$ is the splitting field of $f(X) \in \mathbf{B}_1[X]$, so, by Lemma 2.6.3, τ extends to $\sigma : \mathbf{B}_1\mathbf{B}_2 \to \mathbf{B}_1\mathbf{B}_2$, and then $\sigma \mid \mathbf{B}_2 = \tau$. □

Under our hypothesis we may strengthen Lemma 2.3.4.

Corollary 3.4.4. *Suppose that $\mathbf{B}_2$ is a finite Galois extension of $\mathbf{B}_1 \cap \mathbf{B}_2$. Then*

$$(\mathbf{B}_1\mathbf{B}_2/\mathbf{B}_1) = (\mathbf{B}_2/\mathbf{B}_1 \cap \mathbf{B}_2),$$

$$(\mathbf{B}_1\mathbf{B}_2/\mathbf{B}_2) = (\mathbf{B}_1/\mathbf{B}_1 \cap \mathbf{B}_2),$$

and

$$(\mathbf{B}_1\mathbf{B}_2/\mathbf{B}_1 \cap \mathbf{B}_2) = (\mathbf{B}_1/\mathbf{B}_1 \cap \mathbf{B}_2)(\mathbf{B}_2/\mathbf{B}_1 \cap \mathbf{B}_2).$$

Proof. The first equality is immediate from Theorem 3.4.3 and then the second and third follow as

$$\begin{aligned}(\mathbf{B}_1\mathbf{B}_2/\mathbf{B}_1 \cap \mathbf{B}_2) &= (\mathbf{B}_1\mathbf{B}_2/\mathbf{B}_1)(\mathbf{B}_1/\mathbf{B}_1 \cap \mathbf{B}_2)\\ &= (\mathbf{B}_1\mathbf{B}_2/\mathbf{B}_2)(\mathbf{B}_2/\mathbf{B}_1 \cap \mathbf{B}_2).\end{aligned}$$ □

We also make the following observations.

Corollary 3.4.5. *Let $\mathbf{B}_1$ and $\mathbf{B}_2$ be finite extensions of $\mathbf{F}$ and suppose that $\mathbf{B}_2$ is a Galois extension of $\mathbf{F}$. Then*

$$(\mathbf{B}_1\mathbf{B}_2/\mathbf{F}) = (\mathbf{B}_1/\mathbf{F})(\mathbf{B}_2/\mathbf{F})$$

if and only if $\mathbf{B}_1$ and $\mathbf{B}_2$ are disjoint extensions of $\mathbf{F}$.

Proof. By Corollary 3.4.4,

$$\begin{aligned}(\mathbf{B}_1/\mathbf{F})(\mathbf{B}_2/\mathbf{F}) &= (\mathbf{B}_1/\mathbf{B}_1 \cap \mathbf{B}_2)(\mathbf{B}_1 \cap \mathbf{B}_2/\mathbf{F})(\mathbf{B}_2/\mathbf{B}_1 \cap \mathbf{B}_2)(\mathbf{B}_1 \cap \mathbf{B}_2/\mathbf{F})\\ &= (\mathbf{B}_1\mathbf{B}_2/\mathbf{B}_1 \cap \mathbf{B}_2)(\mathbf{B}_1 \cap \mathbf{B}_2/\mathbf{F})^2\\ &= (\mathbf{B}_1\mathbf{B}_2/\mathbf{F})(\mathbf{B}_1 \cap \mathbf{B}_2/\mathbf{F})\end{aligned}$$

so we have the equality in the statement of the corollary if and only if $(\mathbf{B}_1 \cap \mathbf{B}_2/\mathbf{F}) = 1$, i.e., if and only if $\mathbf{B}_1 \cap \mathbf{B}_2 = \mathbf{F}$. □

Corollary 3.4.6 (Theorem on Natural Irrationalities). *Let $f(X) \in \mathbf{F}[X]$ be a separable polynomial and let $\mathbf{B} \supseteq \mathbf{F}$. Let $\mathbf{E}$ be a splitting field for $f(X)$ over $\mathbf{F}$. Then $\mathbf{BE}$ is a splitting field for $f(X)$ over $\mathbf{B}$ and* $\mathrm{Gal}(\mathbf{BE}/\mathbf{B})$ *is isomorphic to* $\mathrm{Gal}(\mathbf{E}/\mathbf{E} \cap \mathbf{B})$*, a subgroup of* $\mathrm{Gal}(\mathbf{E}/\mathbf{F})$*, with the isomorphism given by $\sigma \mapsto \sigma \mid \mathbf{E}$.*

Proof. Let $f(X)$ have roots $\alpha_1, \ldots, \alpha_k$, so $\mathbf{E} = \mathbf{F}(\alpha_1, \ldots, \alpha_k)$ (Remark 2.5.6); then $\mathbf{BE} = \mathbf{B}(\alpha_1, \ldots, \alpha_k)$ is a splitting field for $f(X) \in \mathbf{B}[X]$. Now $\mathbf{BE}$ is a Galois extension of $\mathbf{B}$ as it is the splitting field of a separable polynomial, and $\mathbf{E}$ is a Galois extension of $\mathbf{F}$ for the same reason. Then, setting $\mathbf{D} = \mathbf{B} \cap \mathbf{E}$, $\mathbf{F} \subseteq \mathbf{D} \subseteq \mathbf{E}$, $\mathbf{E} = \mathbf{D}(\alpha_1, \ldots, \alpha_k)$, and so $\mathbf{E}$ is a Galois extension of $\mathbf{D}$, also for the same reason. Then, by Theorem 3.4.3, $\mathrm{Gal}(\mathbf{BE}/\mathbf{B})$ is isomorphic, via the restriction map, to $\mathrm{Gal}(\mathbf{E}/\mathbf{D})$, which is a subgroup of $\mathrm{Gal}(\mathbf{E}/\mathbf{F})$. □

Theorem 3.4.7. *(1) Let $\mathbf{B}_1$ and $\mathbf{B}_2$ be disjoint Galois extensions of $\mathbf{F}$. Then $\mathbf{E} = \mathbf{B}_1\mathbf{B}_2$ is a Galois extension of $\mathbf{F}$ with* $\mathrm{Gal}(\mathbf{E}/\mathbf{F}) = \mathrm{Gal}(\mathbf{B}_1/\mathbf{F}) \times \mathrm{Gal}(\mathbf{B}_2/\mathbf{F})$.

(2) Let $\mathbf{E}$ be a Galois extension of $\mathbf{F}$ and suppose that $\mathrm{Gal}(\mathbf{E}/\mathbf{F}) = G_1 \times G_2$*. Let $\mathbf{B}_1 = \mathrm{Fix}(G_1)$ and $\mathbf{B}_2 = \mathrm{Fix}(G_2)$. Then $\mathbf{B}_1$ and $\mathbf{B}_2$ are disjoint Galois extensions of $\mathbf{F}$ and $\mathbf{E} = \mathbf{B}_1\mathbf{B}_2$.*

Proof. (1) By Theorem 3.4.3, we have isomorphisms

$$\mathrm{Gal}(\mathbf{E}/\mathbf{B}_i) \to \mathrm{Gal}(\mathbf{B}_j/\mathbf{F})$$

given by $\sigma \mapsto \sigma \mid \mathbf{B}_j$, for $i = 1$ and $j = 2$, and vice versa. In other words, any $\sigma_j \in \mathrm{Gal}(\mathbf{B}_j/\mathbf{F})$ has a unique extension to $\tilde{\sigma}_j \in \mathrm{Gal}(\mathbf{E}/\mathbf{B}_i) \subseteq \mathrm{Gal}(\mathbf{E}/\mathbf{F})$, $j = 1, 2$. Now any $e \in \mathbf{E}$ may be written as $e = \sum b_i^1 b_i^2$ with $b_i^1 \in \mathbf{B}_1$ and $b_i^2 \in \mathbf{B}_2$. Then $\tilde{\sigma}_1\tilde{\sigma}_2(e) = \sum \tilde{\sigma}_1\tilde{\sigma}_2(b_i^1 b_i^2) = \sum \tilde{\sigma}_1\tilde{\sigma}_2(b_i^1)\tilde{\sigma}_1\tilde{\sigma}_2(b_i^2) = \sum \tilde{\sigma}_1(b_i^1)\tilde{\sigma}_2(b_i^2) = \sum \tilde{\sigma}_2\tilde{\sigma}_1(b_i^1)\tilde{\sigma}_2\tilde{\sigma}_1(b_i^2) = \sum \tilde{\sigma}_2\tilde{\sigma}_1(b_i^1 b_i^2) = \tilde{\sigma}_2\tilde{\sigma}_1(e)$, so $\tilde{\sigma}_1$ and $\tilde{\sigma}_2$ commute and we have a map

$$\mathrm{Gal}(\mathbf{B}_1/\mathbf{F}) \times \mathrm{Gal}(\mathbf{B}_2/\mathbf{F}) \mapsto \mathrm{Gal}(\mathbf{E}/\mathbf{F})$$

defined by

$$(\sigma_1, \sigma_2) \mapsto \tilde{\sigma}_1\tilde{\sigma}_2.$$

Next, this map is an injection: Suppose $\tilde{\sigma}_1\tilde{\sigma}_2 = \mathrm{id} : \mathbf{E} \to \mathbf{E}$. Then in particular $\tilde{\sigma}_1\tilde{\sigma}_2 \mid \mathbf{B}_2 = \mathrm{id}$, but $\tilde{\sigma}_2 \mid \mathbf{B}_2 = \mathrm{id}$ by definition, so $\sigma_1 = \tilde{\sigma}_1 \mid \mathbf{B}_1 = \mathrm{id}$, and similarly $\sigma_2 = \tilde{\sigma}_2 \mid \mathbf{B}_2 = \mathrm{id}$.

Finally, on the one hand, as $\mathrm{Gal}(\mathbf{E}/\mathbf{F})$ contains $\mathrm{Gal}(\mathbf{B}_1/\mathbf{F}) \times \mathrm{Gal}(\mathbf{B}_2/\mathbf{F})$,

$$\begin{aligned}
|\mathrm{Gal}(\mathbf{E}/\mathbf{F})| &\geq |\mathrm{Gal}(\mathbf{B}_1/\mathbf{F}) \times \mathrm{Gal}(\mathbf{B}_2/\mathbf{F})| \\
&= |\mathrm{Gal}(\mathbf{B}_1/\mathbf{F})|\,|\mathrm{Gal}(\mathbf{B}_2/\mathbf{F})| \\
&= (\mathbf{B}_1/\mathbf{F})(\mathbf{B}_2/\mathbf{F}) \\
&= (\mathbf{B}_1/\mathbf{F})(\mathbf{E}/\mathbf{B}_1) \\
&= (\mathbf{E}/\mathbf{F}),
\end{aligned}$$

and, on the other hand, $|\mathrm{Gal}(\mathbf{E}/\mathbf{F})| = (\mathbf{E}/\mathbf{D}) \leq (\mathbf{E}/\mathbf{F})$, where $\mathbf{D} \supseteq \mathbf{F}$ is the fixed field of $\mathrm{Gal}(\mathbf{E}/\mathbf{F})$, so $|\mathrm{Gal}(\mathbf{E}/\mathbf{F})| = (\mathbf{E}/\mathbf{F})$. Thus this map is an injection from one group to another group of the same order, and hence an isomorphism; furthermore $\mathbf{E}$ is a Galois extension of $\mathbf{F}$.

(2) Since $G = G_1G_2$ and $\{1\} = G_1 \cap G_2$, this follows directly from Proposition 2.8.14. □

Corollary 3.4.8. *Let* $\mathbf{B}_1$ *and* $\mathbf{B}_2$ *be finite Galois extensions of* $\mathbf{F}$ *and let* $\mathbf{E} = \mathbf{B}_1\mathbf{B}_2$, $\mathbf{B}_0 = \mathbf{B}_1 \cap \mathbf{B}_2$. *Then* $\mathbf{E}$ *is a Galois extension of* $\mathbf{F}$ *and* $\mathrm{Gal}(\mathbf{E}/\mathbf{F}) = \{(\sigma_1, \sigma_2) \in \mathrm{Gal}(\mathbf{B}_1/\mathbf{F}) \times \mathrm{Gal}(\mathbf{B}_2/\mathbf{F}) \mid \sigma_1 \mid \mathbf{B}_0 = \sigma_2 \mid \mathbf{B}_0\}$.

Proof. Since $\mathbf{B}_1$ is a Galois extension of $\mathbf{F}$, it is the splitting field of a separable polynomial $f_1(X) \in \mathbf{F}[X]$, by Theorem 2.7.14. Similarly, $\mathbf{B}_2$ is the splitting field of a separable polynomial $f_2(X) \in \mathbf{F}[X]$. Then $\mathbf{E}$ is the splitting field of the separable polynomial $f(X) = f_1(X)f_2(X) \in \mathbf{F}[X]$, so, by Theorem 2.7.14, it is a Galois extension of $\mathbf{F}$.

As in the proof of Theorem 3.4.3, $\sigma(\mathbf{B}_i) = \mathbf{B}_i$ for each $\sigma \in \mathrm{Gal}(\mathbf{E}/\mathbf{F})$, $i = 1, 2$ (as $\mathbf{B}_i$ is the splitting field of a separable polynomial whose roots are permuted by σ). Thus we have a map $\sigma \mapsto (\sigma_1, \sigma_2) = (\sigma \mid \mathbf{B}_1, \sigma \mid \mathbf{B}_2)$. Again this map is an injection, as $\mathbf{E} = \mathbf{B}_1\mathbf{B}_2$, and clearly any (σ_1, σ_2) in image of this map satisfies $\sigma_1 \mid \mathbf{B}_0 = \sigma_2 \mid \mathbf{B}_0$. Then

$$\begin{aligned}
(\mathbf{E}/\mathbf{F}) = (\mathbf{E}/\mathbf{B}_0)(\mathbf{B}_0/\mathbf{F}) &= (\mathbf{E}/\mathbf{B}_2)(\mathbf{B}_2/\mathbf{B}_0)(\mathbf{B}_0/\mathbf{F}) \\
&= (\mathbf{B}_1/\mathbf{B}_0)(\mathbf{B}_2/\mathbf{B}_0)(\mathbf{B}_0/\mathbf{F}) \\
&= (\mathbf{B}_1/\mathbf{F})(\mathbf{B}_2/\mathbf{F})/(\mathbf{B}_0/\mathbf{F}),
\end{aligned}$$

and these are the orders of the groups in the statement of the corollary. □

Corollary 3.4.9. *Let* **B** *and* **E** *be disjoint extensions of* **F**, *with* **E** *a finite Galois extension of* **F**. *Let* $r : \mathrm{Gal}(\mathbf{BE}/\mathbf{B}) \to \mathrm{Gal}(\mathbf{E}/\mathbf{F})$ *be the restriction isomorphism,* $r(\sigma) = \sigma \mid \mathbf{E}$. *Let* $H \subseteq \mathrm{Gal}(\mathbf{BE}/\mathbf{B})$ *be a subgroup. Then, for the fixed fields, we have*

$$\mathrm{Fix}(H) = \mathbf{B}\,\mathrm{Fix}(r(H)).$$

Proof. First observe that r is an isomorphism by Theorem 3.4.3. Now let $K = r(H)$ and $\mathbf{D} = \mathrm{Fix}(K)$. Then certainly $\mathrm{Fix}(H) \supseteq \mathbf{BD}$. Now

$$(\mathbf{BE}/\mathbf{B}) = (\mathbf{BE}/\mathbf{BD})(\mathbf{BD}/\mathbf{B})$$

and

$$(\mathbf{E}/\mathbf{F}) = (\mathbf{E}/\mathbf{D})(\mathbf{D}/\mathbf{F}).$$

Now $(\mathbf{BE}/\mathbf{B}) = (\mathbf{E}/\mathbf{F})$ as the Galois groups of these two extensions are isomorphic, again by Theorem 3.4.3. On the other hand, by Lemma 2.3.4,

$$(\mathbf{BE}/\mathbf{BD}) \leq (\mathbf{E}/\mathbf{D}) \text{ and } (\mathbf{BD}/\mathbf{B}) \leq (\mathbf{D}/\mathbf{F}),$$

so both inequalities must be equalities. But, since r is an isomorphism between the groups H and K,

$$(\mathbf{BE}/\mathrm{Fix}(H)) = |H| = |K| = (\mathbf{E}/\mathrm{Fix}(K)) = (\mathbf{E}/\mathbf{D})$$

so $(\mathbf{BE}/\mathrm{Fix}(H)) = (\mathbf{BE}/\mathbf{BD})$ and hence $\mathrm{Fix}(H) = \mathbf{BD}$. □

Theorem 3.4.10. *Let* **B** *and* **E** *be disjoint extensions of* **F**.

(1) Assume that both **B** *and* **E** *are finite Galois extensions of* **F**. *Let* $N_1 = \mathrm{Gal}(\mathbf{BE}/\mathbf{B})$ *and* $N_2 = \mathrm{Gal}(\mathbf{BE}/\mathbf{E})$. *Then* **BE** *is a Galois extension of* **F**, *and* N_1 *and* N_2 *are normal subgroups of* $\mathrm{Gal}(\mathbf{BE}/\mathbf{F})$. *Furthermore,* $\mathrm{Gal}(\mathbf{BE}/\mathbf{F})$ *is the direct product* $N_1 \times N_2$ *of the subgroups* N_1 *and* N_2.

(2) Assume that **E** *is a finite Galois extension of* **F**. *Let* $H = \mathrm{Gal}(\mathbf{BE}/\mathbf{B})$ *and* $N = \mathrm{Gal}(\mathbf{BE}/\mathbf{E})$. *Then* **BE** *is a Galois extension of* **F** *and* N *is a normal subgroup of* $\mathrm{Gal}(\mathbf{BE}/\mathbf{F})$. *Furthermore,* $\mathrm{Gal}(\mathbf{BE}/\mathbf{F})$ *is the semidirect product* HN *of the subgroups* H *and* N.

Proof. (1) This follows immediately from Theorem 3.4.3 and Corollary 3.4.6.

(2) **E** is a Galois extension of **F**, by hypothesis, so $N = \mathrm{Gal}(\mathbf{BE}/\mathbf{E})$ is a normal subgroup of $\mathrm{Gal}(\mathbf{BE}/\mathbf{F})$, with quotient $H_0 = \mathrm{Gal}(\mathbf{E}/\mathbf{F})$. By Theorem 3.4.3 the subgroup $H = \mathrm{Gal}(\mathbf{BE}/\mathbf{B})$ maps isomorphically onto H_0 under the projection $G \to G/N$. But then $G = HN$. □

Lemma 3.4.11. *Let $\alpha \in \mathbf{A}$ with separable minimum polynomial $m_\alpha(X) \in \mathbf{F}[X]$. Let $\mathbf{E}$ be a splitting field of $m_\alpha(X)$. If $\mathbf{B}$ is an extension of $\mathbf{F}$ such that $\mathbf{B}$ and $\mathbf{E}$ are disjoint, then $m_\alpha(X)$ is irreducible in $\mathbf{B}[X]$.*

Proof. Let $\tilde{m}_\alpha(X)$ be the minimum polynomial of α over $\mathbf{B}$. Of course, $\tilde{m}_\alpha(X)$ is irreducible in $\mathbf{B}[X]$ (as is $m_\alpha(X)$ in $\mathbf{F}[X]$). Let $\{\alpha_1, \ldots, \alpha_d\}$ be the roots of $m_\alpha(X)$ in $\mathbf{E}$. $\mathbf{E}$ is a Galois extension of $\mathbf{F}$ by Theorem 2.7.14. Now $\mathrm{Gal}(\mathbf{E}/\mathbf{F})$ acts transitively on $\{\alpha_i\}$, by Proposition 2.8.15, and every $\sigma_0 \in \mathrm{Gal}(\mathbf{E}/\mathbf{F})$ extends to $\sigma \in \mathrm{Gal}(\mathbf{EB}/\mathbf{B})$, by Theorem 3.4.3, so $\mathrm{Gal}(\mathbf{EB}/\mathbf{B})$ acts transitively on $\{\alpha_i\}$. Then, by Lemma 2.7.12, $\tilde{m}_\alpha(X) = \prod_{i=1}^{d}(X - \alpha_i) = m_\alpha(X)$. □

Remark 3.4.12. The converse of Lemma 3.4.11 is false in general. Here is an example. Let $\mathbf{F} = \mathbf{Q}$ and $\mathbf{A} = \mathbf{Q}(\omega, \sqrt[3]{2})$ (ω a primitive cube root of 1). Let $\alpha = \sqrt[3]{2}$, so $m_\alpha(X) = X^3 - 2 \in \mathbf{F}[X]$ with splitting field $\mathbf{E} = \mathbf{A}$. Let $\mathbf{B} = \mathbf{Q}(\omega)$. Then $m_\alpha(X)$ is irreducible in $\mathbf{B}[X]$, but $\mathbf{B} \cap \mathbf{E} = \mathbf{B} \supset \mathbf{F}$. ◇

However, we do have the following partial converse.

Lemma 3.4.13. *Let $\alpha \in \mathbf{A}$ with separable minimum polynomial $m_\alpha(X) \in \mathbf{F}[X]$ and suppose that $\mathbf{E} = \mathbf{F}(\alpha)$ is a splitting field of $m_\alpha(X)$. If $\mathbf{B}$ is an extension of $\mathbf{F}$ such that $m_\alpha(X)$ is irreducible in $\mathbf{B}[X]$, then $\mathbf{B}$ and $\mathbf{E}$ are disjoint extensions of $\mathbf{F}$.*

Proof. Since $m_\alpha(X)$ is irreducible in $\mathbf{B}[X]$, $\tilde{m}_\alpha(X) = m_\alpha(X)$. By Corollary 3.4.4, $(\mathbf{B}(\alpha)/\mathbf{B}) = (\mathbf{BF}(\alpha)/\mathbf{B}) = (\mathbf{F}(\alpha)/\mathbf{B} \cap \mathbf{F}(\alpha)) = (\mathbf{E}/\mathbf{B} \cap \mathbf{E})$. But

$$(\mathbf{B}(\alpha)/\mathbf{B}) = \deg \tilde{m}_\alpha(X) = \deg m_\alpha(X) = (\mathbf{F}(\alpha)/\mathbf{F}) = (\mathbf{E}/\mathbf{F}),$$

so $\mathbf{B} \cap \mathbf{E} = \mathbf{F}$. □

Let $\mathbf{B}_1$ and $\mathbf{B}_2$ be extensions of $\mathbf{F}$. In Lemma 2.3.4 we showed that $(\mathbf{B}_1\mathbf{B}_2/\mathbf{F}) \leq (\mathbf{B}_1/\mathbf{F})(\mathbf{B}_2/\mathbf{F})$. In Example 2.3.7 we gave an example with $\mathbf{B}_1$ and $\mathbf{B}_2$ disjoint extensions of $\mathbf{F}$ where we had strict inequality. We now formulate a condition that ensures that we will have equality. Note that in Lemma 3.4.14 and Definition 3.4.15 we do not restrict ourselves to finite extensions.

Lemma 3.4.14. *Let $\mathbf{B}_1$ and $\mathbf{B}_2$ be extensions of $\mathbf{F}$. The following are equivalent:*

(a) If $S_1 = \{s_i'\}$ is a subset of $\mathbf{B}_1$ that is $\mathbf{F}$-linearly independent, then S_1 is $\mathbf{B}_2$-linearly independent.

(b) If $S_2 = \{s_j''\}$ is a subset of $\mathbf{B}_2$ that is $\mathbf{F}$-linearly independent, then S_2 is $\mathbf{B}_1$-linearly independent.

(c) If $S_1 = \{s_i'\}$ is a subset of $\mathbf{B}_1$ that is $\mathbf{F}$-linearly independent, and $S_2 = \{s_j''\}$ is a subset of $\mathbf{B}_2$ that is $\mathbf{F}$-linearly independent, then $S_1S_2 = \{s_i's_j''\}$ is $\mathbf{F}$-linearly independent.

Proof. We show that (a) and (c) are equivalent. Then (b) and (c) are equivalent simply by interchanging the roles of $\mathbf{B}_1$ and $\mathbf{B}_2$, so all three are equivalent.

(a) implies (c): Suppose $\sum_{i,j} a_{ij}s'_i s''_j = 0$ with $a_{ij} \in \mathbf{F}$. Then $(\sum_i s'_i(\sum_j a_{ij}s''_j)) = 0$. As S_1 is $\mathbf{B}_2$-linearly independent, each inner sum is zero; as S_2 is $\mathbf{F}$-linearly independent, each a_{ij} is zero.

(c) implies (a): Suppose $\sum_i \beta_i s'_i = 0$ with $\beta_i \in \mathbf{B}_2$. Choose $S_2 = \{s''_j\}$ to be a basis of $\mathbf{B}_2$ as an $\mathbf{F}$-vector space. Then, for each i, we have $\beta_i = \sum_j a_{ij}s''_j$ with $a_{ij} \in \mathbf{F}$. Then $0 = \sum_i \beta_i s'_i = \sum_i s'_i(\sum_j a_{ij}s''_j) = \sum_{i,j} a_{ij}s'_i s''_j$. As $S_1S_2 = \{s'_i s''_j\}$ is $\mathbf{F}$-linearly independent, each a_{ij} is zero, and hence each β_i is zero. □

Definition 3.4.15. Let $\mathbf{B}_1$ and $\mathbf{B}_2$ be extensions of $\mathbf{F}$. If $\mathbf{B}_1$ and $\mathbf{B}_2$ satisfy the equivalent conditions of Lemma 3.4.14, then $\mathbf{B}_1$ and $\mathbf{B}_2$ are *linearly disjoint* extensions of $\mathbf{F}$. ◇

Lemma 3.4.16. *Let* $\mathbf{B}_1$ *and* $\mathbf{B}_2$ *be finite extensions of* $\mathbf{F}$. *Then* $(\mathbf{B}_1\mathbf{B}_2/\mathbf{F}) = (\mathbf{B}_1/\mathbf{F})(\mathbf{B}_2/\mathbf{F})$ *if and only if* $\mathbf{B}_1$ *and* $\mathbf{B}_2$ *are linearly disjoint extensions of* $\mathbf{F}$.

Proof. Let S_1 be a basis for $\mathbf{B}_1$ and S_2 be a basis for $\mathbf{B}_1$. Then S_1S_2 spans $\mathbf{B}_1\mathbf{B}_2$ as an $\mathbf{F}$-vector space. If $\mathbf{B}_1$ and $\mathbf{B}_2$ are linearly disjoint then S_1S_2 is linearly independent and hence a basis for $\mathbf{B}_1\mathbf{B}_2$ over $\mathbf{F}$. Thus in this case $(\mathbf{B}_1\mathbf{B}_2/\mathbf{F}) = (\mathbf{B}_1/\mathbf{F})(\mathbf{B}_2/\mathbf{F})$. If $\mathbf{B}_1$ and $\mathbf{B}_2$ are not linearly disjoint then S_1S_2 is not linearly independent and hence some proper subset of S_1S_2 is a basis for $\mathbf{B}_1\mathbf{B}_2$ over $\mathbf{F}$. Thus in this case $(\mathbf{B}_1\mathbf{B}_2/\mathbf{F}) < (\mathbf{B}_1/\mathbf{F})(\mathbf{B}_2/\mathbf{F})$. □

Corollary 3.4.17. *Let* $\mathbf{B}_1$ *and* $\mathbf{B}_2$ *be disjoint finite extensions of* $\mathbf{F}$ *and suppose that* $\mathbf{B}_2$ *is a Galois extension of* $\mathbf{F}$. *Then* $\mathbf{B}_1$ *and* $\mathbf{B}_2$ *are linearly disjoint extensions of* $\mathbf{F}$.

Proof. Immediate from Corollary 3.4.5 and Lemma 3.4.16. □

3.5 Simple Extensions

Let $\mathbf{E}$ be a finite extension of $\mathbf{F}$. Then, choosing a basis $\{\alpha_1, \ldots, \alpha_k\}$ for $\mathbf{E}$ as an $\mathbf{F}$-vector space, $\mathbf{E} = \mathbf{F}(\alpha_1, \ldots, \alpha_k)$. We may think of $\mathbf{E}$ as obtained from $\mathbf{F}$ in (at most) k steps, successively adjoining α_1, then $\alpha_2, \ldots,$ and finally α_k. It is natural to ask when we can obtain $\mathbf{E}$ in one step.

Definition 3.5.1. Let $\mathbf{E}$ be an algebraic extension of $\mathbf{F}$ such that $\mathbf{E} = \mathbf{F}(\alpha)$ for some α in $\mathbf{E}$. Then $\mathbf{E}$ is a *simple* extension of $\mathbf{F}$ and α is a *primitive element* of $\mathbf{E}$. ◇

We derive a criterion for $\mathbf{E}$ to be a simple extension of $\mathbf{F}$.

Theorem 3.5.2. *Let* $\mathbf{E}$ *be a finite extension of* $\mathbf{F}$*. Then* $\mathbf{E}$ *is a simple extension of* $\mathbf{F}$ *if and only if there are a finite number of intermediate fields* $\mathbf{F} \subseteq \mathbf{B} \subseteq \mathbf{E}$.

Proof. First, we consider the case when $\mathbf{F}$ is finite. Then $\mathbf{E}$ is finite, so on the one hand there are (trivially) only finitely many intermediate fields, and on the other hand, the multiplicative group of $\mathbf{E}$ is cyclic (Corollary 2.2.2), so it consists of the powers of some $\alpha \in \mathbf{E}$, and then $\mathbf{E} = \mathbf{F}(\alpha)$.

Now we consider the case when $\mathbf{F}$ is infinite. First, assume that $\mathbf{E}$ is simple, $\mathbf{E} = \mathbf{F}(\alpha)$. Then α has minimum polynomial $m_\alpha(X) \in \mathbf{F}[X]$. Let $\mathbf{F} \subseteq \mathbf{B} \subseteq \mathbf{E}$ and consider $\tilde{m}_\alpha(X) \in \mathbf{B}[X]$, the minimum polynomial of α over $\mathbf{B}$. Then $\mathbf{E} = \mathbf{B}(\alpha)$, so $(\mathbf{E}/\mathbf{B}) = \deg \tilde{m}_\alpha(X)$. Let $\mathbf{D}$ be the field obtained from $\mathbf{F}$ by adjoining the coefficients of $\tilde{m}_\alpha(X)$. Then $\mathbf{F} \subseteq \mathbf{D} \subseteq \mathbf{B}$. As $\tilde{m}_\alpha(X)$ is irreducible in $\mathbf{B}[X]$, it is certainly irreducible in $\mathbf{D}[X]$, so $\tilde{m}_\alpha(X)$ is the minimum polynomial of α over $\mathbf{D}$, and $\mathbf{E} = \mathbf{D}(\alpha)$, so $(\mathbf{E}/\mathbf{D}) = \deg \tilde{m}_\alpha(X) = (\mathbf{E}/\mathbf{B})$ and hence $\mathbf{B} = \mathbf{D}$. Thus the intermediate field $\mathbf{B}$ is determined by the polynomial $\tilde{m}_\alpha(X)$. But $\tilde{m}_\alpha(X)$ divides $m_\alpha(X)$ in $\mathbf{B}[X]$ (as $m_\alpha(\alpha) = 0$) and hence $\tilde{m}_\alpha(X)$ divides $m_\alpha(X)$ in $\mathbf{E}[X]$. But $m_\alpha(X)$ has only finitely many factors in $\mathbf{E}[X]$, so there are only finitely many possibilities for $\tilde{m}_\alpha(X)$ and hence for $\mathbf{B}$.

Conversely, suppose there are only finitely many fields between $\mathbf{F}$ and $\mathbf{E}$. We show that given any α and β in $\mathbf{E}$ there is a γ_1 in $\mathbf{E}$ with $\mathbf{F}(\alpha, \beta) = \mathbf{F}(\gamma_1)$. Since $\mathbf{E}$ is a finite extension of $\mathbf{F}$, it is obtained by adjoining a finite number k of elements, and then the theorem follows by induction on k.

Consider the fields $\mathbf{F}(\alpha + a\beta)$ for $a \in \mathbf{F}$. Since there are infinitely many elements of $\mathbf{F}$ and only finitely many intermediate fields, there must exist distinct elements a_1 and a_2 of $\mathbf{F}$ with $\mathbf{F}(\alpha + a_1\beta) = \mathbf{F}(\alpha + a_2\beta)$. Set $\gamma_1 = \alpha + a_1\beta$, $\gamma_2 = \alpha + a_2\beta$. Then $\mathbf{F}(\gamma_1) = \mathbf{F}(\gamma_2)$, so $\gamma_2 \in \mathbf{F}(\gamma_1)$, and certainly $\gamma_1 \in \mathbf{F}(\gamma_1)$, so $\gamma_2 - \gamma_1 = (a_2 - a_1)\beta \in \mathbf{F}(\gamma_1)$, and hence $\beta \in \mathbf{F}(\gamma_1)$ and $\gamma_1 - a_1\beta = \alpha \in \mathbf{F}(\gamma_1)$. Thus $\mathbf{F}(\gamma_1) \subseteq \mathbf{F}(\alpha, \beta) \subseteq \mathbf{F}(\gamma_1)$ and they are equal. □

While Theorem 3.5.2 does not look easy to apply, in fact it readily gives a strong general result.

Corollary 3.5.3. *If* $\mathbf{E}$ *is a finite separable extension of* $\mathbf{F}$*, then* $\mathbf{E}$ *is a simple extension of* $\mathbf{F}$.

Proof. Let $\mathbf{E} = \mathbf{F}(\alpha_1, \ldots, \alpha_k)$. By the definition of a separable extension (Definition 2.7.3) each $m_{\alpha_i}(X)$ is a separable polynomial and hence (also by Definition 2.7.3) the product $f(X) = m_{\alpha_1}(X) \cdots m_{\alpha_k}(X)$ is also a separable polynomial. Let $\mathbf{E}' \supseteq \mathbf{E}$ be an extension of $\mathbf{F}$ that is a splitting field of the polynomial $f(X)$. Then $\mathbf{E}'$ is a Galois extension of $\mathbf{F}$ (Theorem 2.7.14), so by the Fundamental Theorem of Galois Theory there are only finitely many fields intermediate between $\mathbf{F}$ and $\mathbf{E}'$ (corresponding to the subgroups of $\mathrm{Gal}(\mathbf{E}'/\mathbf{F})$), and hence there are certainly only finitely many fields between $\mathbf{F}$ and $\mathbf{E}$. Then, by Theorem 3.5.2, $\mathbf{E}$ is a simple extension of $\mathbf{F}$. □

In case $\mathbf{E}$ is a Galois extension of $\mathbf{F}$, we have the following criterion for an element of $\mathbf{E}$ to be primitive.

Lemma 3.5.4. *Let* $\mathbf{E}$ *be a finite Galois extension of* $\mathbf{F}$ *and let* $\alpha \in \mathbf{E}$. *Then* α *is primitive, i.e.,* $\mathbf{E} = \mathbf{F}(\alpha)$, *if and only if* $\sigma(\alpha) \neq \alpha$ *for every* $\sigma \in \mathrm{Gal}(\mathbf{E}/\mathbf{F})$, $\sigma \neq \mathrm{id}$.

First proof. Let $d = (\mathbf{F}(\alpha)/\mathbf{F})$ and $n = (\mathbf{E}/\mathbf{F})$, so $\mathbf{E} = \mathbf{F}(\alpha)$ if and only if $d = n$. Now $n = |\mathrm{Gal}(\mathbf{E}/\mathbf{F})|$ (Theorem 2.8.5) and $d = \deg m_\alpha(X)$ (Lemma 2.4.6 (2)). But, by Lemma 2.7.12, $m_\alpha(X) = \prod_{i=1}^{d}(X - \alpha_i)$ where $\alpha_1 = \alpha, \alpha_2, \ldots, \alpha_d$ are the Galois conjugates of α in $\mathbf{E}$. Hence $\mathbf{E} = \mathbf{F}(\alpha)$ if and only if α has n conjugates, and this is true if and only if $\sigma(\alpha) = \alpha$, $\sigma \in \mathrm{Gal}(\mathbf{E}/\mathbf{F})$, implies $\sigma = \mathrm{id}$.

Second Proof. Let $\mathbf{B} = \mathbf{F}(\alpha)$ so $\mathbf{B}$ is intermediate between $\mathbf{F}$ and $\mathbf{E}$. By the *FTGT*, $\mathbf{F} = \mathrm{Fix}(H)$ for H a subgroup of $\mathrm{Gal}(\mathbf{E}/\mathbf{F})$, and furthermore $H = \{\sigma \in \mathrm{Gal}(\mathbf{E}/\mathbf{F}) \mid \sigma \mid \mathbf{B} = \mathrm{id}\} = \{\sigma \in \mathrm{Gal}(\mathbf{E}/\mathbf{F}) \mid \sigma(\alpha) = \alpha\}$. Then, also by the *FTGT*, $\mathbf{B} = \mathbf{E}$ if and only if $H = \{\mathrm{id}\}$. □

The following proposition is very useful in constructing primitive elements.

Proposition 3.5.5. *Let* $\mathbf{F}$ *be a field of characteristic* 0 *and let* $\mathbf{B}_1$ *and* $\mathbf{B}_2$ *be disjoint finite Galois extensions of* $\mathbf{F}$. *Set* $\mathbf{B} = \mathbf{B}_1\mathbf{B}_2$. *Let* α_1 *be a primitive element of* $\mathbf{B}_1$ *and let* α_2 *be a primitive element of* $\mathbf{B}_2$. *Then* $\alpha = \alpha_1 + \alpha_2$ *is a primitive element of* $\mathbf{B}$.

Proof. We first observe that $\mathbf{B}$ is a Galois extension of $\mathbf{F}$ (Theorem 3.4.7), so, by Lemma 3.5.4, it suffices to show that the Galois conjugates of α are all distinct. In order to show this, we need only show that if $\sigma(\alpha) = \alpha$ for some $\sigma \in G = \mathrm{Gal}(\mathbf{B}/\mathbf{F})$, then $\sigma = \mathrm{id}$.

Thus, let $\sigma \in G$ with $\sigma(\alpha) = \alpha$. By Theorem 3.4.7, $\sigma = (\sigma_1, \sigma_2)$ with $\sigma_i \in \mathrm{Gal}(\mathbf{B}_i/\mathbf{F})$, $i = 1, 2$. Then

$$\alpha_1 + \alpha_2 = \alpha = \sigma(\alpha) = \sigma_1(\alpha_1) + \sigma_2(\alpha_2),$$

so we see that $\sigma_1(\alpha_1) - \alpha_1 = \sigma_2(\alpha_2) - \alpha_2$. But $\sigma_i(\alpha_i) - \alpha_i \in \mathbf{B}_i$, $i = 1, 2$, and $\mathbf{B}_1$ and $\mathbf{B}_2$ are disjoint extensions of $\mathbf{F}$, so $\sigma_1(\alpha_1) - \alpha_1 = \sigma_2(\alpha_2) - \alpha_2 = a \in \mathbf{F}$. Thus $\sigma_1(\alpha) = \alpha + a$, from which we see that $\sigma_1^k(\alpha) = \alpha + ka$ for every k.

But σ_1 has finite order m, so $\alpha = \sigma_1^m(\alpha) = \alpha + ma$, so $ma = 0$ and hence $a = 0$. Thus $\sigma_1(\alpha_1) = \alpha_1$, and then $\sigma_2(\alpha_2) = \alpha_2$ as well. But α_1 is a primitive element of $\mathbf{B}_1$, so $\sigma_1(\alpha_1) = \alpha_1$ implies $\sigma_1 = \mathrm{id}$ on $\mathbf{B}_1$; similarly $\sigma_2 = \mathrm{id}$ on $\mathbf{B}_2$ as well, so $\sigma = \mathrm{id}$ on $\mathbf{B}$ as required. □

Remark 3.5.6. In order to effectively apply Proposition 3.5.5 we need to develop some more background, which we do in the next chapter. But to show its applicability, as well as to provide some interesting examples of primitive elements, we shall quote an example from the next chapter here.

> **Example 4.6.21.** (1) Let $\mathbf{E} = \mathbf{Q}(\sqrt{2}, \sqrt[3]{2}, \sqrt[5]{2}, \sqrt[7]{2})$. Then $\mathbf{E}$ is an extension of $\mathbf{Q}$ of degree 210 with primitive element $\sqrt{2}+\sqrt[3]{2}+\sqrt[5]{2}+\sqrt[7]{2}$.
> (2) Let $\mathbf{E} = \mathbf{Q}(\sqrt{2}, \sqrt{3}, \sqrt{5}, \sqrt{7})$. Then $\mathbf{E}$ is an extension of $\mathbf{Q}$ of degree 16 with primitive element $\sqrt{2}+\sqrt{3}+\sqrt{5}+\sqrt{7}$. Similarly, if $\mathbf{E} = \mathbf{Q}(\sqrt[3]{2}, \sqrt[3]{3}, \sqrt[3]{5}, \sqrt[3]{7})$, then $\mathbf{E}$ is an extension of $\mathbf{Q}$ of degree 81 with primitive element $\sqrt[3]{2}+\sqrt[3]{3}+\sqrt[3]{5}+\sqrt[3]{7}$.
> (3) Let $\mathbf{E} = \mathbf{Q}(\sqrt{2}, \sqrt[3]{2}, \sqrt{3}, \sqrt[3]{3})$. Then $\mathbf{E}$ is an extension of $\mathbf{Q}$ of degree 36 with primitive element $\sqrt{2}+\sqrt[3]{2}+\sqrt{3}+\sqrt[3]{3}$. ◇

Example 3.5.7. Here is an example that shows that the conclusion of Proposition 3.5.5 may be false if $\mathbf{B}_1$ and $\mathbf{B}_2$ are not Galois extensions of $\mathbf{F}$. Let $\mathbf{F} = \mathbf{Q}$, $\mathbf{B}_1 = \mathbf{Q}(\sqrt[3]{2})$, and $\mathbf{B}_2 = \mathbf{Q}(\omega\sqrt[3]{2})$. Then $\mathbf{B} = \mathbf{Q}(\omega, \sqrt[3]{2})$. Let $\alpha_1 = -\sqrt[3]{2}$, a primitive element of $\mathbf{B}_1$, and let $\alpha_2 = -\omega\sqrt[3]{2}$, a primitive element of $\mathbf{B}_2$. Then $\alpha = \alpha_1 + \alpha_2 = \omega^2\sqrt[3]{2}$ is not a primitive element of $\mathbf{B}$. ◇

Example 3.5.8. Here is an example of an extension that is not simple. Let $\mathbf{F} = \mathbf{F}_p(s, t)$, the field of rational functions in the variables s and t over $\mathbf{F}_p$. Let $\mathbf{E}$ be the splitting field of the polynomial $(X^p - s)(X^p - t)$. More simply, $\mathbf{E} = \mathbf{F}(\beta, \gamma)$ where $\beta^p = s$ and $\gamma^p = t$. Then $[\mathbf{E}/\mathbf{F}] = p^2$. On the other hand, it is easy to check that for any $\alpha \in \mathbf{E}$, $\alpha^p \in \mathbf{F}$, so $[\mathbf{F}(\alpha) : \mathbf{F}] = 1$ or p, and in particular $\mathbf{F}(\alpha) \subset \mathbf{E}$. ◇

3.6 The Normal Basis Theorem

Let $\mathbf{E}$ be a Galois extension of $\mathbf{F}$ of degree n. Then, as an $\mathbf{F}$-vector space, $\mathbf{E}$ has a basis of n elements. On the other hand, $\mathrm{Gal}(\mathbf{E}/\mathbf{F})$ has n elements, and so it natural to ask if there is some element α of $\mathbf{E}$ whose conjugates $\{\sigma(\alpha) \mid \sigma \in \mathrm{Gal}(\mathbf{E}/\mathbf{F})\}$ form a basis for $\mathbf{E}$ over $\mathbf{F}$. In fact, this is always the case, as we shall show in this section.

Definition 3.6.1. Let $\mathbf{E}$ be a Galois extension of $\mathbf{F}$. A basis $\{\alpha_i\}_{i=1,\dots,n}$ of $\mathbf{E}$ as a vector space over $\mathbf{F}$ is a *normal basis* for $\mathbf{E}$ over $\mathbf{F}$ if $\{\alpha_i\} = \{\sigma_i(\alpha)\}$ for some $\alpha \in \mathbf{E}$, where $\{\sigma_i\}_{i=1,\dots,n} = \mathrm{Gal}(\mathbf{E}/\mathbf{F})$. ◇

There is one case of this theorem that follows directly from linear algebra.

Theorem 3.6.2. *Let* **E** *be a finite Galois extension of* **F**. *Suppose* Gal(**E**/**F**) *is cyclic. Then* **E** *has a normal basis.*

Proof. Let σ be a generator of Gal(**E**/**F**) and let $T : \mathbf{E} \to \mathbf{E}$ be given by $T(\alpha) = \sigma(\alpha)$. Then T is a linear transformation that satisfies $T^n - I = 0$. We claim the minimum polynomial $m_T(X) = X^n - 1$. Certainly $m_T(X)$ divides $X^n - 1$. Suppose $m_T(X) \neq X^n - 1$. Then $m_T(X) = \sum_{i=0}^{d} a_i X^i$ with $d < n$. In other words,

$$a_0 \, \mathrm{id} + \cdots + a_{d-1}\sigma^{d-1} = 0$$

as a map from **E** to **E**, and hence as a map from $\mathbf{E}^*$ to **E**.

But $\{\mathrm{id}, \sigma, \ldots, \sigma^{d-1}\}$ are a set of distinct characters of $\mathbf{E}^*$ in **E**, so by Theorem 2.8.4 are linearly independent, so $a_i = 0$ for each i, a contradiction.

Now order the elements of Gal(**E**/**F**) as $\{\mathrm{id}, \sigma, \ldots, \sigma^{d-1}\}$.

Then (by [AW, Theorem 3.7.1]) there is an element α of **E** with $\mathrm{Ann}(\alpha) = \langle m_T(X) \rangle$ (i.e., where the annihilator ideal of α in $\mathbf{F}[X]$ is the ideal generated by $m_T(X)$). Then, since $\deg m_T(X) = n$, $\{\alpha, T(\alpha), \ldots, T^{n-1}(\alpha)\} = \{\alpha, \sigma(\alpha), \ldots, \sigma^{n-1}(\alpha)\}$ are linearly independent, hence form a basis, and this is a normal basis.

(Alternatively phrased, $m_T(X) = m_\alpha(X)$, by Proposition 2.4.6 (3). But **E** has a basis over **F** in which the matrix of T is the companion matrix of $m_T(X)$. But then if α_1 is the first element of the basis, $\sigma^j(\alpha_1) = T^j(\alpha_1) = \alpha_{1+j}$ for $j = 0, \ldots, n-1$ ([AW, Corollary IV.4.15]), so $\{\alpha_i\}$ is a normal basis.) □

The argument for the general case also uses linear algebra, but is more involved.

Lemma 3.6.3. *Let* **E** *be a Galois extension of* **F** *of degree* n, *and let* σ_i, $i = 1, \ldots, n$, *be the elements of* Gal(**E**/**F**). *Let* $\alpha_1, \ldots, \alpha_n$ *be elements of* **E**. *Then* $\{\alpha_1, \ldots, \alpha_n\}$ *is a basis for* **E** *over* **F** *if and only if the matrix* $A = (\alpha_{ij})$, $\alpha_{ij} = \sigma_i(\alpha_j)$, *is nonsingular.*

Proof. Let $\{\alpha_1, \ldots, \alpha_n\}$ be a basis for **E** over **F**, and suppose some linear combination of the rows of A is zero, say $\sum_{i=1}^n c_i \alpha_{ij} = 0$ for each j, i.e., $\sum_{i=1}^n c_i \sigma_i(\alpha) = 0$, $\alpha = \alpha_1, \ldots, \alpha_j$. But each σ_i is an **F**-linear map, so $\sum_{i=1}^n c_i \sigma_i(\alpha) = 0$ for every $\alpha \in \mathbf{E}$. But the σ_i are distinct characters of $\mathbf{E}^*$ in **E**, so by Theorem 2.8.4 are linearly independent; so $c_1 = \cdots = c_n = 0$. Thus the rows of A are linearly independent and A is nonsingular.

On the other, suppose $\{\alpha_1, \ldots, \alpha_n\}$ is not a basis. Then it is not linearly independent, so there is a nontrivial linear combination $\sum_{j=1}^n d_j \alpha_j = 0$. Then

$\sum_{j=1}^{n} d_j \alpha_{ij} = \sum_{j=1}^{n} d_j \sigma_i(\alpha_j) = \sigma_i(\sum_{j=1}^{n} d_j \alpha_j) = \sigma_i(0) = 0$ for each i, so the columns of A are linearly dependent, and A is singular. □

Lemma 3.6.4. *Let* **F** *be an infinite field, let* **E** *be an extension of* **F**, *and let* $f(X_1, \ldots, X_k) \in \mathbf{E}[X_1, \ldots, X_k]$ *be a polynomial. If* $f(a_1, \ldots, a_k) = 0$ *for every* $(a_1, \ldots, a_k) \in \mathbf{F}^k$, *then* $f(X_1, \ldots, X_k)$ *is the zero polynomial.*

Proof. By induction on k. For $k = 1$, let $f(X_1) \in \mathbf{E}[X_1]$ be a nonzero polynomial with $f(a_1) = 0$ for every $a_1 \in \mathbf{F}$. Then $f(X_1)$ has more than $\deg f(X_1)$ roots, contradicting Lemma 2.2.1. Now assume the lemma is true for $k-1$ and consider $f(X_1, \ldots, X_k)$. Regard this as a polynomial in X_k with coefficients in $X_1, \ldots, X_{k-1}$, so

$$f(X_1, \ldots, X_k) = \sum_{i=0}^{n} g_i(X_1, \ldots, X_{k-1}) X_k^i.$$

For any fixed elements $a_1, \ldots, a_{k-1}$ of **F**, $\bar{f}(X_k) = f(a_1, \ldots, a_{k-1}, X_k)$ has more than d roots, so is identically zero by the $k = 1$ case, so each coefficient $g_i(X_1, \ldots, X_{k-1})$ is zero for every $(a_1, \ldots, a_{k-1})$, and hence is identically zero by the $k - 1$ case. □

Theorem 2.8.4 gives us linear independence of characters. In fact, when **F** is infinite and **E** is a finite Galois extension of **F**, we may derive a stronger kind of independence, algebraic independence.

Theorem 3.6.5. *Let* **F** *be an infinite field and let* Gal(**E**/**F**) $= \{\sigma_i\}$, $i = 1, \ldots, n$. *Then* $\{\sigma_i\}$ *are algebraically independent, i.e., if* $f(X_1, \ldots, X_k) \in \mathbf{E}[X_1, \ldots, X_k]$ *is a polynomial with* $f(\sigma_1(\alpha), \ldots, \sigma_k(\alpha)) = 0$ *for every* $\alpha \in \mathbf{E}$, *then* $f(X_1, \ldots, X_k)$ *is the zero polynomial.*

Proof. Fix a basis $\{\alpha_1, \ldots, \alpha_n\}$ for **E** over **F**. By Lemma 3.6.3, the matrix $A = A(\alpha_1, \ldots, \alpha_n)$ is nonsingular, where $A = (\alpha_{ij}) = (\sigma_i(\alpha_j))$.

In this proof we shall identify an n-tuple $(\beta_1, \ldots, \beta_n)$ of elements of **E** with the column vector $[\beta_1, \ldots, \beta_n]^t$. (Purely for typographical convenience, we write this column vector $[\beta_1, \ldots, \beta_n]^t$ as the transpose of the row vector $[\beta_1, \ldots, \beta_n]$.)

For an arbitrary n-tuple $(c_1, \ldots, c_n)$ of elements of **F**, let $\alpha = \sum_{j=1}^{n} c_j \alpha_j$. Then $\sigma_i(\alpha) = \sum_{j=1}^{n} c_j \sigma_i(\alpha_j) = \sum_{j=1}^{n} \alpha_{ij} c_j$ for each $i = 1, \ldots, n$, i.e.,

$$[\sigma_1(\alpha), \ldots, \sigma_n(\alpha)]^t = A[c_1, \ldots, c_n]^t.$$

Now suppose that $f(\sigma_1(\alpha), \ldots, \sigma_n(\alpha)) = 0$ for every $\alpha \in \mathbf{E}$. Then, in our notation,

$$0 = f([\sigma_1(\alpha), \ldots, \sigma_n(\alpha)]^t) = f(A[c_1, \ldots, c_n]^t)$$

for every $(c_1, \ldots, c_n) \in \mathbf{F}^n$. Thus the polynomial

$$g(X_1, \ldots, X_n) = f(A[X_1, \ldots, X_n]^t)$$

is zero for every $(c_1, \ldots, c_n) \in \mathbf{F}^n$, so, by Lemma 3.6.4, it must be the zero polynomial in $\mathbf{E}[X_1, \ldots, X_n]$, and $g(\gamma_1, \ldots, \gamma_n) = 0$ for every $(\gamma_1, \ldots, \gamma_n) \in \mathbf{E}^n$. We claim also that $f(X_1, \ldots, X_n)$ is the zero polynomial, and, again by Lemma 3.6.4, that will follow if we show $f(b_1, \ldots, b_n) = 0$ for every $(b_1, \ldots, b_n) \in \mathbf{F}^n$. Thus, let $(b_1, \ldots, b_n) \in \mathbf{F}^n$. Since A is invertible, we may define $(\gamma_1, \ldots, \gamma_n)$ by $[\gamma_1, \ldots, \gamma_n]^t = A^{-1}[b_1, \ldots, b_n]^t$, and then $f(b_1, \ldots, b_n) = f([b_1, \ldots, b_n]^t) = f(A[\gamma_1, \ldots, \gamma_n]^t) = g(\gamma_1, \ldots, \gamma_n) = 0$, as claimed. □

Remark 3.6.6. Theorem 3.6.5 is false for finite fields. Here is a counterexample. Let $\mathbf{F} = \mathbf{F}_2 = \{0, 1\}$ and $\mathbf{E} = \mathbf{F}_4 = \{0, 1, \alpha_0, \alpha_0 + 1\}$ with $\alpha_0^2 = \alpha_0 + 1$. Let $\mathrm{Gal}(\mathbf{E}/\mathbf{F}) = \{\sigma_1, \sigma_2\}$ with $\sigma_1 = \mathrm{id}$ and σ_2 given by $\sigma_2(\alpha_0) = \alpha + 1$. Let $f(X_1, X_2)$ be the polynomial $f(X_1, X_2) = (X_2 - X_1)(X_2 - X_1 - 1)$. Then $p(\sigma_1(\alpha)\sigma_2(\alpha)) = (\sigma_2(\alpha) - \sigma_1(\alpha))(\sigma_2(\alpha) - \sigma_1(\alpha) - 1) = 0$ for every $\alpha \in \mathbf{E}$. ◊

Theorem 3.6.7. *Let* $\mathbf{E}$ *be a finite Galois extension of* $\mathbf{F}$. *Then* $\mathbf{E}$ *has a normal basis.*

Proof. If $\mathbf{F}$ is finite, then $\mathbf{E}$ is a cyclic extension of $\mathbf{F}$ (Theorem 3.3.1) so the result is a special case of Theorem 3.6.2.

Assume henceforth that $\mathbf{F}$ is infinite. Let $\mathrm{Gal}(\mathbf{E}/\mathbf{F}) = \{\sigma_1, \ldots, \sigma_n\}$. Let $B = B(X_1, \ldots, X_n)$ be the matrix whose entries are indeterminates X_k, with $b_{ij} = X_k$ if $\sigma_i\sigma_j = \sigma_k$. Then $\det(B)$ is a polynomial in $X_1, \ldots, X_k$, $\det(B) = d(X_1, \ldots, X_k)$. Note each X_i appears exactly once in every row and column. Thus, setting $X_1 = 1$ and $X_i = 0$ for $i > 1$, we see $\det(B) = \pm 1$, i.e., $d(1, 0, \ldots, 0) = \pm 1$. Thus d is not the zero polynomial. Hence by Theorem 3.6.5 there is an $\alpha \in \mathbf{E}$ with $d(\sigma_1(\alpha), \ldots, \sigma_n(\alpha)) \neq 0$. Let $\alpha_j = \sigma_j(\alpha)$, $j = 1, \ldots, n$. Now let $A = B(\sigma_1(\alpha), \ldots, \sigma_n(\alpha))$. If $A = (a_{ij})$, then $a_{ij} = \alpha_k$ where $\sigma_i\sigma_j(\alpha) = \sigma_k(\alpha)$, or, in other words, $a_{ij} = \sigma_i\sigma_j(\alpha) = \sigma_i(\alpha_j)$. Thus A is a matrix as in Lemma 3.6.3, and $\det(A) = d(\sigma_1(\alpha), \ldots, \sigma_n(\alpha)) \neq 0$, so A is nonsingular, and then, by Lemma 3.6.3, $\{\alpha_1, \ldots, \alpha_n\} = \{\sigma_1(\alpha), \ldots, \sigma_n(\alpha)\}$ is a basis for $\mathbf{E}$ over $\mathbf{F}$. □

Remark 3.6.8. Let $\mathbf{E}$ be a Galois extension of $\mathbf{F}$ and let $\alpha \in \mathbf{E}$. Then α is a primitive element if the conjugates of α are all distinct (Lemma 3.5.4), but for α to be part of a normal basis, the conjugates of α must be linearly independent. This latter is definitely a stronger condition, as the following simple example shows. Let $\mathbf{F} = \mathbf{Q}$ and $\mathbf{E} = \mathbf{Q}(\sqrt{2})$. Let $\alpha = \sqrt{2}$. Then the conjugates of α are α and $-\alpha$, so the conjugates are distinct (and $\mathbf{E} = \mathbf{F}(\alpha)$) but they are certainly not linearly independent (so $\{\alpha, -\alpha\}$ is not a normal basis for $\mathbf{E}$ over $\mathbf{F}$). ◇

3.7 Abelian Extensions and Kummer Fields

Let $\mathbf{E}$ be a Galois extension of $\mathbf{F}$. Then $\mathbf{E}$ is called an *abelian* extension of $\mathbf{F}$ if the Galois group $\mathrm{Gal}(\mathbf{E}/\mathbf{F})$ is abelian. Since abelian groups are easier to understand than arbitrary groups, we may hope that abelian extensions are easier to understand than arbitrary Galois extensions. As we shall see in this section, under the appropriate conditions on $\mathbf{F}$, we can indeed describe abelian extensions very concretely.

Recall that for a field $\mathbf{F}$, $\mathbf{F}_0$ denotes the prime field: $\mathbf{F}_0 = \mathbf{Q}$ if $\mathrm{char}(\mathbf{F}) = 0$ and $\mathbf{F}_0 = \mathbf{F}_p$ if $\mathrm{char}(\mathbf{F}) = p$.

Definition 3.7.1. An element $\zeta \in \mathbf{F}$ is a *primitive* n^{th} root of 1 in $\mathbf{F}$ if $\zeta^n = 1$, but $\zeta^m \neq 1$ for any $1 \leq m < n$. ◇

Lemma 3.7.2. *(1) If* $\mathbf{F}$ *has a primitive* n^{th} *root of* 1, *then* $\mathrm{char}(\mathbf{F}) = 0$ *or is prime to* n.

(2) The following are equivalent:

(a) $\mathbf{F}$ *has a primitive* n^{th} *root of* 1.

(b) $\mathbf{F}$ *has* n *distinct* n^{th} *roots of* 1.

Proof. (1) Suppose $\mathrm{char}(\mathbf{F}) = p$ and p divides n. Let $\zeta^n = 1$. Then ζ is a root of $X^n - 1$. But $X^n - 1 = (X^{n/p})^p - 1 = (X^{n/p} - 1)^p$ so ζ is a root of $X^{n/p} - 1$, i.e., $\zeta^{n/p} = 1$ and ζ is not primitive.

(2) If ζ is a primitive n^{th} root of 1, then $1, \zeta, \ldots, \zeta^{n-1}$ are distinct and are all n^{th} roots of 1. On the other hand, the group of nth roots of 1 in $\mathbf{F}$ is a subgroup of the multiplicative group of $\mathbf{F}$, so is cyclic (Corollary 2.2.2). Thus if $\mathbf{F}$ has n distinct n^{th} roots of 1, a generator of this group is a primitive n^{th} root of 1. □

Henceforth we assume that $\mathrm{char}(\mathbf{F}) = 0$ or that $p = \mathrm{char}(\mathbf{F})$ is relatively prime to n and we let ζ_n denote an arbitrary but fixed primitive n^{th} root of 1. Observe that $\zeta_n = \exp(2\pi i/n)$ is a primitive n^{th} root of 1 in $\mathbf{C}$.

Lemma 3.7.3. $\mathbf{F}_0(\zeta_n)$ *is a Galois extension of* $\mathbf{F}_0$, *and* $\mathrm{Gal}(\mathbf{F}_0(\zeta_n)/\mathbf{F}_0)$ *is isomorphic to a subgroup of* $(\mathbb{Z}/n\mathbb{Z})^*$. *In particular this group is abelian.*

Proof. $\mathbf{F}(\zeta_n)$ is the splitting field of the separable polynomial $X^n - 1$, so it is a Galois extension of $\mathbf{F}_0$.

If $\sigma_1 \in \text{Gal}(\mathbf{F}_0(\zeta_n)/\mathbf{F}_0)$, then $\sigma_1(\zeta_n)$ is a primitive n^{th} root of 1, so $\sigma_1(\zeta_n) = \zeta_n^{k_1}$ for some k_1 with $(k_1, n) = 1$.

Similarly, $\sigma_2 \in \text{Gal}(\mathbf{F}(\zeta_0)/\mathbf{F})$ satisfies $\sigma_2(\zeta_n) = \zeta_n^{k_2}$ where $(k_2, n) = 1$. Then $\sigma_1\sigma_2(\zeta_n) = \zeta_n^{k_1k_2} = \sigma_2\sigma_1(\zeta_n)$, and $\sigma \in \text{Gal}(\mathbf{F}_0(\zeta_n)/\mathbf{F}_0)$ is determined by $\sigma(\zeta_n)$; so we see that we have an injection $\sigma_i \mapsto k_i$ from $\text{Gal}(\mathbf{F}_0(\zeta_n)/\mathbf{F}_0)$ into $(\mathbb{Z}/n\mathbb{Z})^*$. □

Remark 3.7.4. We will see in Corollary 4.2.7 below that if $\mathbf{F}_0 = \mathbf{Q}$, then $\text{Gal}(\mathbf{F}_0(\zeta_n)/\mathbf{F}_0) = (\mathbb{Z}/n\mathbb{Z})^*$. This is false in general. In the most extreme case, if $p-1$ is a multiple of n, then $\zeta_n \in \mathbf{F}_0$, so in this case $\text{Gal}(\mathbf{F}_0(\zeta_n)/\mathbf{F}_0) = \{\text{id}\}$. ◇

We now introduce some nonstandard but very useful language.

Definition 3.7.5. Let $\alpha \in \mathbf{F}$. Then α is *n-powerless* in $\mathbf{F}$ if α is not an mth power in $\mathbf{F}$ for any m dividing n, $m > 1$. ◇

Proposition 3.7.6. *Let* $\mathbf{F} \supseteq \mathbf{F}_0(\zeta_n)$. *Let* $\mathbf{E}$ *be the splitting field of* $X^n - a \in \mathbf{F}[X]$. *Then* $\text{Gal}(\mathbf{E}/\mathbf{F})$ *is isomorphic to a subgroup of* $\mathbb{Z}/n\mathbb{Z}$. *In particular, this group is cyclic. Furthermore, the following are equivalent:*

(1) a *is n-powerless in* $\mathbf{F}$.

(2) $\text{Gal}(\mathbf{E}/\mathbf{F}) \cong \mathbb{Z}/n\mathbb{Z}$.

(3) $X^n - a$ *is irreducible in* $\mathbf{F}[X]$.

Proof. Let α be a root of $X^n - a$ in $\mathbf{E}$. Then $X^n - a = \prod_{i=0}^{n-1}(X - \zeta_n^i\alpha)$ so $\mathbf{E} = \mathbf{F}(\alpha)$, and $m_\alpha(X)$ divides $X^n - a$ in $\mathbf{F}[X]$ and hence in $\mathbf{E}[X]$ (Lemma 2.2.7) so $m_\alpha(X) = \prod_{k=0}^{m-1}(X - \zeta_n^{i_k}\alpha)$ for some subset $\{i_k\}$ of $\{0, \ldots, n-1\}$.

Since $\zeta_n^{i_k}\alpha$ and α are both roots of the irreducible polynomial $m_\alpha(X)$, by Lemma 2.6.1 there is an element $\sigma_k \in \text{Gal}(\mathbf{E}/\mathbf{F})$ with $\sigma_k(\alpha) = \zeta_n^{i_k}\alpha$, and this equation determines σ_k uniquely. Furthermore, if $\sigma_\ell(\alpha) = \zeta_n^{i_\ell}\alpha$, then $\sigma_k\sigma_\ell(\alpha) = \zeta_n^{i_k+i_\ell}\alpha$, so we see that $\{i_k\}$ form a subgroup of $\mathbb{Z}/n\mathbb{Z}$ isomorphic to $\text{Gal}(\mathbf{E}/\mathbf{F})$.

Now to the second part of the proposition. Since $|\text{Gal}(\mathbf{E}/\mathbf{F})| = (\mathbf{E}/\mathbf{F}) = (\mathbf{F}(\alpha)/\mathbf{F}) = \deg m_\alpha(X)$,

$$
\begin{aligned}
\text{Gal}(\mathbf{E}/\mathbf{F}) = \mathbb{Z}/n\mathbb{Z} &\Leftrightarrow |\text{Gal}(\mathbf{E}/\mathbf{F})| = n \\
&\Leftrightarrow \deg m_\alpha(X) = n \\
&\Leftrightarrow m_\alpha(X) = X^n - a \\
&\Leftrightarrow X^n - a \text{ is irreducible,}
\end{aligned}
$$

showing that (2) $\Leftrightarrow$ (3).

Suppose that (1) is false, so $a = b^m$ for some $b \in \mathbf{F}$. Let $n = jm$. Then $X^n - a = X^{jm} - b^m$ has $X^j - b$ as a proper factor, so it is not irreducible.

On the other hand, suppose that (3) is false. Then, as we have observed, $m_\alpha(X) = \prod_{k=0}^{m-1}(X - \zeta_n^{i_k}\alpha)$ and $\{\zeta^{i_k}\}$ form a group of order m isomorphic to a subgroup of $\mathbb{Z}/n\mathbb{Z}$, so it must be $\{\zeta_m^k \mid k = 0, \ldots, m-1\}$, and then $m_\alpha(X) = \prod_{k=0}^{m-1}(X - \zeta_m^k\alpha) = X - \alpha^m = X - b \in \mathbf{F}[X]$. But if $n = mj$, then $a = \alpha^n = \alpha^{mj} = (\alpha^m)^j = b^j$, $b \in \mathbf{F}$, so α is not n-powerless in $\mathbf{F}$. □

This proposition has a partial converse.

Proposition 3.7.7. *Let* $\mathbf{F} \supseteq \mathbf{F}_0(\zeta_n)$. *Let* $\mathbf{E}$ *be an extension of* $\mathbf{F}$. *The following are equivalent:*

(1) $\mathbf{E}$ *is the splitting field of* $X^n - a \in \mathbf{F}[X]$, *for some a that is n-powerless in* $\mathbf{F}$.

(2) $\mathbf{E}$ *is a Galois extension of* $\mathbf{F}$ *with* $\mathrm{Gal}(\mathbf{E}/\mathbf{F}) \cong \mathbb{Z}/n\mathbb{Z}$.

Proof. (1) implies (2) is (1) implies (2) of Proposition 3.7.6.

Now suppose (2) is true. Let σ be a generator of $\mathrm{Gal}(\mathbf{E}/\mathbf{F})$. By Theorem 3.6.2, $\mathbf{E}$ has a normal basis $\{\sigma^i(\alpha) \mid i = 0, \ldots, n-1\}$ for some $\alpha \in \mathbf{E}$. Let

$$\beta = \alpha + \zeta_n^{-1}\sigma(\alpha) + \cdots + \zeta_n^{-(n-1)}\sigma^{n-1}(\alpha).$$

Since $\{\sigma^i(\alpha)\}$ are independent, $\beta \neq 0$; then direct calculation shows

$$\sigma(\beta) = \sigma(\alpha) + \zeta_n^{-1}\sigma^2(\alpha) + \cdots + \zeta_n^{-(n-1)}\alpha = \zeta_n\beta,$$

and so $\sigma^i(\beta) = \zeta_n^i\beta$ for each i. Thus $\{\sigma^i(\beta)\}$ are all distinct, so $\mathbf{E} = \mathbf{F}(\beta)$ by Lemma 3.5.4. Furthermore,

$$m_\beta(X) = \prod_{i=0}^{n-1}(X - \zeta_n^i\beta) = X^n - \beta^n \in \mathbf{F}[X].$$

Thus, if $a = \beta^n$, $a \in \mathbf{F}[X]$ and $\mathbf{E}$ is the splitting field of $X^n - a$. Finally, a is n-powerless in $\mathbf{F}$ as in the proof of Proposition 3.7.6: If $a = b^m$, $n = mj$, then $m_\beta(X)$ would have $X^j - b$ as a factor, and $m_\beta(X)$ is irreducible so $j = n$ and $m = 1$. □

Remark 3.7.8. Consider the case $\mathbf{F}_0 = \mathbf{Q}$. In order to apply Propositions 3.7.6 and 3.7.7, we must do arithmetic in the field $\mathbf{Q}(\zeta_n)$, which may not be so easy. It would be a lot simpler if we only had to do arithmetic in $\mathbf{Q}$, which is easy. Fortunately, this is often, and indeed usually, the case. We have the following results, which we shall prove later on.

Lemma 4.6.5. *(1) Let n be an odd integer, and let a be n-powerless in $\mathbf{Q}$. Then a is a n-powerless in $\mathbf{Q}(\zeta_N)$ for every N.*
(2) Let n be an even integer, and let a be n-powerless in $\mathbf{Q}$. If a is negative, assume also that $-a$ is not a square in $\mathbf{Q}$. Then one of the following two alternatives holds:
(a) a is n-powerless in $\mathbf{Q}(\zeta_N)$ for every N.
(b) $a = b^2$ for some $b \in \mathbf{Q}(\zeta_N)$, for some N, and b is $n/2$-powerless in $\mathbf{Q}(\zeta_{N'})$ for every multiple N' of N.
Corollary 4.5.5. *Let a be a square-free integer. If a is positive and odd, set $a' = a$. Otherwise set $a' = 4|a|$. Then $a = b^2$ for some $b \in \mathbf{Q}(\zeta_n)$ if and only if a' divides n.* ◇

Definition 3.7.9. Let $\mathbf{F} \supseteq \mathbf{F}_0(\zeta_n)$. A *Kummer field* $\mathbf{E}$ is the splitting field of a polynomial of the form

$$f(X) = (X^n - a_1)(X^n - a_2)\cdots(X^n - a_t),\ a_i \in \mathbf{F},\ i = 1,\ldots,t. \quad \diamond$$

Recall that the *exponent* e of a finite group G is the smallest positive integer with the property that $g^e = 1$ for every $g \in G$, and that the exponent of a group divides its order.

Theorem 3.7.10. *Let $\mathbf{F} \supseteq \mathbf{F}_0(\zeta_n)$, and let $\mathbf{E}$ be an extension of $\mathbf{F}$. Then $\mathbf{E}$ is a Kummer field if and only if:*
(1) $\mathbf{E}$ is a Galois extension of $\mathbf{F}$.
(2) $\mathrm{Gal}(\mathbf{E}/\mathbf{F})$ is abelian.
(3) exponent $(\mathrm{Gal}(\mathbf{E}/\mathbf{F}))$ divides n.

Proof. First suppose that $\mathbf{E}$ is a Kummer field, the splitting field of $f(X) = (X^n - a_1)\cdots(X^n - a_t)$. Let $\alpha_i \in \mathbf{E}$ with $\alpha_i^n = a_i$ for each i. Then $X^n - a_i = \prod_{j=1}^n (X - \zeta_n^j \alpha_i)$, so $X^n - a_i$ splits into a product of distinct linear factors and is separable; then $f(X)$, the product of separable polynomials, is separable as well. Thus $\mathbf{E}$ is the splitting field of a separable polynomial in $\mathbf{F}[X]$, so $\mathbf{E}$ is a Galois extension of $\mathbf{F}$, proving (1). We prove (2) and (3) by induction on t. The $t = 1$ case follows immediately from Proposition 3.7.6. Assume the theorem is true for $t-1$. Let $\mathbf{B}_1 \subseteq \mathbf{E}$ be the splitting field of $f_1(X) = (X^n - a_1)\cdots(X^n - a_{t-1})$ over $\mathbf{F}$ and let $\mathbf{B}_2 \subseteq \mathbf{E}$ be the splitting field of $f_2(X) = X^n - a_t$ over $\mathbf{F}$. Then $\mathbf{E} = \mathbf{B}_1\mathbf{B}_2$. Then, by Corollary 3.4.7, $\mathrm{Gal}(\mathbf{E}/\mathbf{F})$ is a subgroup of $\mathrm{Gal}(\mathbf{B}_1/\mathbf{F}) \times \mathrm{Gal}(\mathbf{B}_2/\mathbf{F})$, so by the inductive hypothesis (for $f_1(X)$) and the $t = 1$ case (for $f_2(X)$) $\mathrm{Gal}(\mathbf{E}/\mathbf{F})$ is a subgroup of an abelian group of exponent dividing n, so $\mathrm{Gal}(\mathbf{E}/\mathbf{F})$ is itself an abelian group of exponent dividing n, and by induction the result follows.

Conversely, suppose $G = \mathrm{Gal}(\mathbf{E}/\mathbf{F})$ is as claimed. Write G as a direct sum of cyclic groups $G = H_1 \oplus \cdots \oplus H_s$. Since the exponent of G divides n, for each i the order m_i of H_i divides n.

We prove the theorem by induction on s. If $s = 1$, then $G = H_1$ is cyclic of order m, so, by Proposition 3.7.7, $\mathbf{E}$ is the splitting field of $X^m - b_1$ for some $b_1 \in \mathbf{F}$ or equivalently (since $\mathbf{F} \supseteq \mathbf{F}_0(\zeta_n)$) of $X^n - a_1$ where $a_1 = b_1^{n/m} \in \mathbf{F}$. Now suppose the theorem is true for $s - 1$ and let G be written as above. Set $\mathbf{B}_1 = \mathrm{Fix}(H_1 \oplus \cdots \oplus H_{s-1})$ and $\mathbf{B}_2 = \mathrm{Fix}(H_s)$. Then, by the inductive hypothesis, $\mathbf{B}_1$ is the splitting field of a polynomial $g_1(X)$ of the form $g_1(X) = (X^n - a_1)(X^n - a_2)\cdots(X^n - a_r)$ and, by the $s = 1$ case, $\mathbf{B}_2$ is the splitting field of a polynomial $g_2(X)$ of the form $g_2(X) = X^n - a_{r+1}$. Then, by Proposition 2.8.14 (1), $\mathbf{B}_1\mathbf{B}_2 = \mathrm{Fix}((H_1 \oplus \cdots \oplus H_{s-1}) \cap H_s) = \mathrm{Fix}(\{1\}) = \mathbf{E}$ and $\mathbf{E} = \mathbf{B}_1\mathbf{B}_2$ is the splitting field of $g_1(X)g_2(X) = (X^n - a_1)(X^n - a_2)\cdots(X^n - a_{r+1})$, and by induction the result follows. □

Let us now explicitly determine the Galois group of a Kummer extension. For a field $\mathbf{F}$, recall that $\mathbf{F}^*$ denotes the group (under multiplication) of nonzero elements of n. We let $(\mathbf{F}^*)^n$ denote those elements that are n^{th} powers, and observe that $(\mathbf{F}^*)^n$ is a subgroup of $\mathbf{F}^*$.

Theorem 3.7.11. *Let $\mathbf{F} \supseteq \mathbf{F}_0(\zeta_n)$ and let $\mathbf{E}$ be the splitting field of*

$$f(X) = (X^n - a_1)\cdots(X^n - a_t)$$

with each $a_i \neq 0$. Then $\mathrm{Gal}(\mathbf{E}/\mathbf{F})$ is isomorphic to $\langle a_1, \ldots, a_t\rangle \subseteq \mathbf{F}^*/(\mathbf{F}^*)^n$ (i.e., to the subgroup of $\mathbf{F}^*/(\mathbf{F}^*)^n$ generated by $a_1, \ldots, a_t$).

Proof. Let $\alpha_i \in \mathbf{E}^*$ with $\alpha_i^n = a_i$, $i = 1, \ldots, n$. Let $G = \langle \alpha_1, \ldots, \alpha_t\rangle \subset \mathbf{E}^*/\mathbf{F}^*$ and let $G' = \langle a_1, \ldots, a_t\rangle \subseteq \mathbf{F}^*/(\mathbf{F}^*)^n$. Let $\varphi : G \to G'$ by $\varphi(\alpha) = \alpha^n$. We claim φ is an isomorphism. Clearly, φ is an epimorphism, so we need only show it is a monomorphism. Suppose $\varphi(\alpha) = 1$. Then $\alpha^n \in (\mathbf{F}^*)^n$, so $\alpha^n = f^n$ for some $f \in \mathbf{F}^*$; then $\alpha = \zeta_n^k f \in \mathbf{F}^*$ for some k (as $\zeta_n \in \mathbf{F}$ by hypothesis). Thus instead of considering G' we may consider G.

Now write G as a direct sum of cyclic subgroups $G = H_1 \oplus \cdots \oplus H_s$ and let $\beta_i \in \mathbf{E}$ be a generator of H_i. Let H_i have order m_i dividing n. Set $b_i = \beta_i^{m_i}$, so $b_i \in \mathbf{F}$ and b_i is m_i-powerless in $\mathbf{F}$. (Then also $G' = H_1' \oplus \cdots \oplus H_s'$ with H_i' the subgroup generated by b_i.)

We prove the corollary by induction on s. We include in the inductive hypothesis that $\mathrm{Gal}(\mathbf{E}/\mathbf{F})$ is generated by elements η_i with $\eta_i(\beta_i) = \zeta_{m_i}\beta_i$ and $\eta_i(\beta_j) = \beta_j$ for $j \neq i$. If $s = 1$, this hypothesis is just the conclusion of Proposition 3.7.7.

Now assume the hypothesis is true for $s - 1$. Let $\mathbf{B} = \mathbf{F}(\beta_1, \ldots, \beta_{s-1})$ and $\mathbf{D} = \mathbf{F}(\beta_s)$.

Claim: **B** and **D** are disjoint extensions of **F** with composite **BD** = **E**.

Assuming this claim, the inductive step, and hence the corollary then follows immediately from Theorem 3.4.7.

Proof of claim: $\mathbf{E} = \mathbf{F}(\alpha_1, \ldots, \alpha_t)$ and $\langle \alpha_1, \ldots, \alpha_t \rangle = \langle \beta_1, \ldots, \beta_s \rangle \subset \mathbf{E}^*/\mathbf{F}^*$, so certainly $\mathbf{BD} = \mathbf{E}$.

Let $\mathbf{D}_0 = \mathbf{B} \cap \mathbf{D}$. By the *FTGT*, $\mathbf{B}_0 = \mathrm{Fix}(K)$ for some subgroup K of $\mathrm{Gal}(\mathbf{B}/\mathbf{F})$, which is a cyclic group of order m_s. From this we see that $\mathbf{D}_0 = \mathbf{F}((\beta_s)^{m'}) = \mathbf{F}(\sqrt[m]{b_s})$, where $mm' = m_s$, for some m dividing m_s. We want to show $m = 1$.

Let us set $\beta = \sqrt[m]{b_s} = (\beta_s)^{m'}$. Then $\beta \in \mathbf{B}$, so by the inductive hypothesis we may write uniquely

$$\beta = (\beta_s)^{m'} = \sum_{i_1,\ldots,i_{s-1}} c_{i_1,\ldots,i_{s-1}} \beta_1^{i_1} \cdots \beta_{s-1}^{i_{s-1}}$$

with $i_j = 0, \ldots, m_j - 1$, for each $j = 1, \ldots, s-1$, and with each $c_{i_1,\ldots,i_{s-1}} \in \mathbf{F}$.

Now let $\sigma \in \mathrm{Gal}(\mathbf{B}/\mathbf{F})$. Then $\sigma(\beta) = \zeta\beta$ for some root of unity ζ, and, similarly, $\sigma(\beta_1^{i_1} \cdots \beta_{s-1}^{i_{s-1}}) = \zeta' \beta_1^{i_1} \cdots \beta_{s-1}^{i_{s-1}}$ for some root of unity ζ'. Thus we see that we must have $\zeta' = \zeta$ for every term with $c_{i_1,\ldots,i_{s-1}} \neq 0$. But, by the inductive hypothesis, we may choose σ with $\sigma(\beta_j) = \zeta_j \beta_j$ where each ζ_j is an arbitrary m_j^{th} root of 1. Hence, if this summation has more than one nonzero term, we may find a σ which has the effect of multiplying one term by ζ and another term by ζ' with $\zeta' \neq \zeta$, which is impossible. Hence we can conclude there is only one term in the summation, and that

$$\beta = (\beta_s)^{m'} = c_{i_1,\ldots,i_{s-1}} \beta_1^{i_1} \cdots \beta_{s-1}^{i_{s-1}};$$

so we see that $(\beta_s)^{m'} \in H_1 \oplus \ldots \oplus H_{s-1} \subset \mathbf{E}^*/\mathbf{F}^*$. But by assumption $G = H_1 \oplus \cdots \oplus H_s$; in particular $(H_1 \oplus \ldots \oplus H_{s-1}) \cap H_s = \{1\}$, and hence we must have $m' = m_s$ and hence $m = 1$, as claimed. □

Corollary 3.7.12. *In the situation of, and in notation of the proof of Theorem 3.7.11, the isomorphism* $\oplus_{i=1}^{s}(\mathbf{Z}/m_i\mathbf{Z}) \to \mathrm{Gal}(\mathbf{E}/\mathbf{F})$ *is given by*

$$(j_1, \ldots, j_s) \mapsto \sigma \text{ where } \sigma(\beta_i) = \zeta_n^{j_i n/m_i} \beta_i \; (= \zeta_{m_i}^{j_i} \beta_i),\; i = 1, \ldots, s.$$

Proof. This follows directly from the proof of Theorem 3.7.11. □

Example 3.7.13. (1) Let **E** be the splitting field of $(X^2-2)(X^2-3)(X^2-5)$ over **Q**. Then $\mathrm{Gal}(\mathbf{E}/\mathbf{Q}) \cong (\mathbb{Z}/2\mathbb{Z})^3$ as $\langle 2, 3, 5 \rangle \subseteq \mathbf{Q}^*/(\mathbf{Q}^*)^2$ is isomorphic to $(\mathbb{Z}/2\mathbb{Z})^3$.

(2) Let $\mathbf{E}$ be the splitting field of $(X^2-2)(X^2-3)(X^2-6)$ over $\mathbf{Q}$. Then $\mathrm{Gal}(\mathbf{E}/\mathbf{Q}) \cong (\mathbb{Z}/2\mathbb{Z})^2$ as $\langle 2,3,6\rangle \subseteq \mathbf{Q}^*/(\mathbf{Q}^*)^2$ is isomorphic to $(\mathbb{Z}/2\mathbb{Z})^2$.

(3) Let $\mathbf{E}$ be the splitting field of $X^{12}-5$ over $\mathbf{Q}(\zeta_{12})$. Then $\mathrm{Gal}(\mathbf{E}/\mathbf{Q}(\zeta_{12})) \cong \mathbb{Z}/12\mathbb{Z}$ as $\langle 5\rangle \subseteq \mathbf{Q}(\zeta_{12})^*/(\mathbf{Q}(\zeta_{12})^*)^{12}$ is isomorphic to $\mathbb{Z}/12\mathbb{Z}$. The polynomial $X^{12}-5$ is irreducible over $\mathbf{Q}$ and also over $\mathbf{Q}(\zeta_{12})$.

(4) Let $\mathbf{E}$ be the splitting field of $X^{12}-3$ over $\mathbf{Q}(\zeta_{12})$. Then $\mathrm{Gal}(\mathbf{E}/\mathbf{Q}(\zeta_{12})) \cong \mathbb{Z}/6\mathbb{Z}$ as $\langle 3\rangle \subseteq \mathbf{Q}(\zeta_{12}^*)/(\mathbf{Q}(\zeta_{12})^*)^{12}$ is isomorphic to $\mathbb{Z}/6\mathbb{Z}$. Note $\sqrt{3} \in \mathbf{Q}(\zeta_{12})$. The polynomial $X^{12}-3$ is irreducible over $\mathbf{Q}$. Over $\mathbf{Q}(\zeta_{12})$, it factors as the product of two irreducible factors $(X^6-\sqrt{3})(X^6+\sqrt{3})$.

(5) Let $\mathbf{E}$ be the splitting field of X^4+1 over $\mathbf{Q}(\zeta_4)$. Then $\mathbf{E} = \mathbf{Q}(\zeta_8)$, $\mathrm{Gal}(\mathbf{E}/\mathbf{Q}(\zeta_4)) \cong \mathbf{Z}/2\mathbb{Z}$, and $\langle -1\rangle \subseteq \mathbf{Q}(\zeta_4)^*/(\mathbf{Q}(\zeta_4)^*)^4 \cong \mathbb{Z}/2\mathbb{Z}$ as -1 is a square but not a 4th power in $\mathbf{Q}(\zeta_4)$. The polynomial X^4+1 is irreducible over $\mathbf{Q}$. Over $\mathbf{Q}(\zeta_4)$, it factors as the product of two irreducible polynomials $(X^2+i)(X^2-i)$.

(6) Let $\mathbf{E}$ be the splitting field of X^4+4 over $\mathbf{Q}(\zeta_4)$. In fact $\mathbf{E} = \mathbf{Q}(\zeta_4)$ as $X^4+4 = (X-(1+i))(X-(1-i))(X-(-1+i))(X-(-1-i))$; thus $\mathrm{Gal}(\mathbf{E}/\mathbf{Q}) = \{\mathrm{id}\}$, and $\langle -4\rangle \subseteq \mathbf{Q}(\zeta_4)^*/(\mathbf{Q}(\zeta_4)^*)^4 \cong \{\mathrm{id}\}$ as $-4 = (1+i)^4$, an equation in $\mathbf{Q}(\zeta_4)$.

In this example we have made a number of claims about arithmetic in cyclotomic fields (i.e., in fields $\mathbf{Q}(\zeta_n)$). They will be verified in Section 4.2. ◇

3.8 The Norm and Trace

Let $\mathbf{E}$ be a finite extension of $\mathbf{F}$. We shall define two important functions from $\mathbf{E}$ to $\mathbf{F}$, the norm and the trace. We shall begin by defining them in case $\mathbf{E}$ is a Galois extension of $\mathbf{F}$, which is the most important case, and the case to which we denote most of our attention. At the end, we will show how to generalize the definition.

Definition 3.8.1. Let $\mathbf{E}$ be a finite Galois extension of $\mathbf{F}$ with Galois group $G = \mathrm{Gal}(\mathbf{E}/\mathbf{F})$. For $\alpha \in \mathbf{E}$, the *norm* (from $\mathbf{E}$ to $\mathbf{F}$) of α is

$$\mathrm{N}_{\mathbf{E}/\mathbf{F}}(\alpha) = \prod_{\sigma \in G} \sigma(\alpha)$$

and the *trace* (from $\mathbf{E}$ to $\mathbf{F}$) of α is

$$\mathrm{Tr}_{\mathbf{E}/\mathbf{F}}(\alpha) = \sum_{\sigma \in G} \sigma(\alpha). \qquad \diamond$$

Lemma 3.8.2. *For any $\alpha \in \mathbf{E}$, $\mathrm{N}_{\mathbf{E}/\mathbf{F}}(\alpha)$ and $\mathrm{Tr}_{\mathbf{E}/\mathbf{F}}(\alpha)$ are elements of $\mathbf{F}$.*

Proof. $\mathrm{N}_{\mathbf{E}/\mathbf{F}}(\alpha)$ and $\mathrm{Tr}_{\mathbf{E}/\mathbf{F}}(\alpha)$ are obviously invariant under G. □

Lemma 3.8.3. *Let $\alpha \in \mathbf{E}$ with $\deg m_\alpha(X) = r$, and write $m_\alpha(X) = \sum_{i=0}^{r} c_i X^i$. Let $(\mathbf{E}/\mathbf{F}) = n$. Then*

$$\mathrm{N}_{\mathbf{E}/\mathbf{F}}(\alpha) = (-1)^n c_0^{n/r} \text{ and } \mathrm{Tr}_{\mathbf{E}/\mathbf{F}}(\alpha) = -(n/r)c_{r-1}.$$

Proof. If α has distinct conjugates $\alpha_1, \ldots, \alpha_r$ (with $\alpha_1 = \alpha$), then $m_\alpha(X) = \prod_{i=1}^{r}(X - \alpha_i)$ by Lemma 2.7.12, so

$$c_0 = (-1)^r \alpha_1 \cdots \alpha_r \quad \text{and} \quad c_{r-1} = -(\alpha_1 + \cdots + \alpha_r). \qquad \square$$

Lemma 3.8.4. *For $\alpha \in \mathbf{E}$, let $T_\alpha : \mathbf{E} \to \mathbf{E}$ be the linear transformation of $\mathbf{F}$-vector spaces given by $T_\alpha(\beta) = \alpha\beta$. Then*

$$\mathrm{N}_{\mathbf{E}/\mathbf{F}}(\alpha) = \mathrm{Det}(T_\alpha) \text{ and } \mathrm{Tr}_{\mathbf{E}/\mathbf{F}}(\alpha) = \mathrm{Trace}(T_\alpha).$$

Proof. Let $\mathbf{B} = \mathbf{F}(\alpha)$. Then $\mathbf{B}$ has $\mathbf{F}$-basis $B = \{1, \alpha, \ldots, \alpha^{r-1}\}$ where $r = \deg m_\alpha(X)$. If $(\mathbf{E}/\mathbf{F}) = n$, then $(\mathbf{E}/\mathbf{B}) = n/r$. Choose a $\mathbf{B}$-basis $\{1 = \epsilon_1, \ldots, \epsilon_{n/r}\}$ for $\mathbf{E}$ over $\mathbf{B}$.

Then

$$C = \{1, \alpha, \ldots, \alpha^{r-1}, \ldots, \epsilon_{n/r}, \epsilon_{n/r}\alpha, \ldots, \epsilon_{n/r}\alpha^{r-1}\}$$

is a basis for $\mathbf{E}$ over $\mathbf{F}$.

Let $S_\alpha = T_\alpha \mid \mathbf{F}(\alpha)$. Now $S_\alpha : \mathbf{F}(\alpha) \to \mathbf{F}(\alpha)$ and the matrix of S_α in the basis B is just $C(m_\alpha(X))$, the companion matrix of the polynomial $m_\alpha(X)$ (see the proof of Proposition 2.4.6(3)). Then the matrix of T_α in the basis C is a block diagonal matrix consisting of n/r diagonal blocks, each equal to $C(m_\alpha(X))$, and the result then follows from Lemma 3.8.3. □

Theorem 3.8.5. *Let $\mathbf{E}$ be a finite Galois extension of $\mathbf{F}$ of degree n. Let $\mathbf{B}$ be a field intermediate between $\mathbf{F}$ and $\mathbf{E}$. (Then $\mathbf{E}$ is a Galois extension of $\mathbf{B}$.) Assume further that $\mathbf{B}$ is a Galois extension of $\mathbf{F}$. Then for any element a of $\mathbf{F}$ and any elements α, β of $\mathbf{E}$:*

(1) $\mathrm{N}_{\mathbf{E}/\mathbf{F}}(a) = a^n$ *and* $\mathrm{Tr}_{\mathbf{E}/\mathbf{F}}(a) = na$.

(2) $\mathrm{N}_{\mathbf{E}/\mathbf{F}}(\alpha\beta) = \mathrm{N}_{\mathbf{E}/\mathbf{F}}(\alpha)\,\mathrm{N}_{\mathbf{E}/\mathbf{F}}(\beta)$ *and* $\mathrm{Tr}_{\mathbf{E}/\mathbf{F}}(\alpha + \beta) = \mathrm{Tr}_{\mathbf{E}/\mathbf{F}}(\alpha) + \mathrm{Tr}_{\mathbf{E}/\mathbf{F}}(\beta)$.

(3) $\mathrm{N}_{\mathbf{E}/\mathbf{F}}(\alpha) = \mathrm{N}_{\mathbf{B}/\mathbf{F}}(\mathrm{N}_{\mathbf{E}/\mathbf{B}}(\alpha))$ *and* $\mathrm{Tr}_{\mathbf{E}/\mathbf{F}}(\alpha) = \mathrm{Tr}_{\mathbf{B}/\mathbf{F}}(\mathrm{Tr}_{\mathbf{E}/\mathbf{B}}(\alpha))$.

Proof. (1) and (2) are clear. We prove (3). Let $H = \mathrm{Gal}(\mathbf{E}/\mathbf{B}) = \{\tau_i\}$ have left coset representatives $\{\rho_j\}$. Then we may identify G/H with $\{\rho_j\}$. Then $\mathrm{N}_{\mathbf{B}/\mathbf{F}}(\mathrm{N}_{\mathbf{E}/\mathbf{B}}(\alpha)) = \prod_j \rho_j(\prod_i \tau_i(\alpha)) = \prod_{i,j} \rho_j\tau_i(\alpha) = \prod_{\sigma \in G} \sigma(\alpha) = \mathrm{N}_{\mathbf{E}/\mathbf{F}}(\alpha)$, and similarly for $\mathrm{Tr}_{\mathbf{E}/\mathbf{F}}(\alpha)$. □

Lemma 3.8.6. *There is an element $\alpha \in \mathbf{E}$ with $\mathrm{Tr}_{\mathbf{E}/\mathbf{F}}(\alpha) \neq 0$.*

Proof. If char(**F**) = 0 or $n = (\mathbf{E}/\mathbf{F})$ is relatively prime to $p = \mathrm{char}(\mathbf{F})$, this is completely trivial, as we may simply choose $\alpha = 1$.

In any case, for each $\sigma \in \mathrm{Gal}(\mathbf{E}/\mathbf{F})$, $\sigma : \mathbf{E}^* \to \mathbf{E}^*$ is a character of the group $\mathbf{E}^*$ in **E** in the sense of Definition 2.8.1, so the result follows immediately from Dirichlet's theorem on the independence of characters (Theorem 2.8.4). □

Theorem 3.8.7 (Hilbert's Theorem 90). *Let* **E** *be a cyclic extension of* **F** *of degree n and let σ be a generator of $G = \mathrm{Gal}(\mathbf{E}/\mathbf{F})$.*

(1) If β is any element of **E** *with $\mathrm{N}_{\mathbf{E}/\mathbf{F}}(\beta) = 1$, then $\beta = \alpha/\sigma(\alpha)$ for some $\alpha \in \mathbf{E}$.*

(2) If β is any element of **E** *with $\mathrm{Tr}_{\mathbf{E}/\mathbf{F}}(\beta) = 0$, then $\beta = \alpha - \sigma(\alpha)$ for some $\alpha \in \mathbf{E}$.*

Proof. (1) By Theorem 2.8.4, $\{1, \sigma, \ldots, \sigma^{n-1}\}$ are independent characters of $\mathbf{E}^*$ in **E**, so for any elements $\epsilon_0, \ldots, \epsilon_{n-1}$ of **E**, there is an element δ of **E** with $\alpha = \sum_{i=0}^{n-1} \epsilon_i \sigma^i(\delta) \neq 0$. Now let

$$\epsilon_i = \sigma^o(\beta)\sigma^1(\beta)\cdots\sigma^{i-1}(\beta), \quad i = 1, \ldots, n-1.$$

Note that $\beta\sigma(\epsilon_i) = \epsilon_{i+1}$ for $i = 1, \ldots, n-2$, and $\epsilon_{n-1} = \mathrm{N}_{\mathbf{E}/\mathbf{F}}(\beta) = 1$ so also $\beta\sigma(\epsilon_{n-1}) = \beta = \sigma^o(\beta) = \epsilon_1$. Let α be as above and note then that $\beta\sigma(\alpha) = \alpha$ so $\beta = \alpha/\sigma(\alpha)$.

(2) By Lemma 3.8.6, there is an element of γ of **E** with $\mathrm{Tr}_{\mathbf{E}/\mathbf{F}}(\gamma) \neq 0$. Set $\delta = \beta\sigma(\gamma) + (\beta + \sigma(\beta))\sigma^2(\gamma) + \cdots + (\beta + \cdots + \sigma^{n-2}(\beta))\sigma^{n-1}(\gamma)$. Then, by direct calculation,

$$\begin{aligned}\delta - \sigma(\gamma) &= \beta(\sigma(\gamma) + \cdots + \sigma^{n-1}(\gamma)) - (\sigma(\beta) + \cdots + \sigma^{n-1}(\beta))\gamma \\ &= \beta\,\mathrm{Tr}_{\mathbf{E}/\mathbf{F}}(\gamma) - \gamma\,\mathrm{Tr}_{\mathbf{E}/\mathbf{F}}(\beta) = \beta Tr_{\mathbf{E}/\mathbf{F}}(\gamma),\end{aligned}$$

so if $\alpha = \delta/\mathrm{Tr}_{\mathbf{E}/\mathbf{F}}(\gamma)$, we have that $\beta = \alpha - \sigma(\alpha)$. □

Now we extend the definitions of norm and trace. First we consider separable extensions. Here we assume familiarity with Section 5.2 below.

Definition 3.8.8. Let **E** be a finite separable extension of **F**. Then

$$\mathrm{N}_{\mathbf{E}/\mathbf{F}} = \mathrm{N}_{\mathbf{D}/\mathbf{F}} \text{ and } \mathrm{Tr}_{\mathbf{E}/\mathbf{F}} = \mathrm{Tr}_{\mathbf{D}/\mathbf{F}}$$

where **D** is any normal closure of **E**. ◇

Remark 3.8.9. (1) It is easy to see that $\mathbf{N}_{\mathbf{E}/\mathbf{F}}$ and $\mathrm{Tr}_{\mathbf{E}/\mathbf{F}}$ are well defined, i.e., independent of the choice of **D**.

(2) The results of this section through Theorem 3.8.5 go through unchanged in this more general situation. ◇

Now we consider arbitrary finite extensions. Here we further assume familiarity with Section 5.1 below.

Definition 3.8.10. Let **E** be a finite extension of **F**. Then

$$\mathrm{N}_{\mathbf{E}/\mathbf{F}} = (\mathrm{N}_{\mathbf{D}/\mathbf{F}})^d \text{ and } \mathrm{Tr}\,\mathbf{E}/\mathbf{F} = d\,\mathrm{Tr}_{\mathbf{D}/\mathbf{F}},$$

where **D** is any normal closure of **E** and $d = (\mathrm{Fix}(\mathrm{Gal}(\mathbf{D}/\mathbf{F}))/\mathbf{F})$. ◇

Remark 3.8.11. (1) Let $\mathbf{F}_i = \mathrm{Fix}(\mathrm{Gal}(\mathbf{D}/\mathbf{F}))$. Then **D** is a Galois extension of $\mathbf{F}_i$ and $\mathbf{F}_i$ is a purely inseparable extension of **F**. If **E** is a separable extension of **F**, then **D** is a Galois extension of **F**, $\mathbf{F}_i = \mathbf{F}$, and $d = 1$. In particular this is always the case for a perfect field **F**.

(2) If $\mathrm{char}(\mathbf{F}) = p$ and **E** is not a separable extension of **F**, then d is a power of p and so in this case $\mathrm{Tr}_{\mathbf{E}/\mathbf{F}} = 0$.

(3) Again the results of this section through Theorem 3.8.5 continue to hold. ◇

3.9 Exercises

Exercise 3.9.1. Give an example of an extension **E** of **F** of degree n and a divisor d of n for which there does *not* exist a field **B** intermediate between **F** and **E** with $(\mathbf{B}/\mathbf{F}) = d$.

Exercise 3.9.2. Let **F** be a field with $\mathrm{char}(\mathbf{F}) \neq 2$. Call a rational function $f(X_1, \ldots, X_n) \in \mathbf{F}(X_1, \ldots, X_n)$ *alternating* if $f(\sigma(X_1), \ldots, \sigma(X_n)) = \mathrm{sign}(\sigma) f(X_1, \ldots, X_n)$ for every $\sigma \in S_n$. Let $s_1, \ldots, s_n$ denote (as usual) the elementary symmetric functions in $X_1, \ldots, X_n$. Let δ be as in Definition 4.8.1.
(a) Show that every alternating rational function $f(X_1, \ldots, X_n)$ can be written as a rational function in $s_1, \ldots, s_n$ and δ in a unique way.
(b) Show that in fact every alternating rational function $f(X_1, \ldots, X_n)$ can be written as $f(X_1, \ldots, X_n) = g(s_1, \ldots, s_n)\delta$ for some rational function $g(s_1, \ldots, s_n)$ of $s_1, \ldots, s_n$.
(c) Show that every alternating polynomial $f(X_1, \ldots, X_n)$ can be written as a polynomial in $s_1, \ldots, s_n$ and δ in a unique way.
(d) Show that in fact every alternating polynomial $f(X_1, \ldots, X_n)$ can be

written as $f(X_1, \dots, X_n) = g(s_1, \dots, s_n)\delta$ for some polynomial function $g(s_1, \dots, s_n)$ in $s_1, \dots, s_n$.
(e) Show that every rational function $f(X_1, \dots, X_n)$ invariant under the action of A_n (i.e., every rational function $f(X_1, \dots, X_n)$ with $f(\sigma(X_1), \dots, \sigma(X_n)) = f(X_1, \dots, X_n)$ for every $\sigma \in A_n$) can be written in a unique way as $f(X_1, \dots, X_n) = f_s(X_1, \dots, X_n) + f_a(X_1, \dots, X_n)$ where $f_s(X_1, \dots, X_n)$ is symmetric and $f_a(X_1, \dots, X_n)$ is alternating. Furthermore, if $f(X_1, \dots, X_n)$ is a polynomial, then so are $f_s(X_1, \dots, X_n)$ and $f_a(X_1, \dots, X_n)$.

Exercise 3.9.3. Show that (2) implies (1) in Exercise 2.10.6.

Exercise 3.9.4. Show that (2) implies (1) in Exercise 2.10.7.

Exercise 3.9.5. Interpret α in Exercise 2.10.1 as being in an extension of $\mathbf{F}_5$, find an irreducible polynomial in $\mathbf{F}_5[X]$ that has α as a root, and find all roots of that polynomial in a splitting field $\mathbf{E}$. Also, find $G = \mathrm{Gal}(\mathbf{E}/\mathbf{F})$. (In the definition of α, there may be an ambiguity because of the roots. Handle all cases.)

Exercise 3.9.6. Interpret α in Exercise 2.10.1 as being in an extension of $\mathbf{F}_7$, find an irreducible polynomial in $\mathbf{F}_7[X]$ that has α as a root, and find all roots of that polynomial in a splitting field $\mathbf{E}$. Also, find $G = \mathrm{Gal}(\mathbf{E}/\mathbf{F})$. (In the definition of α, there may be an ambiguity because of the roots. Handle all cases.)

Exercise 3.9.7. In this exercise, let $\mathbf{E}$ be a splitting field of the polynomial $f(X) \in \mathbf{F}[X]$ and let $G = \mathrm{Gal}(\mathbf{E}/\mathbf{F})$. Find $(\mathbf{E}/\mathbf{F})$ and find G.
(a) $f(X) = X^3 - 1 \in \mathbf{F}_3[X]$.
(b) $f(X) = X^3 - 1 \in \mathbf{F}_5[X]$.
(c) $f(X) = X^3 - 1 \in \mathbf{F}_7[X]$.
(d) $f(X) = X^3 - 3 \in \mathbf{F}_5[X]$.
(e) $f(X) = X^3 + 4X + 2 \in \mathbf{F}_5[X]$.
(f) $f(X) = X^4 + 2X^3 + 3X^2 + 3X + 4 \in \mathbf{F}_5[X]$.
(g) $f(X) = X^4 + 3X^2 + 4 \in \mathbf{F}_5[X]$.
(h) $f(X) = X^4 + 3X^2 + 3 \in \mathbf{F}_5[X]$.

Exercise 3.9.8. Let $\mathbf{F} = \mathbf{F}_p$ and let $f(X) = X^p - X - a$ for some $a \neq 0 \in \mathbf{F}$. Let $\mathbf{E}$ be the splitting field of $\mathbf{F}$ and let α be a root of $f(X)$ in $\mathbf{E}$. Show that $\alpha, \alpha + 1, \dots, \alpha + (p - 1)$ are the roots of $f(X)$ in $\mathbf{E}$. Conclude that $f(X) \in \mathbf{F}[X]$ is irreducible and that $\mathrm{Gal}(\mathbf{E}/\mathbf{F})$ is cyclic of order p with generator σ the automorphism of $\mathbf{E}$ given by $\sigma(\alpha) = \alpha + 1$. Then the Frobenius automorphism Φ of $\mathbf{F}$ must be given by $\Phi(\alpha) = \alpha + k$ for some $k \neq 0 \in \mathbf{F}$. Show that a can be chosen so that $\Phi(\alpha) = \alpha + 1$. ($\mathbf{E}$ is called an *Artin–Schreier extension* of $\mathbf{F}$.)

Exercise 3.9.9. Let $f(X) \in \mathbb{Z}[X]$ be a monic polynomial and let $\bar{f}(X)$ be its image in $\mathbf{F}_p[X]$ (identifying $\mathbf{F}_p$ with $\mathbb{Z}/p\mathbb{Z}$). Let $\mathbf{E}$ be a splitting field for $f(X)$ and let $\bar{\mathbf{E}}$ be a splitting field for $\bar{f}(X)$. Show that if, for some p, all the roots of $\bar{f}(X)$ in $\bar{\mathbf{E}}$ are simple, then all the roots of $f(X)$ in $\mathbf{E}$ are simple.

Exercise 3.9.10. Let $\mathbf{B}_1$ and $\mathbf{B}_2$ be Galois extensions of $\mathbf{F}$, both contained in some field $\mathbf{A}$. Let $\mathbf{E}$ be the composite $\mathbf{E} = \mathbf{B}_1\mathbf{B}_2$.
(a) Show that every $\sigma \in \mathrm{Gal}(\mathbf{B}_1/\mathbf{F})$ extends to a (not necessarily unique) automorphism $\tilde{\sigma} \in \mathrm{Gal}(\mathbf{E}/\mathbf{F})$.
(b) Show that $\mathbf{B}_1$ and $\mathbf{B}_2$ are disjoint extensions of $\mathbf{F}$ if and only if every $\sigma \in \mathrm{Gal}(\mathbf{B}_1/\mathbf{F})$ extends to an automorphism $\tilde{\sigma} \in \mathrm{Gal}(\mathbf{E}/\mathbf{F})$ with $\tilde{\sigma} \mid \mathbf{B}_2 = \mathrm{id}$.

Exercise 3.9.11. Find an example of a field $\mathbf{F}$ and pairwise disjoint Galois extensions $\mathbf{B}_0, \mathbf{B}_1, \mathbf{B}_2$, of $\mathbf{F}$ with $\mathbf{B}_0$ and $\mathbf{B}_1\mathbf{B}_2$ *not* disjoint extensions of $\mathbf{F}$.

Exercise 3.9.12. Let $\mathbf{D}$ be a Galois extension of $\mathbf{F}$ and let $\mathbf{B}$ be an extension of $\mathbf{F}$. Suppose that $\mathbf{D}$ and $\mathbf{B}$ are disjoint extensions of $\mathbf{F}$.
(a) Show that $\mathbf{D}$ and $\mathbf{B}'$ are disjoint extensions of $\mathbf{F}$ for any Galois conjugate $\mathbf{B}'$ of $\mathbf{B}$.
(b) Let $\tilde{\mathbf{B}}$ be the Galois closure of $\mathbf{B}$. Give an example to show that $\mathbf{D}$ and $\tilde{\mathbf{B}}$ need not be disjoint.

Exercise 3.9.13. Let $\mathbf{B}_1$ and $\mathbf{B}_2$ be finite Galois extensions of $\mathbf{F}$. Show that $\mathbf{B}_1 \cap \mathbf{B}_2$ is a Galois extension of $\mathbf{F}$.

Exercise 3.9.14. (a) Let $a \in \mathbf{F}$ and suppose that a is not a pth power in $\mathbf{F}$. Let $\mathbf{B}$ be an extension of $\mathbf{F}$ of degree m, and suppose that m and p are relatively prime. Show that a is not a pth power in $\mathbf{B}$.
(b) More generally, let $a \in \mathbf{F}$ and suppose that a is n-powerless in $\mathbf{F}$. Let $\mathbf{B}$ be an extension of $\mathbf{F}$ of degree m, and suppose that m and n are relatively prime. Show that a is n-powerless in $\mathbf{B}$.

Exercise 3.9.15. Suppose that $X^m - a$ and $X^n - a$ are both irreducible polynomials in $\mathbf{F}[X]$.
(a) Suppose that m and n are relatively prime. Show that $X^{mn} - a$ is an irreducible polynomial in $\mathbf{F}[X]$.
(b) More generally, let $\ell = \mathrm{lcm}(m, n)$. Show that $X^\ell - a$ is an irreducible polynomial in $\mathbf{F}[X]$.

Exercise 3.9.16. (a) Let $\mathbf{E} = \mathbf{Q}(\sqrt{2}, \sqrt{3}, \sqrt{5}, \sqrt{7}, \ldots)$. Show that $\mathbf{E}/\mathbf{Q}$ is infinite.
(b) Let $\mathbf{E} = \mathbf{Q}(\sqrt{2}, \sqrt[3]{2}, \sqrt[4]{2}, \sqrt[5]{2}, \ldots)$. Show that $\mathbf{E}/\mathbf{Q}$ is infinite.

Exercise 3.9.17. (a) Let $\mathbf{F} \supseteq \mathbf{F}_0(\zeta_2)$ (and consequently char(**F**) $\neq 2$) and let $f(X) = X^2 - a \in \mathbf{F}[X]$ be irreducible. Let $\mathbf{E} = \mathbf{F}(\alpha)$ where $\alpha^2 = a$. Let $\beta = b_0 + b_1\alpha$ with $b_0, b_1 \in \mathbf{F}$. Show that the conjugates of β form a normal basis for **E** if and only if $b_0 b_1 \neq 0$.
(b) Let $\mathbf{F} \supseteq \mathbf{F}_0(\zeta_3)$ (and consequently char(**F**) $\neq 3$) and let $f(X) = X^3 - a \in \mathbf{F}[X]$ be irreducible. Let $\mathbf{E} = \mathbf{F}(\alpha)$ where $\alpha^3 = a$. Let $\beta = b_0 + b_1\alpha + b_2\alpha^2$ with $b_0, b_1, b_2 \in \mathbf{F}$. Show that the conjugates of β form a normal basis for **E** if and only if $b_0 b_1 b_2 \neq 0$.

Exercise 3.9.18. Let $\mathbf{F} \subseteq \mathbf{B} \subseteq \mathbf{E}$ with $\mathbf{E}/\mathbf{F}$ Galois. Show that there exist irreducible polynomials $f(X) \in \mathbf{F}[X]$ and $g(X) \in \mathbf{B}[X]$ such that **E** is a splitting field of $f(X)$, and **E** is also a splitting field of $g(X)$, and $g(X)$ divides $f(X)$ in $\mathbf{B}[X]$.

Exercise 3.9.19. (a) Let **E** be a separable extension of **F** and suppose there is an integer n such that $\deg m_\alpha(X) \leq n$ for every $\alpha \in \mathbf{E}$. Show that **E** is a finite extension of **F**. Furthermore, show that $(\mathbf{E}/\mathbf{F}) \leq n$.
(b) Give an explicit counterexample to (a) in case **E** is not a separable extension of **F**.

Exercise 3.9.20. Let **E** be an extension of **F** of degree 2, where $\mathbf{F} = \mathbf{F}_q$ is a finite field, with q odd. Let $\mathbf{E} = \mathbf{F}(\alpha)$, where $\alpha^2 \in \mathbf{F}$. Show that, for any $a, b \in \mathbf{F}$,

$$(a + b\alpha)^q = a - b\alpha \quad \text{or, equivalently,} \quad \mathrm{N}_{\mathbf{E}/\mathbf{F}}(\alpha) = (a + b\alpha)^{q+1}.$$

Exercise 3.9.21. Let **E** be an extension of **F** of degree 2, where char(**F**) $\neq 2$. Let $\mathbf{E} = \mathbf{F}(\delta)$, where $\delta^2 = d \in \mathbf{F}$. Let $T : \mathbf{E} \to \mathbf{E}$ be an **F**-linear transformation. Let $\alpha = T(1)$ and $\beta = T(\delta)/\delta$. (Note $\alpha, \beta \in \mathbf{E}$.)
(a) If $\alpha = \pm\beta$, show that there exists an $f \in \mathbf{F}$ with

$$\mathrm{N}_{\mathbf{E}/\mathbf{F}}(T(\epsilon)) = f\,\mathrm{N}_{\mathbf{E}/\mathbf{F}}(\epsilon) \text{ for all } \epsilon \in \mathbf{E}$$

and in fact $f = \mathrm{N}_{\mathbf{E}/\mathbf{F}}(\alpha) = \mathrm{N}_{\mathbf{E}/\mathbf{F}}(\beta)$.
(b) If $\alpha \neq \pm\beta$, show there does not exist such an f.

Exercise 3.9.22. Let **E** be an extension of **F** of degree 2, where char(**F**) $\neq 2$. Let $\mathbf{E} = \mathbf{F}(\delta)$, where $\delta^2 = d \in \mathbf{F}$, and let σ be the nontrivial element of Gal(**E**/**F**). Prove Hilbert's Theorem 90 (Theorem 3.8.7 (1)) directly in this case. That is, if $\gamma = a + b\delta \in \mathbf{E}$ with $\mathrm{N}_{\mathbf{E}/\mathbf{F}}(\gamma) = a^2 - b^2 d = 1$, then there is an element $\theta = x + y\delta \in \mathbf{E}$ with

$$\gamma = \frac{\theta}{\sigma(\theta)} = \frac{x + y\delta}{x - y\delta}.$$

Exercise 3.9.23. Let $\mathbf{F} = \mathbf{F}_{p^s}$ and let $\mathbf{E} = \mathbf{F}_{p^r}$ for some multiple r of s. Set $m = (p^r - 1)/(p^s - 1)$. Recall that $\mathrm{Gal}(\mathbf{E}/\mathbf{F})$ is cyclic of order r/s with generator Ψ given by $\Psi(\alpha) = \alpha^{p^s}$.
(a) Show that $\mathrm{N}_{\mathbf{E}/\mathbf{F}}(\alpha) = \alpha^m$ for every $\alpha \in \mathbf{E}$.
(b) Show that $\mathrm{N}_{\mathbf{E}/\mathbf{F}} : \mathbf{E} \to \mathbf{F}$ is an epimorphism.
(c) Prove Hilbert's Theorem 90 (Theorem 3.8.7 (1)) directly in this case: If $\beta \in \mathbf{E}$ with $\mathrm{N}_{\mathbf{E}/\mathbf{F}}(\beta) = 1$, then $\beta = (\alpha)(\alpha^{-p^s})$ for some $\alpha \in \mathbf{E}$.

Exercise 3.9.24. (a) Fix a polynomial $k(Y) \in \mathbf{F}[Y]$. Let $f(X) \in \mathbf{F}[X]$ be any separable polynomial. Let $f(X)$ split in $\mathbf{E}$ with roots $\alpha_1, \ldots, \alpha_m$. Let $Q_{k(Y)}(f(X))$ be the polynomial

$$Q_{k(Y)}(f(X)) = \prod_i (X - k(\alpha_i)).$$

Evidently $Q_{k(Y)}(f(X)) \in \mathbf{E}[X]$. Show that in fact $Q_{k(Y)}(f(X)) \in \mathbf{F}[X]$.
(b) Fix a polynomial $k(Y, Z) \in \mathbf{F}[Y, Z]$. Let $f(X), g(X) \in \mathbf{F}[X]$ be any separable polynomials. Let $f(X)$ and $g(X)$ split in $\mathbf{E}$ with roots $\alpha_1, \ldots, \alpha_m$ and $\beta_1, \ldots, \beta_n$, respectively. Let $Q_{k(Y,Z)}(f(X), g(X))$ be the polynomial

$$Q_{k(Y,Z)}(f(X), g(X)) = \prod_{i,j} (X - k(\alpha_i, \beta_j)).$$

Evidently $Q_{k(Y,Z)}(f(X), g(X)) \in \mathbf{E}[X]$. Show that in fact

$$Q_{k(Y,Z)}(f(X), g(X)) \in \mathbf{F}[X].$$

(This generalizes to arbitrarily many polynomials.)

Exercise 3.9.25. (a) Let $\{s_i\}$ be the elementary symmetric polynomials in $\{\alpha_i\}$. Show that for any $k(Y) \in \mathbf{F}[Y]$, $Q_{k(Y)}(f(X))$ is a polynomial whose coefficients are polynomials in $\{s_i\}$, and hence polynomials in the coefficients of $f(X)$.
(b) Let $\{s_i\}$ be the elementary symmetric polynomials in $\{\alpha_i\}$ and let $\{s'_j\}$ be the elementary symmetric polynomials in $\{\beta_j\}$. Show that for any $k(Y, Z) \in \mathbf{F}[Y, Z]$, $Q_{k(Y,Z)}(f(X), g(X))$ is a polynomial whose coefficients are polynomials in $\{s_i\}$ and $\{s'_j\}$, and hence polynomials in the coefficients of $f(X)$ and $g(X)$. (This generalizes to arbitrarily many polynomials.)

Exercise 3.9.26. (a) Show that if $k(Y) \in \mathbb{Z}[Y]$ and $f(X) \in \mathbb{Z}[X]$, then $Q_{k(Y)}(f(X)) \in \mathbb{Z}[X]$.
(b) Show that if $k(Y, Z) \in \mathbb{Z}[Y, Z]$ and $f(X), g(X) \in \mathbb{Z}[X]$, then $Q_{k(Y,Z)}(f(X), g(X)) \in \mathbb{Z}[X]$. (This generalizes to arbitrarily many polynomials.)

Exercise 3.9.27. Let $f(X) = X^2 + aX + b$, $g(X) = X^2 + cX + d$, and $h(X) = X^3 + eX^2 + fX + g$ be polynomials in $\mathbf{F}[X]$. Without finding the roots of these polynomials:
(a) Find $Q_{Y+1}(f(X))$, $Q_{2Y}(f(X))$, $Q_{Y^2}(f(X))$, and $Q_{Y^3}(f(X))$.
(b) Find $Q_{Y+1}(h(X))$, $Q_{2Y}(h(X))$, and $Q_{Y^2}(h(X))$.
(c) Find $Q_{Y+Z}(f(X), g(X))$ and $Q_{YZ}(f(X), g(X))$.
(d) Find $Q_{Y+Z}(f(X), h(X))$ and $Q_{YZ}(f(X), h(X))$.

Exercise 3.9.28. Let R be a subring of $\mathbf{F}$ and let $\mathbf{E}$ be an extension of $\mathbf{F}$. Let $S = \{\alpha \in \mathbf{E} \mid \alpha \text{ is integral over } R\}$. Show that S is a subring of $\mathbf{E}$.

Exercise 3.9.29. Let $\mathbf{E}$ be a field and let $R \subseteq \mathbf{E}$. Suppose that $\mathbf{E} = R[\alpha]$ for some $\alpha \in \mathbf{E}$. Show that there is a nonzero element $s \in R$ such that $R[1/s]$ is a field.

4

Extensions of the Field of Rational Numbers

In this chapter, we assume that all fields we consider are of characteristic 0.

4.1 Polynomials in $\mathbf{Q}[X]$

In this section we deal with a number of questions about polynomials in $\mathbf{Q}[X]$ related to factorization and irreducibility.

We begin with a couple of elementary results.

Lemma 4.1.1. *Let $\mathbf{F} \subseteq \mathbf{R}$ and let $f(X) \in \mathbf{F}[X]$. If $\gamma \in \mathbf{C}$ is a root of $f(X)$, then its complex conjugate $\bar{\gamma}$ is also a root of $f(X)$. Consequently, if $f(X)$ has odd degree, then it has an odd number of real roots, counted with multiplicity, and if $f(X)$ has even degree, then it has an even number of real roots, counted with multiplicity.*

Proof. $f(\bar{\gamma}) = \bar{f}(\bar{\gamma}) = \overline{f(\gamma)} = \bar{0} = 0.$ □

Lemma 4.1.2. *Let $f(X) \in \mathbb{Z}[X]$, $f(X) = \sum_{i=0}^{n} a_i X^i$. Suppose that $r \in \mathbf{Q}$ is a root of $f(X)$ (or equivalently, that $X - r$ is a factor of $f(X)$). If $r = s/t$, with s and t relatively prime, then s divides a_0 and t divides a_n. In particular, if $f(X)$ is monic, then r is an integer dividing a_0.*

Proof. $0 = f(s/t) = t^n f(s/t) = \sum_{i=1}^{n} a_i s^i t^{n-i}$. Now 0 is divisible by s, and every term of this summation except possibly $a_0 t^n$ is divisible by s, so that term must be divisible by s as well, and since t is relatively prime to s, this implies that a_0 is divisible by s. By the same logic applied to the term $a_n s^n$, a_n is divisible by t. □

S.H. Weintraub, *Galois Theory*, DOI 10.1007/978-0-387-87575-0_4,

Note that this lemma, used recursively, provides an effective method for finding all linear factors of $f(X)$ in $\mathbf{Q}[X]$.

Corollary 4.1.6 below is a standard result, but one we shall use repeatedly. First, some preliminaries.

Definition 4.1.3. A polynomial $f(X) = a_n X^n + \cdots + a_0 \in \mathbb{Z}[X]$ is *primitive* if $\gcd(a_n, \ldots, a_0) = 1$. ◇

Lemma 4.1.4 (Gauss's Lemma). *If $g(X)$ and $h(X)$ are primitive polynomials in $\mathbb{Z}[X]$, and $f(X) = g(X)h(X)$, then $f(X)$ is a primitive polynomial in $\mathbb{Z}[X]$.*

Proof. Write $g(X) = b_m X^m + \cdots + b_0$ and $h(X) = c_k X^k + \cdots + c_0$. Let $f(X) = g(X)h(X) = a_n X^n + \cdots + a_0$. Suppose $f(X)$ is not primitive and let $d = \gcd(a_n, \ldots, a_0)$. Pick a prime p dividing d. By hypothesis, p does not divide all of the coefficients of $g(X)$, so let i be the smallest index with b_i not divisible by p. Similarly, p does not divide all of the coefficients of $h(X)$, so let j be the smallest index with c_j not divisible by p.

Consider a_{i+j}, the coefficient of X^{i+j} in $f(X)$. Then

$$a_{i+j} = b_i c_j + (b_{i+1}c_{j-1} + b_{i+2}c_{j-2} + \cdots) + (b_{i-1}c_{j+1} + b_{i-2}c_{j+2} + \cdots).$$

By hypothesis, p divides a_{i+j}, and by construction p divides each of the terms in the two parenthesized expressions, so p divides $b_i c_j$ as well, and hence p divides b_i or p divides c_j, a contradiction. □

Proposition 4.1.5. *Let $f(X)$ be a polynomial in $\mathbb{Z}[X]$ and suppose $f(X) = g(X)h(X)$ with $g(X), h(X) \in \mathbf{Q}[X]$. Then $f(X) = g_1(X)h_1(X)$ with $g_1(X)$, $h_1(X) \in \mathbb{Z}[X]$, where $g_1(X)$ is a constant multiple of $g(X)$, and $h_1(X)$ is a constant multiple of $h(X)$.*

Proof. We may write $f(X) = ef_0(X)$ where $f_0(X) \in \mathbb{Z}[X]$ is primitive and $e \in \mathbb{Z}$. We may write $g(X) = cg_0(X)$ where $g_0(X) \in \mathbb{Z}[X]$ is primitive and $c \in \mathbf{Q}$. Similarly we may write $h(X) = dh_0(X)$ where $h_0(X) \in \mathbb{Z}[X]$ is primitive and $d \in \mathbf{Q}$. Then $ef_0(X) = f(X) = g(X)h(X) = cdg_0(X)h_0(X)$. By Lemma 4.1.4, $g_0(X)h_0(X)$ is primitive so we must have $e = \pm cd$, so $f_0(X) = \pm g_0(X)h_0(X)$ and $f(X) = g_1(X)h_1(X)$ where $g_1(X) = \pm eg_0(X) = (\pm e/c)g(X)$ and $h_1(X) = h_0(X) = (1/d)h(X)$. □

Corollary 4.1.6. *Let $f(X)$ be a monic polynomial in $\mathbb{Z}[X]$ and suppose $f(X) = g(X)h(X)$ with $g(X), h(X) \in \mathbf{Q}[X]$ monic polynomials. Then $g(X), h(X) \in \mathbb{Z}[X]$. Consequently, if $f(X)$ is irreducible in $\mathbb{Z}[X]$, then $f(X)$ is irreducible in $\mathbf{Q}[X]$.*

Proof. Apply Proposition 4.1.5, and note that we must have $g_1(X) = g(X)$ and $h_1(X) = h(X)$ (up to a common sign) as $f(X)$, $g(X)$, and $h(X)$ are monic. The second claim follows by contraposition. □

We have the following useful criterion for a polynomial to be irreducible over $\mathbf{Q}$.

Proposition 4.1.7 (Eisenstein's Criterion). *Let $f(X)$ be a polynomial with integer coefficients, $f(X) = a_nX^n + a_{n-1}X^{n-1} + \cdots + a_0$. Suppose there is a prime p with p not dividing a_n, p dividing $a_0, \ldots, a_{n-1}$, and p^2 not dividing a_0. Then $f(X)$ is irreducible in $\mathbf{Q}[X]$.*

Proof. By Corollary 4.1.6, it suffices to show that $f(X)$ is irreducible in $\mathbb{Z}[X]$. Suppose $f(X) = g(X)h(X)$ with $g(X) = b_mX^m + b_{m-1}X^{m-1} + \cdots + b_0$ and $h(X) = c_kX^k + c_{k-1}X^{k-1} + \cdots + c_0$. Then $b_mc_k = a_n$, so neither is divisible by p. Since p divides $a_0 = b_0c_0$ but p^2 does not divide a_0, by hypothesis, p must divide exactly one of b_0 and c_0. Let that one be b_0. Now $a_1 = b_1c_0 + b_0c_1$. Since p divides a_1 and b_0 but not c_0, p must divide b_1. Then $a_2 = b_2c_0 + b_1c_1 + b_0c_2$, so by similar logic p must divide b_2. Proceeding inductively, we find that p divides b_m, contradicting $b_m = 1$. □

Example 4.1.8. We will use Eisenstein's Criterion to show that for a prime p the polynomial $\Phi_p(X) = X^{p-1} + \cdots + X + 1 = (X^p - 1)/(X - 1)$ is irreducible. Of course, Eisenstein's Criterion does not apply directly. But observe that $\Phi_p(X)$ is irreducible if and only if $\Phi_p(X + 1)$ is irreducible, $\Phi_p(X + 1) = ((X + 1)^p - 1)/X$. Expanding by the Binomial Theorem, we see that $\Phi_p(X + 1) = \sum_{i=1}^{p} \binom{p}{i} X^{i-1}$ is a monic polynomial of degree $p - 1$ with the coefficient of every power of X other than X^{p-1} divisible by p, and with constant term p divisible by p but not p^2, so is irreducible. ($\Phi_p(X)$ is the pth cyclotomic polynomial. For n composite, $\Phi_n(X)$ is not defined by the above formula. We deal with cyclotomic polynomials in Section 4.2.) ◇

Let a be an integer of the form $a = pq$, where p is a prime and p and q are relatively prime. Then, by Eisenstein's Criterion (Proposition 4.1.7), the polynomial $X^n - a$ is irreducible over $\mathbf{Q}$ for any n. Thus, for example, $X^3 - 2$ is irreducible over $\mathbf{Q}$. However, Eisenstein's Criterion does not apply to $X^3 - 4$, which is also irreducible. Let us develop a criterion that does.

Lemma 4.1.9. *Let a be an integer of the form $a = p^mq$, where p is a prime and p and q are relatively prime. Let n be any integer that is relatively prime to m. Then the polynomial $X^n - a$ is irreducible over $\mathbf{Q}$.*

Proof. Let $\alpha = \sqrt[n]{a}$ and $\mathbf{E} = \mathbf{Q}(\alpha)$. Then $(\mathbf{E}/\mathbf{Q}) = \deg m_\alpha(X)$ by Proposition 2.4.6 (2). Thus $(\mathbf{E}/\mathbf{Q}) = n$ if and only $\deg m_\alpha(X) = n$. Now $m_\alpha(X)$ divides $X^n - a$ (as α is a root of this polynomial), so $\deg m_\alpha(X) = n$ if and only if $X^n - a$ is irreducible. Thus, putting these together, we see that $(\mathbf{E}/\mathbf{Q}) = n$ if and only if $X^n - a$ is irreducible.

Let j be an integer with $jm \equiv 1 (\bmod n)$, so $jm = nk + 1$ for some k. Clearly $\mathbf{Q}(\alpha) \supseteq \mathbf{Q}(\alpha^j)$ (and in fact they are equal). Now $(\alpha^j)^n = \alpha^{nj} = a^j = p^{mj}q^j = p^{nk+1}q^j$ so $\beta = \alpha^j/p^k$ satisfies $\beta^n = pq^j$. In other words, β is a root of the polynomial $X^n - pq^j$, and this polynomial is irreducible by Eisenstein's Criterion. Thus $(\mathbf{Q}(\alpha^j)/\mathbf{Q}) = (\mathbf{Q}(\beta)/\mathbf{Q}) = n$ so $(\mathbf{E}/\mathbf{Q}) = n$ as required. □

Corollary 4.1.10. *(1) Let p be an odd prime. If $a \in \mathbb{Z}$ is not a pth power, then $X^{p^t} - a$ is irreducible for every $t \geq 0$.*

(2) If $a \in \mathbb{Z}$ is positive and not a square, then $X^{2^t} - a$ is irreducible for every $t \geq 0$.

Proof. In these cases a satisfies the hypotheses of Lemma 4.1.9 (changing notation so that $n = p$ in case (1) and $n = 2$ in case (2).) □

Remark 4.1.11. In the excluded case, $X^{2^t} - a$ may not be irreducible. For example, $X^4 + 4 = (X^2 + 2X + 2)(X^2 - 2X + 2)$. ◇

Lemma 4.1.9 and Corollary 4.1.10 do not apply to the polynomial $X^6 - 72$. This polynomial is also irreducible, but it will require considerably more work to show this. We do this in Section 4.6, where we show that $X^n - a$ is irreducible in considerable generality.

Remark 4.1.12. There is an algorithm, due to Kronecker, for factoring polynomials $f(X) \in \mathbf{Q}[X]$.

It clearly suffices to consider the case that $f(X) \in \mathbb{Z}[X]$ is a primitive polynomial. Then, by Corollary 4.1.6, it suffices to factor $f(X)$ in $\mathbb{Z}[X]$. Let $f(X)$ have degree n.

We begin by recalling the Lagrange Interpolation Formula. If $a_1, \ldots, a_m$ are distinct, and $b_1, \ldots, b_m$ are arbitrary, there is a unique polynomial $g(X)$ of degree at most $m - 1$ with $g(a_i) = b_i$ for each i, given by

$$g(X) = \sum_{i=1}^{m} b_i \prod_{j \neq i} \frac{(X - a_j)}{(a_i - a_j)}.$$

Using this we proceed as follows, looking for a factor $g(X)$ of $f(X)$. If there is no such factor, then $f(X)$ is irreducible. If there is such a factor, write

$f(X) = g(X)f_1(X)$ and proceed recursively (next factoring $f_1(X), \ldots$). To find such a factor, let $m = 2, \ldots, [m/2]+1$. Pick distinct integers $a_1, \ldots, a_m$. If $f(a_i) = 0$ for some i, then we have found a linear factor $X - a_i$. Assume not and let $B_i = \{b_i\}$ be the finite set of divisors of $f(a_i)$, for each i. For each choice of elements $b_i \in B_i$, let $g(X)$ be the polynomial given by the Lagrange Interpolation Formula with $g(a_i) = b_i$. If $g(X)$ has degree less than $m-1$, or does not have integer coefficients, there is nothing to check. If $g(X)$ is a polynomial of degree $m-1$ with integer coefficients, test whether $f(X)$ is divisible by $g(X)$. ◇

Remark 4.1.13. For future reference, we record an extension of this algorithm. Let R be the polynomial ring $R = \mathbb{Z}[X_1, \ldots, X_n]$ and let $\mathbf{F}$ be its quotient field, the field of rational functions in $X_1, \ldots, X_n$ with coefficients in $\mathbf{Q}$. Let $f(X_{n+1}) \in \mathbf{F}[X_{n+1}]$ be a polynomial. Then there is an algorithm for factoring $f(X_{n+1})$.

We proceed by induction. For $n = 0$ this is just the algorithm of Remark 4.1.12. Assume we have an algorithm for $n = k$ and consider $n = k+1$. Now R is a *UFD* (a Unique Factorization Domain), so the analogs of Gauss's Lemma (Lemma 4.1.4), Proposition 4.1.5, and Corollary 4.1.6 continue to hold for R. Thus we may, as in Remark 4.1.12, assume that $f(X_{n+1})$ is a primitive polynomial in $R[X_{n+1}]$. Then proceed as in Remark 4.1.12. The only thing we need to see is that there is an algorithm for finding each of the sets B_i. But that follows by induction, since in factoring $f(a_i)$, we are in the case $n = k$. ◇

4.2 Cyclotomic Fields

In this section we study the splitting fields of the polynomials $X^n - 1$ over $\mathbf{Q}$.

Recall that ζ_n is a primitive n^{th} root of 1 in a field $\mathbf{E}$ if $\zeta_n^n = 1$ but $\zeta_n^m \neq 1$ for any $1 \leq m < n$. If $\mathbf{E}$ is the splitting field of $X^n - 1$ over $\mathbf{Q}$, then $\mathbf{E} = \mathbf{Q}(\zeta_n)$. In this case, we may simply take $\zeta_n = \exp(2\pi i/n)$, and we do this henceforth.

Definition 4.2.1. $\Phi_n(X) = \prod(X - \zeta)$ where the product is taken over all primitive n^{th} roots of 1.

$\Phi_n(X)$ is the nth *cyclotomic polynomial* and $\mathbf{E} = \mathbf{Q}(\zeta_n)$ is the nth *cyclotomic field.* ◇

Remark 4.2.2. Observe that $\Phi_n(X)$ and $X^n - 1 = \prod_{i=0}^{n-1}(X - \zeta_n^i)$ are both products of distinct linear factors in $\mathbf{E}[X]$, and that $\Phi_n(X)$ divides $X^n - 1$ in $\mathbf{E}[X]$. ◇

We let $\varphi(n)$ (as usual) denote the Euler totient function, $\varphi(n)$ being the number of integers between 1 and n that are relatively prime to n.

Lemma 4.2.3. $\deg \Phi_n(X) = \varphi(n)$.

Proof. We observe that ζ is a primitive n^{th} root of 1 if and only if $\zeta = \zeta_n^k$ for some k relatively prime to n, so $\{\text{primitive } n^{\text{th}} \text{ roots of } 1\} = \{\zeta^k \mid 1 \leq k \leq n, (k, n) = 1\}$ and the cardinality of this set, which by Definition 4.2.1 is the degree of $m_\zeta(X)$, is $\varphi(n)$. □

Lemma 4.2.4. *(1) $\Phi_n(X) \in \mathbf{Q}[X]$ for every n.*

(2) If $n_1 \neq n_2$, then $\Phi_{n_1}(X)$ and $\Phi_{n_2}(X)$ are relatively prime in $\mathbf{Q}[X]$.

(3) $X^n - 1 = \prod_{d|n} \Phi_d(X)$.

Proof. (1) Let $\mathbf{E} = \mathbf{Q}(\zeta_n)$. By Remark 4.2.2, $\mathbf{E}$ is the splitting field of the separable polynomial $X^n - 1$. Thus, by Theorem 2.7.14, $\mathbf{E}$ is a Galois extension of $\mathbf{Q}$. If ζ is a primitive n^{th} root of 1 in $\mathbf{E}$, then $\sigma(\zeta)$ is also a primitive n^{th} root of 1 for any $\sigma \in \text{Gal}(\mathbf{E}/\mathbf{Q})$, so $\Phi_n(X)$ is invariant under $\text{Gal}(\mathbf{E}/\mathbf{Q})$, and hence $\Phi_n(X) \in \mathbf{Q}[X]$.

(2) Let $\mathbf{E} = \mathbf{Q}(\zeta_{n_1 n_2})$. From Definition 4.2.1 we see that $\Phi_{n_1}(X)$ and $\Phi_{n_2}(X)$ are relatively prime in $\mathbf{E}[X]$, and hence also in $\mathbf{Q}[X]$ by Lemma 2.2.7.

(3) Every n^{th} root of 1 is a primitive d^{th} root of 1 for a unique d dividing n. □

Corollary 4.2.5. $\Phi_n(X) \in \mathbb{Z}[X]$.

Proof. From Lemma 4.1.4 we see that $\Phi_n(X)$ divides $X^n - 1$ in $\mathbf{Q}[X]$, and hence, since $\Phi_n(X)$ is monic, $\Phi_n(X) \in \mathbb{Z}[X]$ by Corollary 4.1.6. □

Theorem 4.2.6. $\Phi_n(X)$ *is irreducible for every n.*

First Proof. Suppose $\Phi_n(X)$ is not irreducible, and consider a monic irreducible factor $f(X)$ of $\Phi_n(X)$ that has ζ_n as a root. Then $f(X)$ does not have every primitive n^{th} root of unity as a root, so there is some $k > 1$ with $f(\zeta_n^k) \neq 0$. Choose the smallest such k, and let p be any prime factor of k. Since k is relatively prime to n, p is relatively prime to n as well. Set $\zeta = \zeta_n^{k/p}$. Then, by the minimality of k, $f(\zeta) = 0$ but $f(\zeta^p) \neq 0$. Now $f(X)$ is irreducible and monic, and $f(X)$ divides $X^n - 1$ in $\mathbf{Q}[X]$, so in fact $X^n - 1 = f(X)g(X)$ where both $f(X)$ and $g(X)$ are in $\mathbb{Z}[X]$, by Corollary 4.1.6. Then

$$0 = (\zeta^p)^n - 1 = f(\zeta^p)g(\zeta^p) \text{ and } f(\zeta^p) \neq 0, \text{ so } g(\zeta^p) = 0.$$

Writing $g(X) = \sum c_i X^i$, $c_i \in \mathbb{Z}$, we see that $0 = \sum c_i \zeta^{pi} = \sum c_i (\zeta^p)^i$, so $h(\zeta) = 0$ where $h(X) = g(X^p)$. Since $f(\zeta) = 0$ and $f(X)$ is irreducible and monic, $f(X)$ divides $h(X)$ in $\mathbf{Q}[X]$ and hence in $\mathbb{Z}[X]$, again by Corollary 4.1.6. Now let $\pi : \mathbb{Z}[X] \to (\mathbb{Z}/p\mathbb{Z})[X]$ be the map induced by reducing the coefficients mod p. Then $\pi(f(X))$ divides $\pi(h(X)) = \pi(g(X^p)) = \pi(g(X))^p$ in $(\mathbb{Z}/p\mathbb{Z})[X]$, so $\pi(f(X))$ and $\pi(g(X))$ have an irreducible factor $\bar{q}(X)$ in common. But

$$X^n - 1 = \pi(X^n - 1) = \pi(f(X))\pi(g(X)).$$

This would imply that $X^n - 1$ is divisible by $\bar{q}(X)^2$ in $(\mathbb{Z}/p\mathbb{Z})[X]$, and hence that $X^n - 1$ has a repeated linear factor in $\mathbf{E}_p$, its splitting field over $\mathbb{Z}/p\mathbb{Z} = \mathbf{F}_p$. However, $X^n - 1$ has distinct roots in $\mathbf{E}_p$, a contradiction. (This follows immediately from Lemma 3.2.1, but is easy to see directly.)

Second Proof (Landau). Let $m_{\zeta_n}(X)$ have degree d. Since $m_{\zeta_n}(X)$ is monic and divides X^n-1 in $\mathbf{Q}[X]$, $m_{\zeta_n}(X) \in \mathbb{Z}[X]$, by Corollary 4.1.6. We claim that, for any k, $\zeta_n^k = r_k(\zeta)$ for a unique polynomial $r_k(X) \in \mathbb{Z}[X]$ of degree less than d. To see this, note that, by the division algorithm, $X^k = m_{\zeta_n}(X)q_k(X)+r_k(X)$ with $q_k(X), r_k(X) \in \mathbb{Z}[X]$ unique polynomials with $\deg r_k(X) < d$, and then set $X = \zeta_n$. It follows immediately that for any polynomial $h(X) \in \mathbb{Z}[X]$, $h(\zeta_n^k)$ can be written as a unique polynomial in ζ_n of degree less than d with integer coefficients. In particular, we may write $m_{\zeta_n}(\zeta^k) = g_k(\zeta)$ with $g_k(X)$ a polynomial of this form. Since $\zeta_n^{n+k} = \zeta_n^k$, we see that $g_{n+k}(X) = g_k(X)$.

Now it follows immediately from the Multinomial Theorem and Fermat's Little Theorem ($a^p \equiv a(\text{mod}\, p)$ for every $a \in \mathbb{Z}$, for p a prime) that for any prime p and for any polynomial $f(X) \in \mathbb{Z}[X]$, $f(X^p) - f(X)^p$ has all of its coefficients divisible by p. In particular this is true for $f(X) = m_{\zeta_n}(X)$, so $m_{\zeta_n}(X^p) - m_{\zeta_n}(X)^p = pe(X)$ for some polynomial $e(X) \in \mathbb{Z}[X]$. Setting $X = \zeta$, we see that $g_p(\zeta) = m_{\zeta_n}(\zeta^p) = pe(\zeta^p)$. Now $e(\zeta^p)$ can be written as a polynomial in ζ of degree less than d with integer coefficients in a unique way, and $g_p(\zeta)$ is also such a polynomial, so $g_p(\zeta) = pe(\zeta^p)$. Thus we see that the coefficients of $g_p(X)$ are all divisible by p.

There are only finitely many distinct $g_k(X)$, so there is an upper bound A for the absolute values of all of their coefficients. Let p be any prime with $p > A$. Then we see that $g_p(X)$ is the zero polynomial, so $0 = g_p(\zeta) = m_{\zeta_n}(\zeta_n^p)$, and hence ζ_n and ζ_n^p have the same minimum polynomial $m_{\zeta_n}(X)$ (as $m_{\zeta_n}(X)$ is irreducible). We may iterate this argument to conclude that $m_\zeta(\zeta^s) = 0$ for any s not divisible by any prime $p \leq A$.

Let r be any integer relatively prime to n. Let $p_1, \ldots, p_t$ be the primes less than or equal to A that do not divide r, and set $s = r + np_1 \cdots p_t$. Then s is such an integer. (If $p \leq A$ is a prime, then either p is equal to p_i for some i, in which case p divides the second term but not the first, or it is not, in which case p divides the first term but not the second.) Then $s \equiv r(\text{mod}\, n)$, so

$m_{\zeta_n}(\zeta^r) = m_{\zeta_n}(\zeta^s) = 0$. Thus every primitive n^{th} root of 1 is a root of $m_{\zeta_n}(X)$ and hence $\Phi_n(X) = m_{\zeta_n}(X)$ is irreducible. □

Corollary 4.2.7. *(1)* $m_{\zeta_n}(X) = \Phi_n(X)$ *and* $(\mathbf{Q}(\zeta_n)/\mathbf{Q}) = \varphi(n)$.

(2) $\mathrm{Gal}(\mathbf{Q}(\zeta_n)/\mathbf{Q}) \cong (\mathbb{Z}/n\mathbb{Z})^*$. *In particular,* $\mathrm{Gal}(\mathbf{Q}(\zeta_n)/\mathbf{Q})$ *is abelian.*

Proof. (1) Since $\Phi_n(\zeta_n) = 0$, $m_{\zeta_n}(X)$ divides $\Phi_n(X)$. But $\Phi_n(X)$ is irreducible by Theorem 4.2.6 so they are equal. Then $(\mathbf{Q}(\zeta_n)/\mathbf{Q}) = \deg m_{\zeta_n}(X) = \varphi(n)$ where the first equality is Proposition 2.4.6 (2) and the second is Lemma 4.2.3.

(2) For any k with $(k, n) = 1$, ζ_n^k is also a primitive n^{th} root of 1, i.e, a root of $m_{\zeta_n}(X)$, so by Lemma 2.6.1 there is a map $\sigma_k : \mathbf{Q}(\zeta_n) \to \mathbf{Q}(\zeta_n)$ with $\sigma_k(\zeta_n) = \zeta_n^k$ and $\sigma \mid \mathbf{Q} = \mathrm{id}$, and σ_k is determined by its value on ζ_n. □

Corollary 4.2.8. *Let* m *and* n *be positive integers and set* $d = \gcd(m, n)$ *and* $\ell = \mathrm{lcm}(m, n)$. *Then* $\mathbf{Q}(\zeta_m)\mathbf{Q}(\zeta_n) = \mathbf{Q}(\zeta_\ell)$ *and* $\mathbf{Q}(\zeta_m) \cap \mathbf{Q}(\zeta_n) = \mathbf{Q}(\zeta_d)$.

In particular, if m *and* n *are relatively prime,* $\mathbf{Q}(\zeta_m)\mathbf{Q}(\zeta_n) = \mathbf{Q}(\zeta_{mn})$ *and* $\mathbf{Q}(\zeta_m) \cap \mathbf{Q}(\zeta_n) = \mathbf{Q}$, *i.e.,* $\mathbf{Q}(\zeta_m)$ *and* $\mathbf{Q}(\zeta_n)$ *are disjoint extensions of* $\mathbf{Q}$.

Proof. As $\zeta_m = \zeta_\ell^{\ell/m}$ and $\zeta_n = \zeta_\ell^{\ell/n}$, clearly $\mathbf{Q}(\zeta_\ell) \supseteq \mathbf{Q}(\zeta_m)\mathbf{Q}(\zeta_n)$. Now ℓ/m and ℓ/n are relatively prime, so let s and t be integers with $s(\ell/m) + t(\ell/n) = 1$, so $1/\ell = s/m + t/n$. Then $\zeta_\ell = \zeta_m^s\zeta_n^t$ so $\zeta_\ell \in \mathbf{Q}(\zeta_m)\mathbf{Q}(\zeta_n)$ and $\mathbf{Q}(\zeta_\ell) = \mathbf{Q}(\zeta_m)\mathbf{Q}(\zeta_n)$.

Observe that, for any integers r and s with r dividing s, $\mathbf{Q}(\zeta_r) \subseteq \mathbf{Q}(\zeta_s)$, so $(\mathbf{Q}(\zeta_s)/\mathbf{Q}(\zeta_r))(\mathbf{Q}(\zeta_r)/\mathbf{Q}) = (\mathbf{Q}(\zeta_s)/\mathbf{Q})$, and hence, by Corollary 4.2.7,

$$(\mathbf{Q}(\zeta_s)/\mathbf{Q}(\zeta_r)) = \varphi(s)/\varphi(r).$$

As $\zeta_d = \zeta_m^{m/d}$ and $\zeta_d = \zeta_n^{n/d}$, clearly $\mathbf{Q}(\zeta_d) \subseteq \mathbf{Q}(\zeta_m) \cap \mathbf{Q}(\zeta_n)$. Now, by Corollary 3.4.3 and the above observation,

$$\begin{aligned}(\mathbf{Q}(\zeta_n)/(\mathbf{Q}(\zeta_m) \cap \mathbf{Q}(\zeta_n))) &= (\mathbf{Q}(\zeta_m)\mathbf{Q}(\zeta_n)/\mathbf{Q}(\zeta_m)) \\ &= (\mathbf{Q}(\zeta_\ell)/\mathbf{Q}(\zeta_m)) \\ &= \varphi(\ell)/\varphi(m).\end{aligned}$$

Furthermore, again using the above observation,

$$\begin{aligned}(\mathbf{Q}(\zeta_n)/(\mathbf{Q}(\zeta_m) \cap \mathbf{Q}(\zeta_n)))((\mathbf{Q}(\zeta_m) \cap \mathbf{Q}(\zeta_n))/\mathbf{Q}(\zeta_d)) &= (\mathbf{Q}(\zeta_n)/\mathbf{Q}(\zeta_d)) \\ &= \varphi(n)/\varphi(d),\end{aligned}$$

so

$$\begin{aligned}((\mathbf{Q}(\zeta_m) \cap \mathbf{Q}(\zeta_n))/\mathbf{Q}(\zeta_d)) &= (\varphi(n)/\varphi(d))/(\varphi(\ell))/\varphi(m)) \\ &= (\varphi(m)\varphi(n))/(\varphi(d))\varphi(\ell)).\end{aligned}$$

But it is a fact from elementary number theory that for any two positive integers m and n, $\varphi(m)\varphi(n) = \varphi(d)\varphi(\ell)$, so $\mathbf{Q}(\zeta_m) \cap \mathbf{Q}(\zeta_n) = \mathbf{Q}(\zeta_d)$, as claimed. □

We conclude this section by giving a formula for $\Phi_n(X)$.

Proposition 4.2.9. *For any positive integer n, the n-th cyclotomic polynomial $\Phi_n(X)$ is given by*

$$\Phi_n(X) = \prod_{d|n}(X^d - 1)^{\mu(n/d)},$$

where μ is the Möbius μ function, defined by $\mu(j) = 0$ if j is divisible by the square of a prime, and otherwise $\mu(j) = (-1)^i$ where i is the number of distinct prime factors of j.

Proof. Since, by Lemma 4.2.4 (3), $X^n - 1 = \prod_{d|n} \Phi_d(X)$, the result follows immediately from the multiplicative version of the Möbius Inversion Formula. □

4.3 Solvable Extensions and Solvable Groups

In this section, we show that an equation is solvable by radicals if and only if its group is a solvable group. Of course, we must begin by defining these terms. We are most interested in the case $\mathbf{F} = \mathbf{Q}$, but the proofs apply more generally.

We develop the necessary field theory and the connections between field theory and group theory here. We refer the reader to Section A.1 where we develop the necessary group theory.

Definition 4.3.1. (1) A polynomial $f(X) \in \mathbf{F}[X]$ is a *radical polynomial* if $f(X) = X^n - a$ for some positive integer n and element a of $\mathbf{F}$. An extension $\mathbf{E}$ of $\mathbf{F}$ is a *radical extension* (or an *extension by radicals*) if $\mathbf{E} = \mathbf{F}(\alpha)$ for α a root of a radical polynomial.

(2) An equation $f(X) = 0$, $f(X) \in \mathbf{F}[X]$, is *solvable by radicals* if there is a sequence of extensions $\mathbf{E}_0 = \mathbf{F} \subseteq \mathbf{E}_1 \subseteq \cdots \subseteq \mathbf{E}_k$ with $\mathbf{E}_i$ the splitting field of a radical polynomial $f_i(X) \in \mathbf{E}_{i-1}(X)$ and with $\mathbf{E}_k \supseteq \mathbf{E}$, a splitting field of $f(X)$. ◇

(Radical polynomial is not a standard term, but radical extension is.)

Before proving our main theorem, Theorem 4.3.4, let us observe a special case.

Lemma 4.3.2. *Let $f(X) = X^n - a \in \mathbf{F}[X]$, let $\mathbf{E}$ be the splitting field of $f(X)$, and let $G = \mathrm{Gal}(\mathbf{E}/\mathbf{F})$. Then G is a solvable group.*

Proof. $\mathbf{E} = \mathbf{F}(\zeta_n, \sqrt[n]{a})$. Let $\mathbf{B} = \mathbf{F}(\zeta_n)$. Then $\mathbf{F} \subseteq \mathbf{B} \subseteq \mathbf{E}$. Let $G_1 = \mathrm{Gal}(\mathbf{E}/\mathbf{B})$. Now $\mathbf{B}$ is a Galois extension of $\mathbf{F}$, so G_1 is a normal subgroup of G, and G/G_1 is isomorphic to $\mathrm{Gal}(\mathbf{B}/\mathbf{F})$ by the *FTGT*. But G/G_1 is abelian, by Lemma 3.7.3 if $\mathbf{F} = \mathbf{Q}$, and by Lemma 3.7.3 and Corollary 3.4.6 in general, and G_1 is abelian by Proposition 3.7.6. Then G is solvable, as we see immediately from Definition A.1.1. □

Let us record a generalization of Lemma 4.3.2 that will help us in the proof of Theorem 4.3.4.

Lemma 4.3.3. *Let $f(X) = \prod_{j=1}^{t}(X^{n_j} - a_j)$ with $a_1, \ldots, a_j \in \mathbf{F}$, let $\mathbf{E}$ be the splitting field of $f(X)$ over $\mathbf{F}$, and let $G = \mathrm{Gal}(\mathbf{E}/\mathbf{F})$. Then G is a solvable group.*

Proof. Define a sequence of fields $\mathbf{D}_0 \subseteq \cdots \subseteq \mathbf{D}_t$ inductively, as follows. Let $\mathbf{D}_0 = \mathbf{F}$ and, if $\mathbf{D}_{j-1}$ is defined, let $\mathbf{D}_j$ be the splitting field of $(X^{n_j} - a_j) \in \mathbf{D}_{j-1}[X]$. Note that $\mathbf{D}_t = \mathbf{E}$. Let $G_j = \mathrm{Gal}(\mathbf{D}_{t-j}/\mathbf{D}_0)$ for $j = 0, \ldots, t$. Then $G = G_0 \supseteq \cdots \supseteq G_t = \{1\}$. Also, $G_{j-1}/G_j = \mathrm{Gal}(\mathbf{D}_{t-j+1}/\mathbf{D}_{t-j})$. But this group is solvable by Lemma 4.3.2, so, by Lemma A.1.3, G is solvable. □

Now for our main theorem.

Theorem 4.3.4. *Let $f(X) \in \mathbf{F}[X]$ and let $\mathbf{E}$ be a splitting field of $f(X)$. Then $f(X) = 0$ is solvable by radicals if and only if $G = \mathrm{Gal}(\mathbf{E}/\mathbf{F})$ is a solvable group.*

Proof. First, assume $G = \mathrm{Gal}(\mathbf{E}/\mathbf{F})$ is solvable. Let $\deg f(X) = m$ and set $n = m!$.

Case I: $\mathbf{F} \supseteq \mathbf{Q}(\zeta_n)$. Let $G = G_0 \supseteq \cdots \supseteq G_k = \{1\}$ with G_i a normal subgroup of G_{i-1} and G_{i-1}/G_i cyclic (see Lemma A.1.3), say of order m_i. Note $m_i \leq m$ so in particular m_i divides n. Let $\mathbf{B}_i = \mathrm{Fix}(G_i)$, and note that $\mathbf{B}_0 = \mathbf{F}$ and $\mathbf{B}_k = \mathbf{E}$. Then $\mathbf{B}_i$ is a cyclic extension of $\mathbf{B}_{i-1}$ of degree m_i, and $\mathbf{B}_{i-1} \supseteq \mathbf{Q}(\zeta_n) \supseteq \mathbf{Q}(\zeta_{m_i})$, so $\mathbf{B}_i$ is the splitting field of $X^{m_i} - a_i$, for some $a_i \in \mathbf{B}_{i-1}$, by Proposition 3.7.7.

Case II: The general case. The equation $X^n - 1$ is obviously solvable by radicals. Let $\mathbf{B} = \mathbf{F}(\zeta_n)$, the splitting field of this polynomial. Let $\mathbf{E_B}$ be the splitting field of $f(X) \in \mathbf{B}[X]$. By Corollary 3.4.6, $\mathrm{Gal}(\mathbf{E_B}/\mathbf{B})$ is isomorphic to a subgroup of $\mathrm{Gal}(\mathbf{E}/\mathbf{F})$. But, by Lemma A.1.4 (1), a subgroup of a solvable group is solvable, so we are reduced to case I.

Conversely, assume that $f(X) = 0$ is solvable by radicals. Then there is a sequence of fields $\mathbf{F} = \mathbf{B}_0 \subseteq \mathbf{B}_1 \subseteq \cdots \subseteq \mathbf{B}_k$ with $\mathbf{B}_i$ the splitting field of $X^{m_i} - a_i, a_i \in \mathbf{B}_{i-1}$, and with $\mathbf{B}_k \supseteq \mathbf{E}$. Then $\mathrm{Gal}(\mathbf{B}_i/\mathbf{B}_{i-1})$ is a solvable group, by Lemma 4.3.2.

Suppose for the moment that $k = 2$. Then we have that $\mathrm{Gal}(\mathbf{B}_2/\mathbf{B}_1)$ and $\mathrm{Gal}(\mathbf{B}_1/\mathbf{B}_0)$ are solvable groups, and that $\mathbf{B}_2/\mathbf{B}_1$ and $\mathbf{B}_1/\mathbf{B}_0$ are Galois extensions. If we knew that $\mathbf{B}_2/\mathbf{B}_0$ was a Galois extension, we could apply Lemma A.1.4 (3) to conclude that $G = \mathrm{Gal}(\mathbf{B}_2/\mathbf{B}_0)$ was a solvable group. However, a sequence of Galois extensions may not be Galois (and indeed, one can construct such an example precisely in this case). Thus in order to proceed we must enlarge the fields we are dealing with.

We proceed inductively. We begin by setting $\mathbf{B}'_0 = \mathbf{B}_0 = \mathbf{F}$ and $\mathbf{B}'_1 = \mathbf{B}_1$, and observe that $\mathbf{B}'_1$ is a Galois extension of $\mathbf{F}$. Suppose $\mathbf{B}'_{i-1}$ is defined and is a Galois extension of $\mathbf{F}$. We let $\mathbf{B}'_i$ be the splitting field of $\prod_{\sigma \in \mathrm{Gal}(\mathbf{B}'_{i-1}/\mathbf{B}'_0)}(X^{m_i} - \sigma(a_i)) \in \mathbf{B}'_{i-1}[X]$. Observe that this polynomial is invariant under $\mathrm{Gal}(\mathbf{B}'_{i-1}/\mathbf{B}'_0)$ so it is in fact in $\mathbf{B}'_0[X] = \mathbf{F}[X]$. Thus $\mathbf{B}'_i$ is the splitting field of a separable polynomial in $\mathbf{F}[X]$ so it is a Galois extension of $\mathbf{F}$. Thus we obtain a sequence of fields $\mathbf{F} = \mathbf{B}'_0 \subseteq \mathbf{B}'_1 \subseteq \cdots \subseteq \mathbf{B}'_k$ with each $\mathbf{B}'_i$ a Galois extension of $\mathbf{F}$ and with $\mathbf{B}'_k \supseteq \mathbf{B}_k \supseteq \mathbf{E}$.

Let $G = \mathrm{Gal}(\mathbf{B}'_k/\mathbf{F})$ and let $G_i = \mathrm{Gal}(\mathbf{B}'_{k-i}/\mathbf{B}'_0)$ for $i = 0, \ldots, k$. Then $G = G_0 \supseteq \cdots \supset G_k = \{1\}$. Also, $G_{i-1}/G_i = \mathrm{Gal}(\mathbf{B}'_{k-i+1}/\mathbf{B}'_{k-i})$. By Lemma 4.3.3, G_{i-1}/G_i is solvable for each $i = 1, \ldots, k$. Then, by Lemma A.1.3, $G = \mathrm{Gal}(\mathbf{B}'_k/\mathbf{F})$ is solvable. Now $\mathbf{F} \subseteq \mathbf{E} \subseteq \mathbf{B}'_k$ and $\mathbf{E}$ is a Galois extension of $\mathbf{F}$, so $\mathrm{Gal}(\mathbf{E}/\mathbf{F}) = \mathrm{Gal}(\mathbf{B}'_k/\mathbf{F})/\mathrm{Gal}(\mathbf{B}'_k/\mathbf{E})$ is a quotient of a solvable group and hence is solvable, by Lemma A.1.4 (2). □

Corollary 4.3.5 (Abel). *The general equation of degree n is not solvable by radicals for $n \geq 5$.*

Proof. Let $f_n(X) \in \mathbf{F}[X]$ be a polynomial with splitting field $\mathbf{E}$, where $\mathrm{Gal}(\mathbf{E}/\mathbf{F}) = S_n$, the symmetric group on n elements. Such a polynomial exists by Lemma 3.1.2, and in fact we have explicitly exhibited one in Remark 3.1.5. Now, by Corollary A.3.6, S_n is not solvable for $n \geq 5$. Thus, for $n \geq 5$, $f_n(X)$ is not solvable by radicals, by Theorem 4.3.4. □

While we credit Theorem 4.3.4 to Galois, in fact he did not state it that way. We shall now derive Galois's original result. One the one hand, this is of historical interest, but on the other hand, it provides an easy way of exhibiting a polynomial in $\mathbf{Q}[X]$ whose Galois group is not solvable.

Theorem 4.3.6 (Galois). *Let $f(X) \in \mathbf{F}[X]$ be an irreducible polynomial of prime degree p. Then $f(X) = 0$ is solvable by radicals if and only if all of its roots are rational functions of any two of them.*

Proof. By Theorem 4.3.4 we know that $f(X) = 0$ is solvable by radicals if and only if $G = \mathrm{Gal}(\mathbf{E}/\mathbf{F})$ is a solvable group, where $\mathbf{E}$ is a splitting field of $f(X)$. Since $f(X)$ is irreducible, G acts transitively on the roots of $f(X)$.

First, suppose that all of the roots of $f(X)$ are rational functions of any two of them. Regard G as a subgroup of the symmetric group S_p by its action permuting the roots of $f(X)$, and let $\sigma \in G$. If $\alpha_1 \neq \alpha_2$ are roots, then for any $\epsilon \in \mathbf{E}$, $\sigma(\epsilon)$ is determined by $\sigma(\alpha_1)$ and $\sigma(\alpha_2)$. There are at most $p(p-1)$ choices for $\sigma(\alpha_1)$ and $\sigma(\alpha_2)$, so G has order at most $p(p-1)$, and so by Corollary A.1.9 G is solvable.

Conversely, suppose that G is solvable. Let α_1 and α_2 be any two distinct roots of $f(X)$, and let $\mathbf{B} = \mathbf{F}(\alpha_1, \alpha_2) \subseteq \mathbf{E}$. We need to show that $\mathbf{B} = \mathbf{E}$.

We apply the Fundamental Theorem of Galois Theory (*FTGT*), Theorem 2.8.8. $\mathbf{B}$ is a field intermediate between $\mathbf{F}$ and $\mathbf{E}$, so $G' = \mathrm{Gal}(\mathbf{E}/\mathbf{B}) \subseteq G = \mathrm{Gal}(\mathbf{E}/\mathbf{F})$. By definition, $\mathrm{Gal}(\mathbf{E}/\mathbf{B}) = \{\sigma \in G \mid \sigma(\beta) = \beta \text{ for every } \beta \in \mathbf{B}\}$. But the *FTGT* establishes a $1-1$ correspondence between subgroups of G and fields intermediate between $\mathbf{F}$ and $\mathbf{E}$, so $\mathbf{B} = \mathbf{E}$ if and only if $G' = \{\mathrm{id}\}$. We proceed to show this.

Let $\sigma \in G$. Note that $\sigma \in G'$ if and only if $\sigma(\alpha_1) = \alpha_1$ and $\sigma(\alpha_2) = \alpha_2$.

By Proposition A.1.7 (and using the notation there), we may identify G with G_H for some subgroup H of $\mathbf{F}_p^*$, and α_1 and α_2 with $\left[\begin{smallmatrix} i_1 \\ 1 \end{smallmatrix}\right]$ and $\left[\begin{smallmatrix} i_2 \\ 1 \end{smallmatrix}\right]$ for some $i_1 \neq i_2$.

Let $\sigma = \left[\begin{smallmatrix} h & n \\ 0 & 1 \end{smallmatrix}\right] \in G_H$ with $\sigma(\alpha_1) = \alpha_1$ and $\sigma(\alpha_2) = \alpha_2$. Then

$$\begin{bmatrix} i_1 \\ 1 \end{bmatrix} = \begin{bmatrix} h & n \\ 0 & 1 \end{bmatrix} \begin{bmatrix} i_1 \\ 1 \end{bmatrix} = \begin{bmatrix} hi_1 + n \\ 1 \end{bmatrix}, \quad \begin{bmatrix} i_2 \\ 1 \end{bmatrix} = \begin{bmatrix} h & n \\ 0 & 1 \end{bmatrix} \begin{bmatrix} i_2 \\ 1 \end{bmatrix} = \begin{bmatrix} hi_2 + n \\ 1 \end{bmatrix},$$

and these two equations readily imply $h = 1$, $n = 0$, i.e., $\sigma = \mathrm{id}$. Thus we see that $G' = \{\mathrm{id}\}$, completing the proof. □

Corollary 4.3.7. *Let* $\mathbf{F} \subseteq \mathbf{R}$ *and let* $f(X) \in \mathbf{F}[X]$ *be an irreducible polynomial of prime degree* p *with* k *real roots,* $1 < k < p$. *Then* $f(X) = 0$ *is not solvable by radicals.*

Proof. Suppose $f(X)$ has real roots $\alpha_1 \neq \alpha_2$ and a nonreal root β. Then β is not a rational function of α_1 and α_2, so, by Theorem 4.3.6, $f(X) = 0$ is not solvable by radicals. □

Remark 4.3.8. In Example 4.7.5 we will see that, for each odd prime p, there is an irreducible polynomial $f_p(X) \in \mathbf{Q}[X]$ with $p - 2$ real roots. Thus, by Corollary 4.3.7, $f_p(X)$ is not solvable by radicals for $p \geq 5$. But in fact we will show the stronger result that the Galois group of $f_p(X)$ over $\mathbf{Q}$ is isomorphic to S_p, thus providing an example of Corollary 4.3.5 over $\mathbf{Q}$. We will further provide an example of such a polynomial $f_n(X)$ over $\mathbf{Q}$ for every $n \geq 5$ in Proposition 4.7.10, but the construction there is considerably more complicated. ◇

4.4 Geometric Constructions

In this section we will investigate the question of constructibility by straightedge and compass. We will see that the three classical questions of Greek antiquity (trisecting the angle, duplicating the cube, and squaring the circle) are unsolvable by these methods. We will also see when it is possible to construct a regular n-gon by these methods, a result originally due to Gauss. In particular, we will show how to construct a regular 17-gon.

We begin by drawing an arbitrary line L, and choose two distinct points O and A on it. We normalize coordinates by letting O be the origin and A be the point $(1, 0)$, so in particular the line segment OA has length 1. Thus the line L is the x-axis.

We recall that, using straightedge and compass, it is possible to construct a line through a given point parallel to, or perpendicular to, a given line. Thus, in particular, we may construct the y-axis and hence we have the usual Cartesian coordinates in the plane.

There are two ways to regard points in the plane. We may consider that a point P has coordinates (x, y), or that P is represented by the complex number $z = x + iy$. If we have constructed the point P represented by $z = x + iy$, then we may drop perpendiculars to obtain the real and imaginary parts x and y of P; conversely, given real numbers x and y we may certainly obtain the point (x, y) corresponding to $z = x + iy$. Also, all arithmetic operations on complex numbers (addition, subtraction, multiplication, division) may be performed by doing arithmetic operations on their real and imaginary parts. Thus we shall pass freely between these two different descriptions.

Our first main result is the following:

Theorem 4.4.1. *The complex number z can be constructed by straightedge and compass if and only if z is algebraic over $\mathbf{Q}$ and there is a sequence of fields*

$$\mathbf{F}_0 = \mathbf{Q} \subset \mathbf{F}_1 \subset \cdots \subset \mathbf{F}_k$$

with $\mathbf{Q}(z) \subseteq \mathbf{F}_k$ and with $(\mathbf{F}_i/\mathbf{F}_{i-1}) = 2$ for each $i = 1, \ldots, k$.

Proof. First, we observe that we may perform all arithmetic operations with straightedge and compass. Clearly we can add and subtract (positive) real numbers x and y. We may multiply them as follows: Let the line segment AB have length x and extend AB to a line L. Let M be another line through A and let C and D be points on the same side of A with the line segments AC and AD having lengths 1 and y, respectively. Construct the line N through D parallel to the line segment CB and let this line intersect the line L in the point E. Then, by similar triangles, the line segment AE has length xy. To divide we perform a similar construction, this time letting N be the line through C parallel to the line segment DB, and then AE has length x/y.

We also observe that we may take the square root of a (positive) real number x using straightedge and compass. To do so, let line segment AB have length x and extend AB to a line L. Let C be a point on the opposite side of A from B with the length of the line segment AC equal to 1. Construct a semicircle with diameter CB, and a line M perpendicular to CB at A. Let D be the point at which the line M intersects this semicircle. Then, by similar triangles, AD has length $\sqrt{x}$.

With this in hand, we turn to the proof of the theorem.

First, suppose there is a sequence of fields as indicated. By our first observation, every $x \in \mathbf{Q} = \mathbf{F}_0$ can be constructed. Since $(\mathbf{F}/\mathbf{F}_0) = 2$, $\mathbf{F}_1 = \mathbf{F}_0(\alpha_1)$ where α_1 is a root of an irreducible quadratic equation with coefficients in $\mathbf{F}_0$. Then, by the quadratic formula, $\mathbf{F}_0(\alpha_1) = \mathbf{F}_0(\sqrt{a_1})$ for some $a_1 \in \mathbf{F}_0$, and by our second observation, $\sqrt{a_1}$ can be constructed, and hence every $x \in \mathbf{F}_1$ can be constructed. Then proceed by induction.

Conversely, suppose a point can be constructed with a straightedge and compass. Then it is obtained by a succession of the following operations:

(1) Finding the intersection of two lines.
(2) Finding the intersection of a circle and a line.
(3) Finding the intersection of two circles.

In case of operation (1), let the two lines be given by the equations $a_1x + b_1y = c_1$ and $a_2x + b_2y = c_2$ with $a_1, a_2, b_1, b_2, c_1, c_2 \in \mathbf{F}$ for some field $\mathbf{F}$. Then the solution (x, y) has $x \in \mathbf{F}$ and $y \in \mathbf{F}$. In case of operation (2), let the circle be given by $(x - a_1)^2 + (y - b_1)^2 = c_1^2$ and the line be given by $a_2x + b_2y = c_2$ with $a_1, a_2, b_1, b_2, c_1, c_2 \in \mathbf{F}$. Solving for y in terms of x (or vice versa) in the linear equation, substituting in the quadratic equation, and solving, shows $x \in \mathbf{F}(d)$ and $y \in \mathbf{F}(d)$ where $(\mathbf{F}(d)/\mathbf{F}) = 1$ or 2. In case of operation (3), let the two circles have equations $(x - a_1)^2 + (y - b_1)^2 = c_1^2$ and $(x - a_2)^2 + (y - b_2)^2 = c_2^2$ with $a_1, a_2, b_1, b_2, c_1, c_2 \in \mathbf{F}$. Subtracting the second equation from the first gives a linear equation $a_3x + b_3y = c_3$ for some $a_3, b_3, c_3 \in \mathbf{F}$, so combining the equation of the first circle (say) with this linear equation, we are reduced to operation (2). This yields the theorem. □

Corollary 4.4.2. *It is impossible to perform the following constructions using straightedge and compass:*

(1) (Trisecting the angle) Trisecting an arbitrary angle.

(2) (Duplicating the cube) Constructing a cube with volume twice that of a given cube.

(3) (Squaring the circle) Constructing a square with area that of a given circle.

Proof. (1) Given an angle θ, let one side of the angle be the x-axis and let the other side of the angle intersect the unit circle at P. Then P has coordinates $(\cos\theta, \sin\theta)$. Thus we may construct the angle $\theta/3$ if and only if we may construct the real number $\cos\theta/3$. Recall the formula $\cos 3\varphi = 4\cos^4\varphi - 3\cos\varphi$. Letting $\varphi = \theta/3$ and $x = \cos\varphi$, we see that x satisfies the cubic equation $4x^3 - 3x = \cos\theta$. We may certainly construct an angle $\theta = \pi/3$ ($= 60°$), and $\cos(\pi/3) = \frac{1}{2}$. Thus x is a root of the polynomial $f(X) = 8X^3 - 6X - 1$. Using Lemma 4.1.2, it is easy to check that $f(X)$ has no linear factor in $\mathbf{Q}(X)$, and hence is irreducible, so $(\mathbf{Q}(x)/\mathbf{Q}) = 3$. But then we cannot have $\mathbf{Q}(x) \subseteq \mathbf{F}_k$ with $\mathbf{F}_k$ as in Theorem 4.4.1, as $(\mathbf{F}_k/\mathbf{Q})$ is a power of 2 and hence not divisible by 3.

(2) Trivially, we may construct a segment of length the side of a cube of volume 1. But then to construct a segment of length the side of a cube of volume 2 would be to construct $x = \sqrt[3]{2}$. But $(\mathbf{Q}(\sqrt[3]{2})/\mathbf{Q}) = 3$ does not divide any power of 2, so this is impossible as in part (1).

(3) Trivially, we may construct a segment of length the radius of a circle of area π. But then to construct a segment of length the side of a square of area π would be to construct $x = \sqrt{\pi}$. But it is a famous theorem of Lindemann that π is transcendental (i.e., not algebraic, or equivalently that π and hence $\sqrt{\pi}$ does not satisfy any algebraic equation) so $(\mathbf{Q}(\sqrt{\pi})/\mathbf{Q})$ is infinite and hence $\sqrt{\pi}$ is not an element of any finite extension of $\mathbf{Q}$ (regardless of degree). □

We have an equivalent formulation of the condition in Theorem 4.4.1.

Lemma 4.4.3. *For a complex number z, the following are equivalent:*

(1) There is a sequence of fields $\mathbf{F}_0 = \mathbf{Q} \subset \mathbf{F}_1 \subset \cdots \subset \mathbf{F}_k$ *with* $\mathbf{Q}(z) \subseteq \mathbf{F}_k$ *and with* $(\mathbf{F}_i/\mathbf{F}_{i-1}) = 2$ *for each* $i = 1, \ldots, k$.

(2) If $\mathbf{E}$ *is a splitting field of* $m_z(X)$, *then* $(\mathbf{E}/\mathbf{Q})$ *is a power of* 2.

Proof. (2) implies (1): Let $G = \mathrm{Gal}(\mathbf{E}/\mathbf{Q})$ and let G have order 2^k. By Corollary A.2.3 there is a sequence of subgroups $G = G_0 \supset G_1 \supset \cdots \supset G_k = \{1\}$ with $[G : G_i] = 2^i$. Let $\mathbf{F}_i = \mathrm{Fix}(G_i)$. Then $\{\mathbf{F}_i\}$ is as required.

(1) implies (2): We prove this by induction on k. For $k = 0$ it is trivial and for $k = 1$ it is immediate, as every quadratic extension is a splitting field. Let $(\mathbf{F}_k/\mathbf{F}_{k-1}) = 2$. Then $\mathbf{F}_k = \mathbf{F}_{k-1}(\alpha_k)$ for some $\alpha_k \in \mathbf{F}_k$ with

$\deg m_{\alpha_k}(X) = 2$. In the course of the proof we shall have occasion to replace the field $\mathbf{F}_{k-1}$ by another field $\mathbf{E}_{k-1}$. Observe that for any field $\mathbf{E}_{k-1} \supseteq \mathbf{F}_{k-1}$, $(\mathbf{E}_{k-1}(\alpha_k)/\mathbf{E}_{k-1}) = 1$ or 2 by Lemma 2.4.7.

Let $\mathbf{E}_0 = \mathbf{F}_0$ and $\mathbf{E}_1 = \mathbf{F}_1$. If $\alpha_k \in \mathbf{E}_{k-1}$, then $\mathbf{E}_k = \mathbf{E}_{k-1}(\alpha_k) = \mathbf{E}_{k-1}$ and there is nothing to do. Otherwise, let $G = \mathrm{Gal}(\mathbf{E}_{k-1}/\mathbf{E}_0)$ and let

$$f(X) = \prod_{\sigma \in G} \sigma(m_{\alpha_k}(X)).$$

Then $f(X)$ is invariant under G, so $f(X) \in \mathbf{E}_0[X]$. Let $\mathbf{E}_k$ be a splitting field of $f(X)$ containing $\mathbf{F}_k$. Then $f(X)$ has roots $\beta_1 = \alpha_k, \ldots, \beta_m$. Now $m_{\alpha_k}(X)$ has degree 2, so $\sigma(m_\alpha(X))$ has degree 2 for each $\sigma \in G$, and hence $(\mathbf{E}_{k-1}(\beta_i)/\mathbf{E}_{k-1}) = 2$ for each i. Now consider

$$\mathbf{E}_{k-1} \subseteq \mathbf{E}_{k-1}(\beta_1) \subseteq \mathbf{E}_{k-1}(\beta_1, \beta_2) \subseteq \cdots \subseteq \mathbf{E}_{k-1}(\beta_1, \ldots, \beta_m) = \mathbf{E}_k.$$

Set $\mathbf{E}'_{k-1} = \mathbf{E}_{k-1}(\beta_1, \ldots, \beta_{i-1})$. By Remark 2.3.6, $\mathbf{E}_{k-1}(\beta_1, \ldots, \beta_i) = \mathbf{E}_{k-1}(\beta_1, \ldots, \beta_{i-1})(\beta_i) = \mathbf{E}'_{k-1}(\beta_i)$ is the composite $\mathbf{E}'_{k-1}\mathbf{E}_{k-1}(\beta_i)$, so, by Lemma 2.3.8,

$$\begin{aligned}(\mathbf{E}_{k-1}(\beta_1, \ldots, \beta_i)/\mathbf{E}_{k-1}(\beta_1, \ldots, \beta_{i-1})) &= (\mathbf{E}'_{k-1}(\beta_i)/\mathbf{E}'_{k-1})\\ &\leq (\mathbf{E}_{k-1}(\beta_i)/\mathbf{E}_{k-1}) = 2,\end{aligned}$$

so each successive extension is of degree 1 (i.e., is trivial) or of degree 2. Thus $(\mathbf{E}_k/\mathbf{Q})$ is of degree a power of 2, and is a splitting field, hence a Galois extension of $\mathbf{Q}$. Now $z \in \mathbf{E}_k$, so $m_z(X)$ splits in $\mathbf{E}_k$ and hence $m_z(X)$ has a splitting field $\mathbf{E}$ with $\mathbf{Q} \subseteq \mathbf{E} \subseteq \mathbf{E}_k$. Then $(\mathbf{E}/\mathbf{Q})$ divides $(\mathbf{E}_k/\mathbf{Q})$, so $(\mathbf{E}/\mathbf{Q})$ is a power of 2. □

Remark 4.4.4. For an arbitrary prime p, the implication (2) implies (1) of Lemma 4.4.3 holds, with the identical proof, but the implication (1) implies (2) does not. The proof breaks down as Lemma 2.3.8 merely guarantees a degree between 1 and p, but it may be strictly between 1 and p. A counter-example to the implication (1) implies (2) of Lemma 4.4.3 is given by $z = \sqrt[3]{2}$. Then $(\mathbf{Q}(z)/\mathbf{Q}) = 3$ but $m_z(X) = X^3 - 2$, whose splitting field $\mathbf{E}$ over $\mathbf{Q}$ has $(\mathbf{E}/\mathbf{Q}) = 6$, which is not a power of 3. (Compare Example 2.3.7.) ◇

We now do a bit of elementary number theory. Let $n = 2^t + 1$ for some positive integer t. If t has an odd factor $r > 1$, then, writing $t = rs$, n is divisible by $2^s + 1$ (as we see by substituting $X = 2^s$ in the algebraic identity $X^r + 1 = (X+1)(X^{r-1} - X^{r-2} + \cdots + 1)$, valid for any odd r). Thus the only possible primes of this form are when $t = 2^k$ for some k. Set $F_k = 2^{2^k} + 1$.

If F_k is prime, it is called a *Fermat prime*. This terminology is due to the fact that Fermat claimed F_k is prime for every k. Now $F_0 = 3$, $F_1 = 5$, $F_2 = 17$, $F_3 = 257$, and $F_4 = 65537$ are indeed prime, but Euler discovered that $F_5 = 4294967297$ is divisible by 641 and hence is composite. In fact, there is no known value of $k > 4$ for which F_k is prime.

Here is our second main result:

Theorem 4.4.5 (Gauss). *A regular n-gon is constructible by straightedge and compass if and only if n is of the form*

$$n = 2^k p_1 \cdots p_j$$

with $\{p_i\}$ distinct Fermat primes.

Proof. Clearly a regular n-gon is constructible if and only if a regular n-gon inscribed in a circle of radius 1 is constructible, and this is true if and only if its vertices are constructible, and this is true if and only if $\zeta_n = \exp(2\pi i/n)$ is constructible.

By Theorem 4.4.1 and Lemma 4.4.3, this is true if and only if the splitting field of $m_{\zeta_n}(X)$ has degree a power of 2. But this splitting field is simply $\mathbf{Q}(\exp(2\pi i(n))$, the nth cyclotomic field, which has degree $\varphi(n)$ by Corollary 4.2.7.

If $n = 2^a p_1^b p_2^c \cdots$, then $\varphi(n) = 2^{a-1}(p_1 - 1)p_1^{b-1}(p_2 - 1)p_2^{c-1} \cdots$, so we see that $\varphi(n)$ can be a power of 2 if and only if $b = c = \cdots = 1$ and each of $p_1, p_2, \ldots$ is more than a power of 2, and hence a Fermat prime. □

Example 4.4.6. (1) Everyone knows how to construct an equilateral triangle. But in any case, ζ_3 satisfies the equation $X^2 + X + 1 = 0$. In fact, constructing $\zeta_3 = (-1 + i\sqrt{3})/2$ gives a construction of an equilateral triangle that is different from the usual one. It is the following (using the notation at the beginning of this section): Extend the line segment AO so that it intersects the unit circle at the point D. Construct the midpoint E of the line segment OD, and then construct the line M perpendicular to the segment OD at E. Let this line intersect the unit circle at the points B and C. Then ABC is an equilateral triangle.

(2) We essentially showed how to construct a regular pentagon in Example 1.1.4. To review that, let $\theta = \zeta_5 + \zeta_5^4 = \zeta_5 + \zeta_5^{-1}$. Then θ satisfies the equation $X^2 - X - 1 = 0$ and ζ_5 satisfies the equation $X^2 - \theta X + 1 = 0$.

(3) We show how to construct a regular 17-gon. Let $\zeta = \zeta_{17} = \exp(2\pi i/17)$. Let $G = G_0 = \mathrm{Gal}(\mathbf{Q}(\zeta)/\mathbf{Q})$. Then we know that G is cyclic of order 16. Write $G = G_0 \supset G_1 \supset G_2 \supset G_3 \supset G_4 = \{1\}$ where $[G : G_i] = 2^i$, and note that G_i is cyclic of order 2^{4-i}. Let $\mathbf{F}_i = \mathrm{Fix}(G_i)$

so $\mathbf{F}_0 = \mathbf{Q}$ and $\mathbf{F}_4 = \mathbf{Q}(\zeta)$. G is isomorphic to $\mathbf{F}_{17}^*$, and direct calculation shows that 3 is a generator of this group. (In general, a generator of $\mathbf{F}_p^*$ is called a *primitive root* mod p.) Let $\sigma_i(\zeta) = \zeta^{3^i}$, $i = 0, \ldots, 15$. Then $G_0 = \{\sigma_0, \sigma_1, \ldots, \sigma_{15}\}$, $G_1 = \{\sigma_0, \sigma_2, \ldots, \sigma_{14}\}$, $G_2 = \{\sigma_0, \sigma_4, \sigma_8, \sigma_{12}\}$, $G_3 = \{\sigma_0, \sigma_8\}$, and $G_4 = \{\sigma_0\}$. Also, $(\mathbf{F}_i/\mathbf{F}_{i-1}) = 2$ for $i = 1, \ldots, 4$. Write $\mathbf{E} = \mathbf{F}_4 = \mathbf{Q}(\zeta)$.

In order to simplify our computations, we recall the following elementary fact: If r and s are the roots of the quadratic equation $X^2 - aX + b = 0$, then $r + s = -a$ and $rs = b$. (Proof: $(X - r)(X - s) = X^2 - (r + s)X + rs$.)

We work from the bottom up. $(\mathbf{F}_1/\mathbf{F}_0) = 2$ so $\mathbf{F}_1 = \mathbf{F}_0(\alpha)$ for any $\alpha \in \mathbf{E}$ fixed under G_1 but not under G_0, i.e., any α with $\sigma_2(\alpha) = \alpha$ but $\sigma_1(\alpha) \neq \alpha$. Let

$$\alpha_0 = \textstyle\sum_{i=0}^{7} \zeta^{3^{2i}} = \zeta + \zeta^9 + \zeta^{13} + \zeta^{15} + \zeta^{16} + \zeta^8 + \zeta^4 + \zeta^2,$$
$$\alpha_1 = \textstyle\sum_{i=0}^{7} \zeta^{3^{2i+1}} = \zeta^3 + \zeta^{10} + \zeta^5 + \zeta^{11} + \zeta^{14} + \zeta^7 + \zeta^{12} + \zeta^6.$$

If $\alpha = \alpha_0$ or α_1, then $\sigma_2(\alpha) = \alpha$ but $\sigma_1(\alpha) \neq \alpha$. Direct calculation shows $\alpha_0 + \alpha_1 = -1$ and $\alpha_0\alpha_1 = -4$, so α_0 and α_1 are the roots of $X^2 + X - 4 = 0$.

Now $(\mathbf{F}_2/\mathbf{F}_1) = 2$ so $\mathbf{F}_2 = \mathbf{F}_1(\beta)$ for any $\beta \in \mathbf{E}$ fixed under G_2 but not G_1, i.e., any β with $\sigma_4(\beta) = \beta$ but $\sigma_2(\beta) \neq \beta$. Let

$$\beta_0 = \zeta + \zeta^{13} + \zeta^{16} + \zeta^4,$$
$$\beta_1 = \zeta^3 + \zeta^5 + \zeta^{14} + \zeta^{12},$$
$$\beta_2 = \zeta^9 + \zeta^{15} + \zeta^8 + \zeta^2,$$
$$\beta_3 = \zeta^{10} + \zeta^{11} + \zeta^7 + \zeta^6.$$

If $\beta = \beta_0,\ \beta_1,\ \beta_2$, or β_3, then $\sigma_4(\beta) = \beta$ but $\sigma_2(\beta) \neq \beta$. Then $\beta_0 + \beta_2 = \alpha_0$, $\beta_0\beta_2 = -1$, and $\beta_1\beta_3 = \alpha_1$, $\beta_1\beta_3 = -1$, so β_0 and β_2 are the roots of $X^2 - \alpha_0 X - 1 = 0$ and β_1 and β_3 are the roots of $X^2 - \alpha_1 X - 1 = 0$.

Next $(\mathbf{F}_3/\mathbf{F}_2) = 2$ and $\mathbf{F}_3 = \mathbf{F}_2(\gamma)$. Let

$$\gamma_0 = \zeta + \zeta^{16},$$
$$\gamma_4 = \zeta^{13} + \zeta^4.$$

If $\gamma = \gamma_0$ or γ_4, then $\sigma_8(\gamma) = \gamma$ but $\sigma_4(\gamma) \neq \gamma$. Then $\gamma_0 + \gamma_4 = \beta_0$, $\gamma_0\gamma_4 = \beta_1$, so γ_0 and γ_4 are the roots of $X^2 - \beta_0 X + \beta_1 = 0$.

Finally, let $\delta_0 = \zeta$ and $\delta_8 = \zeta^{16}$. If $\delta = \delta_0$ or δ_8, then $\sigma_8(\delta) \neq \delta$. Set $\mathbf{F}_4 = \mathbf{F}_3(\delta_0) = \mathbf{F}_3(\zeta)$. Note that ζ satisfies $\zeta + \zeta^{16} = \gamma_0$, $\zeta\zeta^{16} = 1$, so ζ is a root of $X^2 - \gamma_0 X + 1 = 0$.

Summarizing, we construct $\zeta = \zeta_{17}$ as follows:

(1) Solve the quadratic $X^2 + X - 4 = 0$; call the roots α_0 and α_1.

(2a) Solve the quadratic $X^2 - \alpha_0 X - 1 = 0$; call the roots β_0 and β_2.
(2b) Solve the quadratic $X^2 - \alpha_1 X - 1 = 0$; call the roots β_1 and β_3.
(3) Solve the quadratic $X^2 - \beta_0 X + \beta_1 = 0$; call the roots γ_0 and γ_4.
(4) Solve the quadratic $X^2 - \gamma_0 X + \gamma_4 = 0$; its roots are $\zeta = \delta_0$ and $\zeta^{16} = \delta_8$.

With enough patience (and careful attention to sign) the reader may obtain an explicit formula for ζ which involves square roots nested four deep. (To start off, $\alpha_0 = (-1 + \sqrt{17})/2$ and $\alpha_1 = (-1 - \sqrt{17})/2$, showing that $\mathbf{Q}(\sqrt{17}) \subset \mathbf{Q}(\zeta_{17})$. Compare Corollary 4.5.2.) ◇

4.5 Quadratic Extensions of Q

We recall a bit of elementary number theory. Let p be a prime. Identifying $\mathbf{F}_p$ with $\mathbb{Z}/p\mathbb{Z}$, we know that $G = \mathbf{F}_p^* = \{1, \dots, p-1\}$ is a cyclic group, and a generator r of this group is called a *primitive root* mod p. For example, 3 is a primitive root mod 7 as $3^1 \equiv 3 \pmod 7$, $3^2 \equiv 2 \pmod 7$, $3^3 \equiv 6 \pmod 7$, $3^4 \equiv 4 \pmod 7$, $3^5 \equiv 5 \pmod 7$, and $3^6 \equiv 1 \pmod 7$. On the other hand, 2 is not a primitive root mod 7 as $2^3 \equiv 1 \pmod 7$. (We must be careful not to confuse a primitive root mod p with a primitive p^{th} root of 1. Observe that ζ_p^k is a primitive p^{th} root of 1 for any k not divisible by p.) Observe also that $(\mathbb{Z}/4\mathbb{Z})^* = \{1, 3\}$, so 3 is a primitive root mod 4.

If p is an odd prime, then G is cyclic of even order $p-1$, so it has a unique subgroup H of index 2. Let $\pi : G \to G/H \cong \mathbb{Z}/2\mathbb{Z} = \{0, 1\}$ be the quotient map. For an integer k relatively prime to p, we regard k as an element of $\mathbb{Z}/p\mathbb{Z}$. Then χ_p, the *quadratic residue* character mod p, is defined by $\chi_p(k) = (-1)^{\pi(k)}$. (For fixed p, we shall write χ for χ_p.)

There is a more traditional definition of the quadratic residue character. Recalling that G is a multiplicative group, the subgroup H consists of the (quadratic) residues, i.e., of those elements of G that are squares, and its complement consists of the (quadratic) nonresidues, i.e., of those elements of G that are nonsquares. Then $\chi(k) = 1$ if k is a residue and $\chi(k) = -1$ if k is a nonresidue.

Furthermore, if r is a generator of G, i.e., a primitive root mod p, then $\chi(k) = 1$ if k is an even power of r and $\chi(k) = -1$ if k is an odd power of r (and this is independent of the choice of r).

For example, if $p = 7$, $1 = (\pm 1)^2$, $4 = (\pm 2)^2$, and $2 = (\pm 3)^2$ are the residues and 3, 5, and 6 are the nonresidues. Choosing the primitive root 3, we see that $1 = 3^0$, $2 = 3^2$, and $4 = 3^4$, while $3 = 3^1$, $5 = 3^5$, and $6 = 3^3$.

It is then easy to see that the quadratic residue character has the following properties:

(1) $\chi(jk) = \chi(j)\chi(k)$. ($\chi$ is a homomorphism to the multiplicative group $\{\pm 1\}$.)

(2) $\sum_{k=1}^{p-1} \chi(k) = 0$. (There are $(p-1)/2$ residues and $(p-1)/2$ nonresidues.)

(3) $\chi(-1) = (-1)^{(p-1)/2} = 1$ if $p \equiv 1 \pmod 4$ and $= -1$ if $p \equiv 3 \pmod 4$. (If r is a primitive root mod p, let $a = r^{(p-1)/2}$. Then $a \neq 1$ but $a^2 = 1$. On the other hand, the equation $X^2 - 1 = 0$ has only two solutions in $\mathbf{F}_p$, $X = \pm 1$. Hence $a = -1$. Thus if $p \equiv 1 \pmod 4$, $(p-1)/2$ is even, a is an even power of r, and $\chi(a) = 1$, while if $p \equiv 3 \pmod 4$, $(p-1)/2$ is odd, a is an odd power of r, and $\chi(a) = -1$.)

Theorem 4.5.1. *Let* $\mathbf{E}$ *be a quadratic extension of* $\mathbf{Q}$. *Then* $\mathbf{E}$ *is intermediate between* $\mathbf{Q}$ *and some cyclotomic field. In particular:*
(1) $\mathbf{Q}(\sqrt{-1}) = \mathbf{Q}(\zeta_4)$.
(2) $\mathbf{Q}(\sqrt{2}) \subset \mathbf{Q}(\zeta_8)$ *and* $\mathbf{Q}(\sqrt{-2}) \subset \mathbf{Q}(\zeta_8)$.
(3) If p is a prime congruent to 1 modulo 4, *then* $\mathbf{Q}(\sqrt{p}) \subseteq \mathbf{Q}(\zeta_p)$.
(4) If p is a prime congruent to 3 modulo 4, *then* $\mathbf{Q}(\sqrt{-p}) \subseteq \mathbf{Q}(\zeta_p)$.

Proof. Observe that if $\mathbf{Q}(\sqrt{a}) \subseteq \mathbf{Q}(\zeta_m)$ and $\mathbf{Q}(\sqrt{b}) \subseteq \mathbf{Q}(\zeta_n)$, then $\mathbf{Q}(\sqrt{ab}) \subseteq \mathbf{Q}(\sqrt{a}, \sqrt{b}) \subseteq \mathbf{Q}(\zeta_m, \zeta_n) \subseteq \mathbf{Q}(\zeta_{mn})$, so in order to prove the general statement it suffices to prove the particular statements (1), (2), (3), and (4).

(1) is obvious.

(2) is direct calculation: $\zeta_8 = (1+i)/\sqrt{2}$.

We prove (3) and (4) simultaneously. Actually, we present two proofs.

Let p be an odd prime and let

$$S_p = \sum_{k=1}^{p-1} \chi(k)\zeta_p^k.$$

Then

$$S_p^2 = \sum_{k=1}^{p-1}\sum_{j=1}^{p-1} \chi(k)\zeta_p^k \chi(j)\zeta_p^j = \sum_{k=1}^{p-1}\sum_{j=1}^{p-1} \chi(kj)\zeta_p^{k+j}.$$

Set $j \equiv km \pmod p$ and notice that as j runs over the nonzero congruence classes mod p, so does m. Also, $\chi(kj) = \chi(k^2 m) = \chi(m)$. Thus

$$\begin{aligned} S_p^2 &= \sum_{k=1}^{p-1}\sum_{m=1}^{p-1} \chi(m)\zeta_p^{k+km} \\ &= \sum_{m=1}^{p-1} \chi(m) \sum_{k=1}^{p-1} (\zeta_p^{1+m})^k. \end{aligned}$$

Now $1+\zeta_p+\cdots+\zeta_p^{p-1}=0$, so as long as $m \neq p-1$, the inner sum is -1. If $m=p-1$, then the inner sum is of course $p-1$.

Thus

$$S_p^2=\left(-\sum_{m=1}^{p-2}\chi(m)\right)+(p-1)\chi(p-1).$$

But $\sum_{m=1}^{p-1}\chi(m)=0$, so the first term is $+\chi(-1)$ and we see

$$S_p^2=p\chi(-1),$$

as required.

Our second proof uses more theory and a minimum of calculation. Let $G=\mathrm{Gal}(\mathbf{Q}(\zeta_p)/\mathbf{Q})$ and recall that G is generated by σ_0, where $\sigma_0(\zeta_p)=\zeta_p^r$ with r a primitive root mod p. Since $\mathbf{Q}(\zeta_p)$ is a Galois extension of $\mathbf{Q}$, any $\alpha \in \mathbf{Q}(\zeta_p)$ with $\sigma_0(\alpha)=\alpha$ (and hence with $\sigma(\alpha)=\alpha$ for every $\sigma \in G$) is in fact an element of $\mathbf{Q}$.

Now $(\mathbf{Q}(\zeta_p)/\mathbf{Q})=p-1$ and $\mathbf{Q}(\zeta_p)$ is certainly spanned by $\{1,\zeta_p,\ldots,\zeta_p^{p-1}\}$. We know one relation among these p elements: $1+\zeta_p+\cdots+\zeta_p^{p-1}=0$. Hence $\{\zeta_p,\ldots,\zeta_p^{p-1}\}$ form a basis for $\mathbf{Q}(\zeta_p)$, so in particular this set is $\mathbf{Q}$-linearly independent. Thus if $\alpha \in \mathbf{Q}(\zeta_p)$ with $\sigma_0(\alpha)=\alpha$, then, writing $\alpha=\sum_{i=1}^{p-1}a_i\zeta_p^i$ with $a_i \in \mathbf{Q}$, we must have $a_1=\cdots=a_{p-1}=a$, for some a. (Of course, in this case $\alpha=-a\in\mathbf{Q}$.)

With these observations in mind, let us set

$$\alpha_0=\alpha_0(p)=\sum_{i=1}^{(p-1)/2}\zeta_p^{r^{2i-2}},$$

$$\alpha_1=\alpha_1(p)=\sum_{i=1}^{(p-1)/2}\zeta_p^{r^{2i-1}}.$$

Observe that the exponents in α_0 are all quadratic residues mod p, and that the exponents in α_1 are all quadratic nonresidues mod p. Then

$$\alpha_0+\alpha_1=\zeta_p+\cdots\zeta_p^{p-1}=-1 \text{ and } \sigma(\alpha_0)=\alpha_1,\ \sigma(\alpha_1)=\alpha_0.$$

Let $b=\alpha_0\alpha_1$. Then

$$\sigma_0(b)=\sigma_0(\alpha_0\alpha_1)=\sigma_0(\alpha_0)\sigma_0(\alpha_1)=\alpha_1\alpha_0=b,$$

so $b\in\mathbf{Q}$. We now compute b.

First, suppose $p \equiv 1 \pmod 4$. There are $(p-1)/2$ terms in α_0 and $(p-1)/2$ terms in α_1, so there are $(p-1)^2/4$ terms in the product. Each term is of the form $\zeta_p^j \zeta_p^k$ with j a quadratic residue and k a quadratic nonresidue. Since $p \equiv 1 \pmod 4$, $\chi(-j) = \chi(-1)\chi(j) = \chi(j)$, so j and $-j$ are always either both residues or both nonresidues, thus $\zeta_p^k \neq (\zeta_p^j)^{-1}$ and hence no term $\zeta_p^j \zeta_p^k$ is equal to 1. Then, writing $b = \sum_{i=1}^{p-1} a_i \zeta_p^i$, we see that $\sum a_i = (p-1)^2/4$ and that $a_1 = \cdots = a_{p-1} = a$, for some a, so $\sum a_i = (p-1)a = (p-1)^2/4$ and hence $a = (p-1)/4$. Then

$$b = a \sum_{i=1}^{p-1} \zeta^i = ((p-1)/4) \sum_{i=1}^{p-1} \zeta^i = ((p-1)/4)(-1) = -(p-1)/4.$$

Now suppose $p \equiv 3 \pmod 4$. Then $\chi(-j) = -\chi(j)$, so one of j and $-j$ is always a quadratic residue and the other is a quadratic nonresidue. Hence each term in α_0 "pairs up" with one of the terms in α_1 to give a product of 1. This takes care of $(p-1)/2$ of the $(p-1)^2/4$ terms. Thus we may write $b = (p-1)/2 + b'$ where b' is the sum of the remaining $(p-1)^2/4 - (p-1)/2 = (p-1)(p-3)/4$ terms. We now apply the same logic to b'. Writing $b' = \sum_{i=1}^{p-1} a_i' \zeta_p^i$, we see that $\sum a_i' = (p-1)(p-3)/4$ and that $a_1' = \cdots = a_{p-1}' = a'$, for some a', so $\sum a_i' = (p-1)a' = (p-1)(p-3)/4$, and hence $a' = (p-3)/4$. Then

$$b' = a' \sum_{i=1}^{p-1} \zeta^i = ((p-3)/4) \sum_{i=1}^{p-1} \zeta^i = ((p-3)/4)(-1) = -(p-3)/4,$$

and then

$$b = (p-1)/2 + b' = (p-1)/2 - (p-3)/4 = (p+1)/4.$$

Now

$$\begin{aligned}(X-\alpha_0)(X-\alpha_1) &= X^2 - (\alpha_0+\alpha_1)X + \alpha_0\alpha_1 \\ &= X^2 + X + b,\end{aligned}$$

so

$$\begin{aligned}(X-\alpha_0)(X-\alpha_1) &= X^2 + X - (p-1)/4 \quad \text{if } p \equiv 1 \pmod 4, \\ (X-\alpha_0)(X-\alpha_1) &= X^2 + X + (p+1)/4 \quad \text{if } p \equiv 3 \pmod 4.\end{aligned}$$

In other words, α_0 and α_1 are the roots of this quadratic equation. But we may find these roots by the quadratic formula, and we obtain

$$\{\alpha_0, \alpha_1\} = \{(-1 \pm \sqrt{p})/2\} \qquad \text{if } p \equiv 1 \pmod 4,$$
$$\{\alpha_0, \alpha_1\} = \{(-1 \pm \sqrt{-p})/2\} \qquad \text{if } p \equiv 3 \pmod 4. \qquad \square$$

Corollary 4.5.2. *For p an odd prime, $\mathbf{Q}(\sqrt{\chi_p(-1)p})$ is the unique quadratic field intermediate between $\mathbf{Q}$ and $\mathbf{Q}(\zeta_p)$.*

Proof. $\mathrm{Gal}(\mathbf{Q}/\zeta_p)/\mathbf{Q}) \cong \mathbb{Z}/(p-1)\mathbb{Z}$ contains a unique subgroup of index two, so there is a unique quadratic field intermediate between $\mathbf{Q}$ and $\mathbf{Q}(\zeta_p)$, and we have just identified that field. □

Remark 4.5.3. Let us cite some results from number theory that will enable us to extend Corollary 4.5.2. They are:

(1) If m_1 and m_2 are relatively prime, then $(\mathbb{Z}/(m_1m_2)\mathbb{Z})^* \cong (\mathbb{Z}/m_1\mathbb{Z})^* \oplus (\mathbb{Z}/m_2\mathbb{Z})^*$. (This follows easily from the Chinese Remainder Theorem.)

(2) For p an odd prime and k a positive integer, $(\mathbb{Z}/p^k\mathbb{Z})^*$, a group of order $\varphi(p^k) = (p-1)p^{k-1}$, is cyclic. (This is usually phrased as saying there is a primitive root mod p^k.)

(3) $(\mathbb{Z}/4\mathbb{Z})^* \cong \mathbb{Z}/2\mathbb{Z}$ and for $k > 2$, $(\mathbb{Z}/2^k\mathbb{Z})^*$, a group of order $\varphi(2^k) = 2^{k-1}$, is isomorphic to the direct sum of a cyclic group of order 2 and a cyclic group of order 2^{k-2}. ◇

Corollary 4.5.4. *Let $n > 2$ be an integer. Define a set $A = \{a_i\}_{i=1,\ldots,t}$ as follows: If p is an odd prime factor of n, then $\chi_p(-1)p \in A$. If n is divisible by 4, then $-1 \in A$. If n is divisible by 8, then $2 \in A$. Then $\mathbf{Q}(\zeta_n)$ contains $2^t - 1$ quadratic extensions of $\mathbf{Q}$, and they are $\mathbf{Q}(\sqrt{m})$ for m any nontrivial product of distinct elements of A.*

Proof. By the Fundamental Theorem of Galois Theory, the quadratic extensions of $\mathbf{Q}$ contained in $\mathbf{Q}(\zeta_n)$ are in $1-1$ correspondence with the subgroups of index 2 of $\mathrm{Gal}(\mathbf{Q}(\zeta_n)/\mathbf{Q} \cong (\mathbb{Z}/n\mathbb{Z})^*$. Now if $n = 2^{e_2}3^{e_3}5^{e_5}\ldots$, $(\mathbb{Z}/n\mathbb{Z})^* \cong (\mathbb{Z}/2^{e_2}\mathbb{Z})^* \times (\mathbb{Z}/3^{e_3}\mathbb{Z})^* \times (\mathbb{Z}/5^{e_5}\mathbb{Z})^* \times \cdots$. For p odd and $k \geq 1$, $(\mathbb{Z}/p^k\mathbb{Z})^*$ is a cyclic group of even order. Also, $(\mathbb{Z}/2)^*$ is trivial, $(\mathbb{Z}/4)^* \cong \mathbb{Z}/2\mathbb{Z}$, and $(\mathbb{Z}/2^k\mathbb{Z})^* \cong (\mathbb{Z}/2\mathbb{Z}) \oplus (\mathbb{Z}/2^{k-2}\mathbb{Z})$. Thus $(\mathbb{Z}/n\mathbb{Z})^*$ contains $2^t - 1$ subgroups of index 2, where t is as in the statement of the corollary. (A subgroup of index 2 is the kernel of an epimorphism $\psi\colon \mathrm{Gal}(\mathbf{Q}(\zeta_n)/\mathbf{Q}) \to \mathbb{Z}/2\mathbb{Z}$ and since $\mathrm{Gal}(\mathbf{Q}(\zeta_n)/\mathbf{Q})$ is isomorphic to the direct sum of t cyclic groups of even order, there are 2^t homomorphisms from $\mathrm{Gal}(\mathbf{Q}(\zeta_n)/\mathbf{Q})$ to $\mathbb{Z}/2\mathbb{Z}$, one of which is the trivial one.) Since $\mathbf{Q}(\sqrt{m}) \subseteq \mathbf{Q}(\zeta_n)$ for each of the $2^t - 1$ values of m in the statement of the corollary, by Theorem 4.5.3, we see that these are all the quadratic subfields of $\mathbf{Q}(\zeta_n)$. □

Corollary 4.5.5. *Let m be a square-free integer. If $m \equiv 1 \pmod 4$, set $m' = |m|$. Otherwise set $m' = 4|m|$. Then $\mathbf{Q}(\sqrt{m}) \subseteq \mathbf{Q}(\zeta_n)$ if and only if n is a multiple of m'.*

Proof. Corollary 4.5.4 gives all the quadratic subfields of $\mathbf{Q}(\zeta_n)$. □

Remark 4.5.6. We have shown in the proof of Theorem 4.5.1 that for an odd prime p, $S_p = \pm\sqrt{\chi_p(-1)p}$. It is a famous theorem of Gauss that in every case the sign is positive. It is easy to check that $S_p = 2\alpha_0(p) + 1$, so we see that $\alpha_0(p) = (-1 + \sqrt{\chi_p(-1)p})/2$ in every case as well. ◇

4.6 Radical Polynomials and Related Topics

In this section we investigate radical polynomials $f(X) \in \mathbf{Q}[X]$, i.e., polynomials of the form $X^n - a$, their associated radical extensions $\mathbf{Q}(\sqrt[n]{a})$, and their splitting fields $\mathbf{Q}(\sqrt[n]{a}, \zeta_n)$. The results we obtain are interesting in their own right, and, as we shall see, also have applications to cyclotomic fields and Kummer fields.

First we determine the Galois groups of their splitting fields. We remind the reader of our discussion of primitive roots at the beginning of Section 4.5.

Lemma 4.6.1. *Let p be an odd prime. Let $a \in \mathbf{Q}$ with a not a pth power in $\mathbf{Q}$, and let $\mathbf{E}$ be the splitting field of $X^p - a$. Let $G = \mathrm{Gal}(\mathbf{E}/\mathbf{Q})$. Then*

$$G = \langle \sigma, \tau \mid \sigma^p = 1, \tau^{p-1} = 1, \tau\sigma\tau^{-1} = \sigma^r \rangle$$

for some primitive root r mod p.

Proof. Certainly $\mathbf{E} = \mathbf{Q}(\alpha, \zeta_p)$ where $\alpha \in \mathbf{E}$, $\alpha^p = a$. We claim that $\mathbf{Q}(\alpha) \cap \mathbf{Q}(\zeta_p) = \mathbf{Q}$. To see this, let $\mathbf{B} = \mathbf{Q}(\alpha) \cap \mathbf{Q}(\zeta_p)$. Then $\mathbf{Q} \subseteq \mathbf{B} \subseteq \mathbf{Q}(\zeta_p)$, so $(\mathbf{B}/\mathbf{Q})$ divides $(\mathbf{Q}(\zeta_p)/\mathbf{Q}) = p - 1$, and $\mathbf{Q} \subseteq \mathbf{B} \subseteq \mathbf{Q}(\alpha)$, so $(\mathbf{B}/\mathbf{Q})$ divides $(\mathbf{Q}(\alpha)/\mathbf{Q}) = p$. Thus $(\mathbf{B}/\mathbf{Q}) = 1$, i.e., $\mathbf{B} = \mathbf{Q}$. In particular, $\alpha \notin \mathbf{Q}(\zeta_p)$ and $(\mathbf{E}/\mathbf{Q}(\zeta_p)) = p$. Thus, if $\tilde{m}_\alpha(X)$ denotes the minimum polynomial of α over $\mathbf{Q}(\zeta_p)$, and $m_\alpha(X)$ denotes the minimum polynomial of α over $\mathbf{Q}$, we see that

$$\tilde{m}_\alpha(X) = m_\alpha(X) = \prod_{i=0}^{p-1}(X - \zeta_p^i\alpha) \in \mathbf{E}[X].$$

Then, by either Lemma 2.6.1 or Lemma 2.6.3, there is an automorphism σ of $\mathbf{E}$ with

$$\sigma(\alpha) = \zeta_p\alpha, \quad \sigma(\zeta_p) = \zeta_p.$$

On the other hand, consider a generator τ_0 of $\mathrm{Gal}(\mathbf{Q}(\zeta_p)/\mathbf{Q})$. This is given by $\tau_0(\zeta_p) = \zeta_p^r$ for r a primitive root mod p. Again, by either Lemma 2.6.1 or Lemma 2.6.3, there is an automorphism τ_0 of $\mathbf{E}$ extending τ_0. Then $\tau_0(\alpha) = \zeta_p^k\alpha$ for some k. Set $\tau = \sigma^{-k}\tau_0$. Then τ is an automorphism of $\mathbf{E}$ with

$$\tau(\alpha) = \alpha, \quad \tau(\zeta_p) = \zeta_p^r.$$

Then direct calculation shows

$$\begin{aligned}\sigma^r\tau(\alpha) &= \sigma^r(\tau(\alpha)) = \sigma^r(\alpha) = \zeta_p^r\alpha\\ \sigma^r\tau(\zeta_p) &= \sigma^r(\tau(\zeta_p)) = \sigma^r(\zeta_p^r) = \zeta_p^r,\end{aligned}$$

while

$$\begin{aligned}\tau\sigma(\alpha) &= \tau(\sigma(\alpha)) = \tau(\zeta_p\alpha) = \zeta_p^r\alpha\\ \tau\sigma(\zeta_p) &= \tau(\sigma(\zeta_p)) = \tau(\zeta_p) = \zeta_p^r,\end{aligned}$$

so $\sigma^r\tau = \tau\sigma \in \mathrm{Gal}(\mathbf{E}/\mathbf{Q})$, i.e., $\tau\sigma\tau^{-1} = \sigma^r$, as claimed. □

Remark 4.6.2. (1) We see that, in the situation of Lemma 4.6.1, $G = HN$ is the semidirect product of the normal subgroup $N = \langle \sigma \mid \sigma^p = 1\rangle$, isomorphic to $\mathbb{Z}/p\mathbb{Z}$, and the subgroup $H = \langle \tau \mid \tau^{p-1} = 1\rangle$, isomorphic to $\mathbb{Z}/(p-1)\mathbb{Z}$.

(2) The structure of G is independent of the primitive root r. (There is a choice involved in the generators of G. Varying this choice varies r among all the primitive roots.)

(3) Note that G is nonabelian. Note also that Lemma 4.6.1 also holds for $p = 2$, when it recovers the much easier fact that in this case $G \cong \mathbb{Z}/2\mathbb{Z}$. ◊

Lemma 4.6.3. *Let $a \in \mathbf{Q}$ with $a \neq \pm b^2$ for any $b \in \mathbf{Q}$, and let $\mathbf{E}$ be the splitting field of $X^4 - a$. Let $G = \mathrm{Gal}(\mathbf{E}/\mathbf{Q})$. Then*

$$G = \langle \sigma, \tau \mid \sigma^4 = 1, \tau^2 = 1, \tau\sigma\tau^{-1} = \sigma^3\rangle.$$

Proof. Let $\alpha \in \mathbf{E}$ with $\alpha^4 = a$. Then $\mathbf{E} = \mathbf{Q}(\alpha, \zeta_4)$. First we must show that $\mathbf{Q}(\alpha) \cap \mathbf{Q}(\zeta_4) = \mathbf{Q}$. Assume $\mathbf{Q}(\alpha) \cap \mathbf{Q}(\zeta_4) \supset \mathbf{Q}$. Since $(\mathbf{Q}(\zeta_4)/\mathbf{Q}) = 2$, this implies that $\mathbf{Q}(\alpha) \supseteq \mathbf{Q}(\zeta_4)$ and hence that $\mathbf{E} = \mathbf{Q}(\alpha)$. Then $(\mathbf{E}/\mathbf{Q}) = 4$ and $(\mathbf{Q}(\zeta_4)/\mathbf{Q}) = 2$, and so $\mathbf{Q}(\zeta_4)$ is the fixed field of a subgroup $\{\mathrm{id}, \sigma\}$ of $\mathrm{Gal}(\mathbf{E}/\mathbf{Q})$. Since α is a root of the polynomial $X^4 - a$, $\sigma(\alpha)$ must also be a root of this polynomial, so $\sigma(\alpha) = \zeta_4^k\alpha$ for some k. But $\sigma(\zeta_4) = \zeta_4$ and $\sigma \neq \mathrm{id}$ but $\sigma^2 = \mathrm{id}$, so $\zeta_4^k\alpha \neq \alpha$ but $\zeta_4^{2k}\alpha = \alpha$. Hence $k = 2$ and $\sigma(\alpha) = -\alpha$. Now $\mathbf{E} = \{c_0 + c_1\alpha + c_2\alpha^2 + c_3\alpha^3 \mid c_i \in \mathbf{Q}\}$ and $\mathrm{Fix}(\sigma) = \{c_0 + c_2\alpha^2 \mid c_i \in \mathbf{Q}\}$. We claim that $i = \zeta_4 \notin \mathrm{Fix}(\sigma)$.

Suppose $i = c_0 + c_2\alpha^2$. Then $-1 = i^2 = (c_0 + c_2\alpha^2)^2 = (c_0^2 + c_2^2 a) + 2c_0c_2\alpha^2$. Since a is not a square, $\alpha^2 \notin \mathbf{Q}$, so $c_0c_2 = 0$ and either $c_0 = 0$,

$c_2^2 = -1/a$, which is impossible since $-a$ is not a square, or $c_2 = 0$, $c_0^2 = -1$, which is certainly impossible. Thus $\mathbf{Q}(\alpha) \cap \mathbf{Q}(\zeta_4) = \mathbf{Q}$.

The rest of the proof entirely parallels the proof of Lemma 4.6.1. □

Lemma 4.6.1 and 4.6.3 have a considerable generalization in Theorem 4.6.6.

We have seen in Section 4.5 that every quadratic field $\mathbf{Q}(\sqrt{a})$ is contained in some cyclotomic field. We now use the theory we have developed to show that no cubic, quartic, ... field is contained in a cyclotomic field.

Corollary 4.6.4. *Let $a \in \mathbf{Q}$ and let $n > 2$ be an integer. Suppose that a is not a pth power for some odd prime p dividing n or that $\pm a$ is not a square if n is divisible by 4. Then $\mathbf{Q}(\sqrt[n]{a})$ is not contained in any cyclotomic field.*

Proof. Let $k = p$ or $k = 4$ according as n is divisible by the odd prime p or by 4.

Suppose that $\mathbf{Q}(\sqrt[n]{a}) \subseteq \mathbf{Q}(\zeta_m)$. Since $\sqrt[k]{a} = (\sqrt[n]{a})^{n/k}$, $\mathbf{Q}(\sqrt[k]{a}) \subseteq \mathbf{Q}(\sqrt[n]{a})$. Then $\mathbf{E} = \mathbf{Q}(\sqrt[k]{a}, \zeta_k) \subseteq \mathbf{Q}(\zeta_m, \zeta_k) \subseteq \mathbf{Q}(\zeta_{mk})$. Now $\mathbf{E}$ is the splitting field of $X^k - a$, so is a Galois extension of $\mathbf{Q}$. Then, by the Fundamental Theorem of Galois Theory,

$$\mathrm{Gal}(\mathbf{E}/\mathbf{Q}) = \mathrm{Gal}(\mathbf{Q}(\zeta_{mk})/\mathbf{Q})/\,\mathrm{Gal}(\mathbf{Q}(\zeta_{mk})/\mathbf{E}).$$

In particular, $\mathrm{Gal}(\mathbf{E}/\mathbf{Q})$ is a quotient of $\mathrm{Gal}(\mathbf{Q}(\zeta_{mk})/\mathbf{Q})$. But we have that $\mathrm{Gal}(\mathbf{Q}(\zeta_{mk})/\mathbf{Q})$ is abelian, by Corollary 4.2.7, while under our hypothesis $\mathrm{Gal}(\mathbf{E}/\mathbf{Q})$ is nonabelian, by Lemma 4.6.1 or Lemma 4.6.3, which is impossible. □

We now prove a result which allows us to weaken the hypothesis of Lemma 3.7.7 in our discussion of Kummer fields, as well as to further investigate the Galois groups of splitting fields of radical polynomials.

Lemma 4.6.5. *(1) Let n be an odd integer, and let a be n-powerless in $\mathbf{Q}$. Then a is n-powerless in $\mathbf{Q}(\zeta_N)$ for every N.*

(2) Let n be an even integer, and let a be n-powerless in $\mathbf{Q}$. If a is negative, assume also that $-a$ is not a square in $\mathbf{Q}$. Then one of the following two alternatives holds:

(a) a is n-powerless in $\mathbf{Q}(\zeta_N)$ for every N.

(b) $a = b^2$ for some $b \in \mathbf{Q}(\zeta_N)$, for some N, and b is $n/2$-powerless in $\mathbf{Q}(\zeta_{N'})$ for every multiple N' of N.

Proof. If a is n-powerless in $\mathbf{Q}(\zeta_N)$, there is nothing to prove. Otherwise, a is an mth power in $\mathbf{Q}(\zeta_N)$ for some m dividing n, $m > 1$. We shall show that the only possibility is $m = 2$, from which the lemma follows.

Suppose $m > 2$. Then m is divisible by k, where k is either an odd prime or 4. Write $a = b^k$ where $b \in \mathbf{Q}(\zeta_N)$. Then we see that $\mathbf{Q}(\sqrt[k]{a})$ is contained in some cyclotomic field. But, under our hypothesis on a, this is impossible, by Corollary 4.6.4. □

Theorem 4.6.6. *Let n be an integer and let a be n-powerless in $\mathbf{Q}$. Let $\mathbf{E}$ be the splitting field of $X^n - a$ over $\mathbf{Q}$, and let $G = \mathrm{Gal}(\mathbf{E}/\mathbf{Q})$.*

(1) If a is n-powerless in $\mathbf{Q}(\zeta_n)$, then $G = HN$, the semidirect product of a normal subgroup N isomorphic to $\mathbb{Z}/n\mathbb{Z}$ and a subgroup H isomorphic to $(\mathbb{Z}/n\mathbb{Z})^*$.

(2) If $a = b^2$ with b $n/2$-powerless in $\mathbf{Q}(\zeta_n)$, then $G = HN$, the semidirect product of a normal subgroup N isomorphic to $\mathbb{Z}/(n/2)\mathbb{Z}$ and a subgroup H isomorphic to $(\mathbb{Z}/n\mathbb{Z})^*$.

Proof. (1) We have that $\mathbf{E} = \mathbf{Q}(\zeta_n, \alpha)$, where $\alpha^n = a$. Let $\mathbf{B} = \mathbf{Q}(\zeta_n)$. Then $X^n - a$ is irreducible over $\mathbf{B}$, so by Proposition 3.7.6 the Galois group $N = \mathrm{Gal}(\mathbf{E}/\mathbf{B})$ is isomorphic to $\mathbb{Z}/n\mathbb{Z}$. Now N is a subgroup of G and is normal as $\mathbf{B}$ is a Galois extension of $\mathbf{Q}$. Also, $H = \mathrm{Gal}(\mathbf{E}/\mathbf{Q}(\alpha))$ is isomorphic to $\mathrm{Gal}(\mathbf{B}/\mathbf{Q})$ which is isomorphic to $(\mathbb{Z}/n\mathbb{Z})^*$. Also, $(\mathbf{E}/\mathbf{Q}) = (\mathbf{E}/\mathbf{B})(\mathbf{B}/\mathbf{Q}) = (\mathbf{Q}(\alpha)/\mathbf{Q})(\mathbf{B}/\mathbf{Q})$ so $\mathbf{Q}(\alpha)$ and $\mathbf{B}$ are disjoint extensions of $\mathbf{Q}$ by Corollary 3.4.5, so $G = HN$ by Theorem 3.4.10.

(2) Let $\mathbf{F} = \mathbf{Q}(b) \subseteq \mathbf{Q}(\zeta_n)$. Then $(\mathbf{F}/\mathbf{Q}) = 2$, and so $G' = \mathrm{Gal}(\mathbf{E}/\mathbf{F})$ is a subgroup of index 2 of G. The proof of part (1) goes through with $\mathbf{Q}$ replaced by $\mathbf{F}$ and n replaced by m to give that $G' = H'N$ with $H' = \mathrm{Gal}(\mathbf{Q}(\zeta_n)/\mathbf{Q}(b))$ a subgroup of $H = \mathrm{Gal}(\mathbf{Q}(\zeta_n)/\mathbf{Q})$ of index 2. Now $[G : G'] = [H : H'] = 2$. To show $G = HN$ it suffices to show that $\sigma \notin N$ where σ is any element of H not in H'. But for any $\sigma \in H$, except $\sigma = \mathrm{id}$, σ is non-trivial on $\mathbf{B} = \mathbf{Q}(\zeta_n)$, while every element of N acts trivially on $\mathbf{B}$ as $N = \mathrm{Gal}(\mathbf{E}/\mathbf{B})$, so $\sigma \notin N$, as required. □

Remark 4.6.7. As we have seen in our discussion of quadratic fields, both cases in Lemma 4.6.5 (2) may indeed arise. For example, 5 is 6-powerless in $\mathbf{Q}$ and $\sqrt{5} \notin \mathbf{Q}(\zeta_6)$, so we are in case (a) for $n = 6$, while 5 is 10-powerless in $\mathbf{Q}$ but $\sqrt{5} \in \mathbf{Q}(\zeta_{10})$, so we are in case (b) for $n = 10$. ◇

We now wish to determine when the polynomial $X^n - a$ is irreducible. Clearly if a is not n-powerless it is not, so we need only consider the case when a is n-powerless. Recall that in Corollary 4.1.10 we showed that $X^n - a$ is irreducible if a is p-powerless for p an odd prime, or if a is positive and a is 2-powerless.

We shall generalize this result in several stages.

Corollary 4.6.8. *Let a be n-powerless in $\mathbf{Q}$. If n is odd, then the polynomial $X^n - a$ is irreducible in $\mathbf{Q}[X]$.*

Proof. Since a is n-powerless in $\mathbf{Q}$, and n is odd, we have, by Lemma 4.6.5, that a is n-powerless in $\mathbf{Q}(\zeta_n)$. Then, by Proposition 3.7.6, $X^n - a$ is irreducible in $\mathbf{Q}(\zeta_n)[X]$, and hence $X^n - a$ is certainly irreducible in $\mathbf{Q}[X]$. □

Lemma 4.6.9. *Let $n = 2^t$, and assume that a is not a square in $\mathbf{Q}$. If $t \geq 2$, assume also that $-4a$ is not a 4th power in $\mathbf{Q}$. Then the polynomial $X^n - a$ is irreducible in $\mathbf{Q}[X]$.*

Proof. Clearly the lemma holds when $t = 1$; so suppose $t \geq 2$.

Let α be a root of $X^{2^t} - a$ in some extension $\mathbf{E}$ of $\mathbf{Q}$. Consider

$$\mathbf{Q} = \mathbf{Q}(a) = \mathbf{Q}(\alpha^{2^t}) \subseteq \mathbf{Q}(\alpha^{2^{t-1}}) \subseteq \cdots \subseteq \mathbf{Q}(\alpha).$$

Claim: $(\mathbf{Q}(\alpha^{2^{k-1}})/\mathbf{Q}(\alpha^{2^k})) = 2$ for $k = 1, \ldots, t$.

Assuming this claim, we then immediately have that $(\mathbf{Q}(\alpha)/\mathbf{Q}) = 2^t$, and hence that $m_\alpha(X)$ is a polynomial of degree 2^t. But α is a root of $X^{2^t} - a$, so $m_\alpha(X) = X^{2^t} - a$ and hence $X^{2^t} - a$ is irreducible.

Proof of claim: We prove this by downward induction on k.

If $k = t$, then $\alpha^{2^{k-1}}$ is a root of $X^2 - a$, and a is not a square in $\mathbf{Q} = \mathbf{Q}(\alpha^{2^k})$, so $(\mathbf{Q}(\alpha^{2^{k-1}})/\mathbf{Q}(\alpha^{2^k})) = 2$.

Now suppose the claim is true for all integers between k and t and consider $k - 1$. We wish to show $(\mathbf{Q}(\alpha^{2^{k-2}})/\mathbf{Q}(\alpha^{2^{k-1}})) = 2$. We prove this by contradiction. Assume it is not the case. Then $\mathbf{Q}(\alpha^{2^{k-2}}) = \mathbf{Q}(\alpha^{2^{k-1}})$. For simplicity, set $\beta = \alpha^{2^{k-2}}$, $\gamma = \alpha^{2^{k-1}}$, and $\delta = \alpha^{2^k}$. Now $(\mathbf{Q}(\gamma)/\mathbf{Q}(\delta)) = 2$ by the inductive hypothesis, so $\mathbf{Q}(\gamma)$ has basis $\{1, \gamma\}$ as a $\mathbf{Q}(\delta)$-vector space. Hence we may write $\beta = c + d\gamma$ uniquely with $c, d \in \mathbf{Q}(\delta)$. Then $\gamma = \beta^2 = (c + d\gamma)^2 = (c^2 + d^2\delta) + 2cd\gamma$, so $(c^2 + d^2\delta) = 0$ and $2cd = 1$. Solving for δ yields $\delta = -4c^4$ with $c \in \mathbf{Q}(\delta)$. We claim this is impossible.

If $k - 1 = t - 1$, then $k = t$ and $\mathbf{Q}(\delta) = \mathbf{Q}$. But then $\delta = -4c^4$ implies $-4\delta = 16c^4 = (2c)^4$, contradicting the hypotheses of the lemma.

If $k - 1 < t - 1$, then $k < t$ and $(\mathbf{Q}(\delta)/\mathbf{Q}(\epsilon)) = 2$ by the inductive hypothesis, where $\epsilon = \alpha^{2^{k+1}}$. Thus, if φ is the nontrivial element of $\mathrm{Gal}(\mathbf{Q}(\delta)/\mathbf{Q}(\epsilon))$, $\varphi(\delta) = -\delta$. Then $-\delta = \varphi(\delta) = \varphi(-4c^4) = -\varphi(4c^4) = -\varphi((2c)^2) = -(\varphi(2c))^2$, so $\delta = (\varphi(2c))^2$ with $2c \in \mathbf{Q}(\delta)$. But $\delta = \gamma^2$, so $\gamma = \pm\varphi(2c) \in \mathbf{Q}(\delta)$, contradicting the inductive hypothesis. □

Theorem 4.6.10. *Let n be an integer and let a be n-powerless in $\mathbf{Q}$. If n is divisible by 4, assume also that $-4a$ is not a 4th power in $\mathbf{Q}$. Then the polynomial $X^n - a$ is irreducible in $\mathbf{Q}[X]$.*

Proof. Let $n = n_1 n_2$ where n_1 is odd and n_2 is a power of 2. Let α be a root of $X^n - a$ in some extension field $\mathbf{E}$ of $\mathbf{Q}$. Then α^{n_2} is a root of $X^{n_1} - a$, which is irreducible by Lemma 4.6.9, so $(\mathbf{Q}(\alpha^{n_2})/\mathbf{Q}) = n_1$ divides $(\mathbf{Q}(\alpha)/\mathbf{Q})$. Similarly, α^{n_1} is a root of $X^{n_2} - a$, which is irreducible by Lemma 4.6.10, so $(\mathbf{Q}(\alpha^{n_1})/\mathbf{Q}) = n_2$ divides $(\mathbf{Q}(\alpha)/\mathbf{Q})$. But n_1 and n_2 are relatively prime, so $n = n_1 n_2$ divides $(\mathbf{Q}(\alpha)/\mathbf{Q})$. But certainly $(\mathbf{Q}(\alpha)/\mathbf{Q}) \leq n$, so $(\mathbf{Q}(\alpha)/\mathbf{Q}) = n$, and $X^n - a = m_\alpha(X)$ is irreducible. □

Remark 4.6.11. Theorem 4.6.10 is sharp. If $n = 4m$ and $-4a = b^4$, then $b = 2c$ and

$$X^n - a = (X^{2m} + 2cX^m + 2c^2)(X^{2m} - 2cX^m + 2c^2),$$

so $X^n - a$ is not irreducible in this case. ◇

Remark 4.6.12. (1) Theorem 4.6.10 and its proof show that the hypothesis in Lemma 4.6.5 (2) may be weakened to: If a is a negative integer, assume also that $-4a$ is not a 4th power in $\mathbb{Z}$. However, it may not be weakened further. For example, let $n = 4$ and $a = -4$. Then $a = (2i)^2$ in $\mathbf{Q}(\zeta_4)$ and $2i = (1+i)^2$ in $\mathbf{Q}(\zeta_4)$.

(2) A similar observation applies to Theorem 4.6.6. If $\mathbf{E}$ is the splitting field of $X^4 + 5$, then $|\operatorname{Gal}(\mathbf{E}/\mathbf{Q})| = 8$ and we are in case (1) of that theorem. If $\mathbf{E}$ is the splitting field of $X^4 + 9$, then $|\operatorname{Gal}(\mathbf{E}/\mathbf{Q})| = 4$ and we are in case (2) of that theorem. On the other hand, the roots of $X^4 + 4$ are $\pm(1 \pm i)$, so if $\mathbf{E}$ is the splitting field of $X^4 + 4$, then $|\operatorname{Gal}(\mathbf{E}/\mathbf{Q})| = 2$. ◇

We conclude this section by considering multiple radical extensions. First we show they have the degrees we expect, and then we find primitive elements. Again we proceed in stages.

Theorem 4.6.13. *Let $\{n_i\}$, $i = 1, \ldots, t$, be pairwise relatively prime integers and let $\{a_i\}$, $i = 1, \ldots, t$ be integers with a_i n_i-powerless for each i. If some n_i is divisible by 4, assume that $-4a_i$ is not a 4th power in $\mathbb{Z}$. Then $\mathbf{Q}(\sqrt[n_1]{a_1}, \ldots, \sqrt[n_t]{a_t})$ is an extension of $\mathbf{Q}$ of degree $n_1 \cdots n_t$.*

Proof. Immediate from Theorem 4.6.12 and Lemma 3.4.2. □

Lemma 4.6.14. *Let $A = \{a_1, \ldots, a_t\}$ be a set of integers with the property that no product of the elements of any nonempty subset of A is a square. (In particular, this holds if $a_1, \ldots, a_t$ are pairwise relatively prime, none of which is a square, and at most one of which is the negative of a square.) Then $\operatorname{Gal}(\mathbf{Q}(\sqrt{a_1}, \ldots, \sqrt{a_t})/\mathbf{Q}) \cong (\mathbb{Z}/2\mathbb{Z})^t$. In particular, $(\mathbf{Q}(\sqrt{a_1}, \ldots, \sqrt{a_t})/\mathbf{Q}) = 2^t$.*

First proof. This is a special case of Theorem 3.7.11.

Second proof. We proceed by induction on t. For $t = 0$ there is nothing to prove, and for $t = 1$ the result is obvious. By induction, it suffices to prove that $\mathbf{Q}(\sqrt{a_1}, \ldots, \sqrt{a_{t-1}})$ and $\mathbf{Q}(\sqrt{a_t})$ are disjoint extensions of $\mathbf{Q}$. Since $(\mathbf{Q}(\sqrt{a_t})/\mathbf{Q}) = 2$, to show this it suffices to show that $\mathbf{Q}(\sqrt{a_t}) \not\subseteq \mathbf{Q}(\sqrt{a_1}, \ldots, \sqrt{a_{t-1}})$. Suppose $\sqrt{a_t} = x \in \mathbf{Q}(\sqrt{a_1}, \ldots, \sqrt{a_{t-1}})$. Write $x = y + z\sqrt{a_{t-1}}$ with $y, z \in \mathbf{Q}(\sqrt{a_1}, \ldots, \sqrt{a_{t-2}})$. Squaring, $a_t = (y^2 + z^2 a_{t-1}) + 2yz\sqrt{a_{t-1}}$. Since every term but the last is contained in $\mathbf{Q}(\sqrt{a_1}, \ldots, \sqrt{a_{t-2}})$, by induction the last term must be zero, so either $z = 0$, in which case $x \in \mathbf{Q}(\sqrt{a_1}, \ldots, \sqrt{a_{t-1}})$, contradicting the inductive hypothesis, or $y = 0$, in which case $x = z\sqrt{a_{t-1}}$ and $a_t = x^2 = z^2 a_{t-1}$ with $z \in \mathbf{Q}$, which is certainly impossible. □

Theorem 4.6.15. *Let $n_1, \ldots, n_t$ be integers and let $a_1, \ldots, a_t$ be pairwise relatively prime integers with a_i n_i-powerless, $i = 1, \ldots, t$. If some n_i is even, assume that no a_i is the negative of a square. Then*

$$(\mathbf{Q}(\sqrt[n_1]{a_1}, \ldots, \sqrt[n_t]{a_t})/\mathbf{Q}) = n_1 \cdots n_t$$

with basis $\{\prod_{i=1}^{t} a_i^{k_i/n_i} \mid 0 \le k_i < n_i\}$.

Proof. Let $n = \operatorname{lcm}(n_1, \ldots, n_t)$. By hypothesis, each a_i is n_i-powerless. We will first prove the theorem under the stronger hypothesis that each a_i is n-powerless, and then show how to remove this stronger hypothesis.

Thus, assume that each a_i is n-powerless. Observe that in this case it suffices to prove the theorem for $n_1 = \cdots = n_t = n$ as clearly

$$(\mathbf{Q}(\sqrt[n_1]{a_1}, \ldots, \sqrt[n_t]{a_t})/\mathbf{Q}) \le n_1 \cdots n_t$$

and

$$(\mathbf{Q}(\sqrt[n]{a_1}, \ldots, \sqrt[n]{a_t})/\mathbf{Q}(\sqrt[n_1]{a_1}, \ldots, \sqrt[n_t]{a_t})) \le (n/n_1) \cdots (n/n_t),$$

so

$$(\mathbf{Q}(\sqrt[n]{a_1}, \ldots, \sqrt[n]{a_t})/\mathbf{Q}) = n^t$$

forces both inequalities to be equalities.

First, suppose n is odd. We prove the theorem by induction on n. It is trivially true for $n = 1$, so assume it is true for $n' < n$ and consider the case $n' = n$.

Let $\mathbf{F} = \mathbf{Q}(\zeta_N)$ where N is any multiple of n and let $\mathbf{E} = \mathbf{F}(\sqrt[n]{a_1}, \ldots, \sqrt[n]{a_t})$. (We need to consider not only $N = n$ but also N a multiple of n for the inductive step below.) Then, by Theorem 3.7.11, $\mathrm{Gal}(\mathbf{E}/\mathbf{F})$ is isomorphic to $\langle a_1, \ldots, a_t \rangle \subset \mathbf{F}^*/(\mathbf{F}^*)^n$. Suppose $a = a_1^{i_1} \cdots a_t^{i_t} \in (\mathbf{F}^*)^n$, i.e., $a = \alpha^n$ for some $\alpha \in \mathbf{F}$. In other words, a is not n-powerless in $\mathbf{F}$, so, by the contrapositive of Lemma 4.6.5, a is not n-powerless in $\mathbf{Q}$, and under our hypothesis on $a_1, \ldots, a_t$ this is true if and only if each of $i_1, \ldots, i_t$ is a multiple of n'' for some n'' dividing n, $n'' > 1$.

We claim that each $i_1, \ldots, i_t$ is a multiple of n. If $n'' = n$ we are done, so suppose not. Set $n' = n/n''$. Let $a' = a_1^{i_1/n''} \cdots a_t^{i_t/n''}$. Then $(a')^{n''} = a = \alpha^n = (\alpha^{n/n''})^{n''}$, so $a' = \zeta_{n''}^k \alpha^{n/n''}$ for some k, i.e., $a' = (\zeta_n^{n/n''})^k \alpha^{n'}$ for $\alpha' = \zeta_n^k \alpha$, so $a' \in (\mathbf{F}^*)^{n'}$. Now $n' < n$, so by induction each i_k/n'' is a multiple of $n' = n/n''$, i.e., each i_k is a multiple of n, as claimed.

Since, by Lemma 2.3.4,

$$n^t = |\operatorname{Gal}(\mathbf{E}/\mathbf{F})| = (\mathbf{E}/\mathbf{F}) \leq (\mathbf{Q}\sqrt[n]{a_1}, \ldots, \sqrt[n]{a_t})/\mathbf{Q}) \leq n^t,$$

we must have equality.

Next, suppose that n is even and let $m = n/2$. Choose N such that $\mathbf{Q}(\zeta_N) \supseteq \mathbf{Q}(\zeta_n, \sqrt{a_1}, \ldots, \sqrt{a_t})$, which is possible by Theorem 4.5.1, and set $\mathbf{F} = \mathbf{Q}(\zeta_N)$, $\mathbf{E} = \mathbf{F}(\sqrt[n]{a_1}, \ldots, \sqrt[n]{a_t})$. Again, by Theorem 3.7.11, $\operatorname{Gal}(\mathbf{E}/\mathbf{F})$ is isomorphic to $\langle a_1, \ldots, a_t \rangle \subset \mathbf{F}^*/(\mathbf{F}^*)^n$, and now by the same argument as in the case n odd,

$$m^t = |\operatorname{Gal}(\mathbf{E}/\mathbf{F})| = (\mathbf{E}/\mathbf{F}) \leq (\mathbf{Q}(\sqrt[n]{a_1} \ldots, \sqrt[n]{a_t})/\mathbf{Q}(\sqrt{a_1}, \ldots, \sqrt{a_t})) \leq m^t,$$

and we have equality. But $(\mathbf{Q}(\sqrt{a_1}, \ldots, \sqrt{a_t})/\mathbf{Q}) = 2^t$, by Lemma 4.6.14, so $(\mathbf{Q}(\sqrt[n]{a_1} \ldots, \sqrt[n]{a_t})/\mathbf{Q}) = 2^t m^t = n^t$, completing the proof in this case.

Now let us see that we may in fact assume that the a_i are all n-powerless. To do so we must use an argument from elementary number theory. First let $i = 1$, and let a_1 have prime factorization $a_1 = p_1^{e_1} \cdots p_k^{e_k}$, and let $d_1 = \gcd(e_1, \ldots, e_k)$. Since a_1 is n_1-powerless, $\gcd(d_1, n_1) = 1$. Let $\bar{n}_1$ be the product of the primes p that divide n but that do not divide n_1. By the Chinese Remainder Theorem, for each i there is an integer e_i' satisfying the congruences

$$e_i' \equiv e_i (\operatorname{mod} n_1), \qquad e_i' \equiv 1 (\operatorname{mod} \bar{n}_1).$$

Let $d = \gcd(e_1', \ldots, e_k')$. We claim that $\gcd(d, n) = 1$. To see this, consider a prime p dividing n. If p divides n_1, then, since $\gcd(d_1, n_1) = 1$, p does not divide e_i for some i. But $e_i' \equiv e_i (\operatorname{mod} n_1)$ implies $e_i' \equiv e_i (\operatorname{mod} p)$, so p does not divide e_i'. If p does not divide n_1, then p divides $\bar{n}_1$. But $e_i' \equiv 1 (\operatorname{mod} \bar{n}_1)$ for every i, and $e_i' \equiv 1 (\operatorname{mod} \bar{n}_1)$ implies $e_i' \equiv 1 (\operatorname{mod} p)$, so p does not divide e_i'.

Now let $a_1' = p_1^{e_1'} \cdots p_k^{e_k'}$. Then a_1' is n-powerless. Also, writing $e_i' = e_i + x_i n_1$, note that

$$\begin{aligned}\sqrt[n_1]{a_1'} &= \sqrt[n_1]{p_1^{e_1+x_1n_1} \cdots p_t^{e_t+x_tn_1}} \\ &= \sqrt[n_1]{(p_1^{e_1} \cdots p_k^{e_k})(p_1^{x_1} \cdots p_t^{x_t})^{n_1}} \\ &= (p_1^{x_1} \cdots p_t^{x_t}) \sqrt[n_1]{(p_1^{e_1} \cdots p_k^{e_k})} = r_1 \sqrt[n_1]{a_1}\end{aligned}$$

where $r_1 = (p_1^{x_1} \cdots p_t^{x_t}) \in \mathbf{Q}$.

Proceeding in exactly the same fashion, we may obtain $a_1', \ldots, a_t'$ with each a_i' n-powerless, and with $\sqrt[n_i]{a_i'} = r_i \sqrt[n_i]{a_i}$ with $r_i \in \mathbf{Q}$ for each i. But then $\mathbf{Q}(\sqrt[n_1]{a_1'}, \ldots, \sqrt[n_t]{a_t'}) = \mathbf{Q}(\sqrt[n]{a_1}, \ldots, \sqrt[n]{a_t})$, so we may replace a_i by a_i' and argue as above. □

Remark 4.6.16. (1) To illustrate a point in the proof of Theorem 4.6.15, let $a_1 = 8 = 2^3, n_1 = 5, a_2 = 3, n_2 = 3$. Then $n = 15$ and a_1 is not 15-powerless, but we may rechoose $a_1 = 256 = 2^8 = 2^{3+5}$ which is 15-powerless.

(2) Clearly Theorem 4.6.15 is true somewhat more generally, but the best result is clumsy to state. Our hypothesis were chosen so that we could directly appeal to Lemma 4.6.5. ◇

Corollary 4.6.17. *Let n be an integer and let $a_1, \ldots, a_t$ be pairwise relatively prime n-powerless integers. If n is even, suppose that no a_i is the negative of a square and that n is relatively prime to each a_i. Let $\mathbf{E}$ be the splitting field of $f(X) = \prod_{i=1}^{t}(X^n - a_i)$ over $\mathbf{Q}$. Then* Gal($\mathbf{E}/\mathbf{Q}$) *is the semidirect product of a group isomorphic to $(\mathbb{Z}/n\mathbb{Z})^t$ and a normal subgroup isomorphic to $(\mathbb{Z}/n\mathbb{Z})^*$.*

Proof. Let $\mathbf{B} = \mathbf{Q}(\sqrt[n]{a_1}, \ldots, \sqrt[n]{a_t})$ and let $\mathbf{D} = \mathbf{Q}(\zeta_n)$. Under our hypotheses, $\mathbf{B}$ and $\mathbf{D}$ are disjoint extensions of $\mathbf{Q}$, by Corollary 4.6.4 or Corollary 4.5.5. Now $\mathbf{E} = \mathbf{BD}$ and, by Theorem 4.6.15 and Corollary 3.4.4, $|\operatorname{Gal}(\mathbf{E}/\mathbf{D})| = (\mathbf{E}/\mathbf{D}) = (\mathbf{BD}/\mathbf{D}) = (\mathbf{B}/\mathbf{B} \cap \mathbf{D}) = (\mathbf{B}/\mathbf{Q}) = n^t$. Clearly $\operatorname{Gal}(\mathbf{E}/\mathbf{D}) \subseteq (\mathbb{Z}/n\mathbb{Z})^t$, as any automorphism must take $\sqrt[n]{a_i}$ to $\zeta\sqrt[n]{a_i}$ for some ζ with $\zeta^n = 1$, so these two groups are equal. Now $\operatorname{Gal}(\mathbf{D}/\mathbf{Q}) = \operatorname{Gal}(\mathbf{Q}(\zeta_n)/\mathbf{Q}) \cong (\mathbb{Z}/n\mathbb{Z})^*$, so the result follows from Theorem 3.4.10. □

Lemma 4.6.18. *Let $\{n_i\}$, $i = 1, \ldots, t$, be pairwise relatively prime integers and let $\{a_i\}$, $i = 1, \ldots, t$, be integers with a_i n_i-powerless for each i. If some n_i is divisible by 4, assume that $-4a_i$ is not a 4th power in $\mathbb{Z}$. Then $\alpha = \sqrt[n_1]{a_1} + \cdots + \sqrt[n_t]{a_t}$ is a primitive element of $\mathbf{Q}(\sqrt[n_1]{a_1}, \ldots, \sqrt[n_t]{a_t})$.*

Proof. Let $N = n_1 \cdots n_t$ and let $\mathbf{B}_i = \mathbf{Q}(\sqrt[n_i]{a_i})$, for each i. By Theorem 4.6.13, $(\mathbf{B}_i/\mathbf{Q}) = n_i$. Let $\mathbf{B} = \mathbf{B}_1 \cdots \mathbf{B}_t = \mathbf{Q}(\sqrt[n_1]{a_1}, \ldots, \sqrt[n_t]{a_t})$. Let $\mathbf{F} = \mathbf{Q}(\zeta_N)$, let $\mathbf{D}_i = \mathbf{F}\mathbf{B}_i = \mathbf{F}(\sqrt[n_i]{a_i})$, for each i, and let $\mathbf{D} = \mathbf{D}_1 \cdots \mathbf{D}_t$.

First, suppose each n_i is odd. Then $\mathbf{B}_i$ and $\mathbf{F}$ are disjoint extensions of $\mathbf{Q}$ by Corollary 4.6.4, and $\mathbf{F}$ is a Galois extension of $\mathbf{Q}$, so, by Corollary 3.4.4, $(\mathbf{D}_i/\mathbf{F}) = n_i$. Since the $\{n_i\}$ are pairwise relatively prime, we have, applying Lemma 3.4.2 successively, that $\mathbf{D}_2$ is disjoint from $\mathbf{D}_1$ (as extensions of $\mathbf{F}$), and then that $\mathbf{D}_3$ is disjoint from $\mathbf{D}_1\mathbf{D}_2, \ldots$. Consequently, $(\mathbf{D}/\mathbf{F}) = N$. Clearly $\alpha_i = \sqrt[n_i]{a_i}$ is a primitive element of $\mathbf{D}_i$ as an extension of $\mathbf{F}$. Then, applying Proposition 3.5.5 successively, we have that $\alpha_1 + \alpha_2$ is a primitive element of $\mathbf{D}_1\mathbf{D}_2, \ldots$, and finally that α is a primitive element of $\mathbf{D}$ as an extension of $\mathbf{F}$. It remains to show that α is a primitive element of $\mathbf{B}$ as an extension of $\mathbf{Q}$. Let $\mathbf{B}' = \mathbf{Q}(\alpha)$. Then, by Lemma 2.3.4,

$$N = (\mathbf{D}/\mathbf{F}) = (\mathbf{F}\mathbf{B}'/\mathbf{F}) \leq (\mathbf{B}'/\mathbf{Q}) \leq (\mathbf{B}/\mathbf{Q}) = N,$$

so we must have equality and $\mathbf{B}' = \mathbf{B}$.

Now suppose some n_i, say n_1, is even. The same argument as above applies, except that $(\mathbf{D}_1/\mathbf{F}) = n_1$ or $n_1/2$, enabling us to conclude that $(\mathbf{D}/\mathbf{F}) = N$ or $N/2$ and hence that $\mathbf{B} = \mathbf{B}'$ or that $(\mathbf{B}/\mathbf{B}') = 2$. But in the latter case we must have $\mathbf{B}' = \mathbf{Q}(\sqrt[n_1/2]{a_1}, \sqrt[n_2]{a_2}, \ldots, \sqrt[n_t]{a_t})$, which is impossible as α is not an element of this field. □

Lemma 4.6.19. *Let $n_1, \ldots, n_t$ be integers and let $a_1, \ldots, a_t$ be pairwise relatively prime integers with a_i n_i-powerless, $i = 1, \ldots, t$. If some n_i is even, assume that no a_i is the negative of a square. Then $\alpha = \sqrt[n_1]{a_1} + \cdots + \sqrt[n_t]{a_t}$ is a primitive element of $\mathbf{Q}(\sqrt[n_1]{a_1}, \ldots, \sqrt[n_t]{a_t})$.*

Proof. The proof proceeds along the same lines as the proof of Lemma 4.6.18. We need to know, in the notation of that proof, that $(\mathbf{D}/\mathbf{F}) = N$. In the easy case when $n_1 = \cdots = n_t = 2$, $\mathbf{F} = \mathbf{Q}$ and that follows immediately from Lemma 4.6.14. In the general case, we have shown that in the proof of Theorem 4.6.15. We then apply that fact to conclude that $\mathbf{D}_2$ is disjoint from $\mathbf{D}_1$ (as extensions of $\mathbf{F}$), and then that $\mathbf{D}_3$ is disjoint from $\mathbf{D}_1\mathbf{D}_2 \ldots$, (as otherwise $(\mathbf{D}/\mathbf{F}) < N$). The remainder of the argument is the same. □

We assemble these last two results into one theorem.

Theorem 4.6.20. *Let $n_1, \ldots, n_t$ be integers and let $a_1, \ldots, a_t$ be positive integers with a_i n_i-powerless, $i = 1, \ldots, t$. If n_i and n_j are not relatively prime, assume that a_i and a_j are relatively prime. Then $\alpha = \sqrt[n_1]{a_1} + \cdots + \sqrt[n_t]{a_t}$ is a primitive element of $\mathbf{Q}(\sqrt[n_1]{a_1}, \ldots, \sqrt[n_t]{a_t})$.*

Proof. We proceed by induction on t. The $t = 1$ case is trivial.

Now assume the theorem is true for all positive integers at most t. Consider the case $t+1$. Let $A = \{n_i \mid i = 1, \ldots, t\}$, let $A_1 = \{n_i \in A \mid \gcd(n_i, n_t) = 1\}$, and let $A_2 = \{n_i \in A \mid \gcd(n_i, n_t) > 1\}$. If $A_1 = A$, then this is just Lemma 4.6.18, and if $A_2 = A$, this is just Lemma 4.6.19. Thus we must consider the "mixed" case. Reordering if necessary, we may assume $A_1 = \{n_1, \ldots, n_s\}$ and $A_2 = \{n_{s+1}, \ldots, n_t\}$ for some s with $1 \leq s < t$. Then, by Lemma 4.6.18, $\alpha_1 = \sqrt[n_1]{a_1} + \cdots + \sqrt[n_s]{a_s}$ is a primitive element of $\mathbf{B}_1 = \mathbf{Q}(\sqrt[n_1]{a_1}, \ldots, \sqrt[n_s]{a_s})$, and, by Lemma 4.6.19, $\alpha_2 = \sqrt[n_{s+1}]{a_{s+1}}, \ldots, \sqrt[n_{t+1}]{a_{t+1}}$ is a primitive element of $\mathbf{B}_2 = \mathbf{Q}(\sqrt[n_{s+1}]{a_{s+1}}, \ldots, \sqrt[n_{t+1}]{a_{t+1}})$.

Furthermore, if $\mathbf{F} = \mathbf{Q}(\zeta_N)$ for $N = n_1 \cdots n_{t+1}$, $\mathbf{B}_1\mathbf{F}$ and $\mathbf{B}_2\mathbf{F}$ are disjoint Galois extensions of $\mathbf{F}$, so we may apply Proposition 3.5.5 as above to conclude that $\alpha = \alpha_1 + \alpha_2$ is a primitive element of $\mathbf{B}_1\mathbf{B}_2\mathbf{F}$; then we continue to argue as above to conclude that α is a primitive element of $\mathbf{B}_1\mathbf{B}_2 = \mathbf{Q}(\sqrt[n_1]{a_1}, \ldots, \sqrt[n_{t+1}]{a_{t+1}})$, completing the inductive step. □

We conclude this section with an example that illustrates these last several results.

Example 4.6.21. (1) Let $\mathbf{E} = \mathbf{Q}(\sqrt{2}, \sqrt[3]{2}, \sqrt[5]{2}, \sqrt[7]{2})$. Then $\mathbf{E}$ is an extension of $\mathbf{Q}$ of degree 210 with primitive element $\sqrt{2} + \sqrt[3]{2} + \sqrt[5]{2} + \sqrt[7]{2}$.

(2) Let $\mathbf{E} = \mathbf{Q}(\sqrt{2}, \sqrt{3}, \sqrt{5}, \sqrt{7})$. Then $\mathbf{E}$ is an extension of $\mathbf{Q}$ of degree 16 with primitive element $\sqrt{2} + \sqrt{3} + \sqrt{5} + \sqrt{7}$. Similarly, if $\mathbf{E} = \mathbf{Q}(\sqrt[3]{2}, \sqrt[3]{3}, \sqrt[3]{5}, \sqrt[3]{7})$, then $\mathbf{E}$ is an extension of $\mathbf{Q}$ of degree 81 with primitive element $\sqrt[3]{2} + \sqrt[3]{3} + \sqrt[3]{5} + \sqrt[3]{7}$.

(3) Let $\mathbf{E} = \mathbf{Q}(\sqrt{2}, \sqrt[3]{2}, \sqrt{3}, \sqrt[3]{3})$. Then $\mathbf{E}$ is an extension of $\mathbf{Q}$ of degree 36 with primitive element $\sqrt{2} + \sqrt[3]{2} + \sqrt{3} + \sqrt[3]{3}$. ◇

4.7 Galois Groups of Extensions of Q

We showed in Lemma 3.1.2 that every finite group is the Galois group of some Galois extension. It is natural to ask whether every finite group is the Galois group of some Galois extension of $\mathbf{Q}$. The answer to this question is unknown.

In this section we shall investigate two special cases: *abelian extensions*, i.e., extensions with abelian Galois group, and *symmetric extensions*, i.e., the splitting fields of polynomials of degree n with Galois group the symmetric group S_n. (Abelian extension is standard terminology but symmetric extension is not.)

Theorem 4.7.1. *Every finite abelian group is the Galois group of a Galois extension* $\mathbf{E}$ *of* $\mathbf{Q}$.

Proof. Let H be an abelian group and write H as the direct sum of cyclic groups of prime power order, $H = H_1 \oplus \cdots \oplus H_s$. Let H_i have order m_i.

Choose distinct primes $\{p_i\}$ with $p_i \equiv 1 (\bmod m_i)$, $i = 1, \dots, s$. Such primes p_i exist by Lemma C.1.1.

Let $n_0 = 1$ and $n_i = p_1 \cdots p_i$ for $i > 0$. Then $\mathbf{Q}(\zeta_{n_{i-1}})$ and $\mathbf{Q}(\zeta_{p_i})$ are disjoint extensions of $\mathbf{Q}$ for each $i > 0$, by Corollary 4.2.8, so by Theorem 3.4.7, applied inductively, we see that $G = \mathrm{Gal}(\mathbf{Q}(\zeta_{n_s})/\mathbf{Q}) = G_1 \oplus \cdots \oplus G_s$ where $G_i \cong \mathbb{Z}/(p_i - 1)\mathbb{Z}$. Since the order of H_i divides the order of G_i for each i, we see that H is a quotient of G, so by the Fundamental Theorem of Galois Theory there is a field $\mathbf{E}$ intermediate between $\mathbf{Q}$ and $\mathbf{Q}(\zeta_{n_s})$ with $\mathrm{Gal}(\mathbf{E}/\mathbf{Q}) = H$. □

Remark 4.7.2. Instead of using Lemma C.1.1, we may use a deep and famous theorem of Dirichlet: *Every arithmetic progression $ax + b$ with a and b relatively prime contains infinitely many primes.* However, we have chosen to include Lemma C.1.1 in order to give a self-contained and elementary proof of Theorem 4.7.1, rather than to simply quote this deep result. ◇

Remark 4.7.3. We can ask: When can a Galois extension $\mathbf{E}$ of $\mathbf{Q}$ be a subfield of a cyclotomic field $\mathbf{Q}(\zeta_n)$ for some n? Since $\mathrm{Gal}(\mathbf{Q}(\zeta_n)/\mathbf{Q})$ is abelian and $\mathbf{E} \subseteq \mathbf{Q}(\zeta_n)$ implies $\mathrm{Gal}(\mathbf{E}/\mathbf{Q})$ is a quotient of $\mathrm{Gal}(\mathbf{Q}(\zeta_n)/\mathbf{Q})$, an obvious necessary condition is that $\mathbf{E}$ be an abelian extension of $\mathbf{Q}$, i.e., that $\mathrm{Gal}(\mathbf{E}/\mathbf{Q})$ be abelian. This necessary condition turns out to be sufficient. This is the famous Kronecker–Weber Theorem: *Every abelian extension of* $\mathbf{Q}$ *is a subfield of a cyclotomic field.* The proof of this deep theorem is far beyond the confines of this book. (Note that we proved a very special case by direct construction. We showed in Theorem 4.5.1 that every quadratic extension of $\mathbf{Q}$ is a subfield of a cyclotomic field. Of course, our construction in Theorem 4.7.1 produced a subfield of a cyclotomic field as well.) ◇

Now we turn to the case of symmetric extensions. First, we treat the case of a polynomial of prime degree, which is relatively simple. Then we handle the case of a polynomial of arbitrary degree, using a construction of van der Waerden.

Theorem 4.7.4. *Let $f(X) \in \mathbf{Q}[X]$ be an irreducible polynomial of degree p, p a prime, with exactly $p - 2$ real roots. Let $\mathbf{E}$ be the splitting field of $f(X)$. Then* $\mathrm{Gal}(\mathbf{E}/\mathbf{Q}) = S_p$, *the symmetric group on p elements.*

Proof. Recall that since $f(X)$ is irreducible and $\mathrm{char}(\mathbf{Q}) = 0$, $f(X)$ must have p distinct roots in $\mathbf{C}$. Since $f(X) \in \mathbf{R}[X]$, the two nonreal roots of $f(X)$ must be conjugates of each other.

Let the roots of $f(X)$ in $\mathbf{C}$ be $\alpha_1, \dots, \alpha_p$ with α_1 and α_2 nonreal. Let τ be complex conjugation. Then $\tau(\alpha_1) = \alpha_2$, $\tau(\alpha_2) = \alpha_1$, $\tau(\alpha_i) = \alpha_i$ for $i > 2$.

We identify $G = \mathrm{Gal}(\mathbf{E}/\mathbf{Q})$ with a subgroup of S_p by its action permuting $\{\alpha_1, \ldots, \alpha_p\}$. Under this identification, τ is the transposition $(1\,2)$.

Now, since $f(X)$ is irreducible, G acts transitively on $\{\alpha_1, \ldots, \alpha_p\}$. Thus we see that G is a transitive subgroup of S_p, p a prime, that contains a transposition. Then, by Lemma A.3.1, $G = S_p$. □

Example 4.7.5. We now show that there exists a polynomial satisfying the hypotheses of Theorem 4.7.4, for each prime p. For $p = 2$ the result is trivial. Suppose now that p is an odd prime. Fix p and let

$$f(X) = X(X-2)(X-2a)(X-4a)\cdots(X-2(p-2)a) - (b+2).$$

We will show that with the proper choice of a and b, $f(X)$ is irreducible and has exactly $p-2$ real roots.

Let $g(X) = X(X-2)(X-2a)(X-4a)\cdots(X-2(p-2)a)$, so $g(X)$ is a polynomial of degree p with roots $0, 2, 2a, 4a, \ldots, 2(p-2)a$. Then, by elementary calculus, $g'(X) = f'(X)$, a polynomial of degree $p-1$, has roots $s_1, \ldots, s_{p-1}$ with $0 < s_1 < 2 < s_2 < 2a < \cdots < s_{p-1} < 2(p-2)a$. Then we see that $f(X)$ is increasing on the interval $(-\infty, s_1)$, decreasing on the interval (s_1, s_2), increasing on the interval (s_2, s_3), ..., and increasing on the interval (s_{p-1}, ∞). We also see that the maximum value of $f(X)$ on the interval $[0, 2]$ is at most $(2)(2)(2a)(4a)\cdots(2(p-2)a) = 2^p a^{p-2}(p-2)!$, and that $g(3a) \geq a^p$, $g(7a) \geq a^p$, $g(11a) \geq a^p$, and so forth. Thus we choose

$$a = 2^p(p-2)! \text{ and } b = 2^p a^{p-2}(p-2)!.$$

First, with this choice of a and b, $f(X)$ is irreducible by Eisenstein's Criterion (Proposition 4.1.7).

Next, with this choice of a and b, $f(X) < 0$ for $x \leq 2$, and $f(3a) > 0$, $f(7a) > 0$, $f(11a) > 0$, and so forth. Now $s_1 < 2 < s_2 < 2a < s_3$. Thus $f(X)$ is decreasing on the interval $[2, s_2]$, increasing on the interval $[s_2, 2a]$, with $f(2) < 0$, $f(2a) < 0$, so $f(X)$ has no roots on the interval $[2, 2a]$. Then $s_2 < 2a < s_3 < 4a < s_4$. Thus $f(X)$ is increasing on the interval $[2a, s_3]$, decreasing on the interval $[s_3, 4a]$, with $f(2a) < 0$, $f(4a) < 0$, and $f(s_3) \geq f(3a) > 0$, so $f(X)$ has exactly two roots r_1 and r_2 on the interval $[2a, 4a]$. By the same argument, $f(X)$ has no roots on the interval $[4a, 6a]$ and exactly two roots r_3 and r_4 on the interval $[6a, 8a]$. Proceeding in this fashion we find roots $r_1, r_2, \ldots, r_{p-3}$ of $f(X)$ on the interval $(-\infty, 2(p-2)]$. But $s_{p-1} < 2(p-2)a$, so $f(X)$ is increasing on the interval $[2(p-2)a, \infty)$, and $f(2(p-2)a) < 0$, so there is exactly one more root r_{p-2} of $f(X)$ in that interval, and lastly $f(X)$ has a total of $p-2$ real roots, as claimed. ◇

Now we turn to the case of arbitrary degree, which is considerably harder. Let $f(X) \in \mathbf{F}[X]$ be a separable polynomial and let $\mathbf{E}$ be a splitting field of $f(X)$. Let $G = \mathrm{Gal}(\mathbf{E}/\mathbf{F})$. If the irreducible factors of $f(X)$ are all distinct, then $f(X)$ has distinct roots, and eliminating repeated irreducible factors will not change $\mathbf{E}$, so we assume $f(X)$ has distinct roots $\{\alpha_1, \ldots, \alpha_n\}$. Then G permutes $\{\alpha_1, \ldots, \alpha_n\}$ (and this determines the action of G on $\mathbf{E}$). Let $\{Y_1, \ldots, Y_n\}$ be a set of indeterminates and observe that the symmetric group S_n acts in a natural way as a group of permutations of this set. Then there is an isomorphism $G \to G' \subseteq S_n$ given by $\sigma \mapsto \sigma'$ where $\sigma'(Y_i) = Y_j$ if $\sigma(\alpha_i) = \alpha_j$.

For brevity, let $\tilde{\mathbf{F}}$ denote the field of rational functions $\tilde{\mathbf{F}} = \mathbf{F}(Y_1, \ldots, Y_n)$ and let $\tilde{\mathbf{E}}$ denote the field of rational functions $\tilde{\mathbf{E}} = \mathbf{E}(Y_1, \ldots, Y_n)$.

Definition 4.7.6. In the above situation, let

$$\theta = \alpha_1 Y_1 + \cdots + \alpha_n Y_n \in \tilde{\mathbf{E}}$$

and let

$$F(Z) = \prod_{\tau' \in S_n} (Z - \tau'(\theta)) \in \tilde{\mathbf{E}}[Z]. \qquad \diamond$$

The polynomial $F(Z)$ is a symmetric function of its roots, so by Lemma 3.1.12 its coefficients are functions of the elementary symmetric functions of its roots and hence of $Y_1, \ldots, Y_n$ and the coefficients of $f(X)$. Thus, $F(Z) \in \tilde{\mathbf{F}}[Z]$. Now we may factor $F(Z)$ into a product of irreducibles in $\tilde{\mathbf{F}}[Z]$,

$$F(Z) = F_1(Z) \cdots F_t(Z).$$

One of these is divisible by $Z - \theta$ in $\tilde{\mathbf{E}}(Z)$. Renumbering if necessary we may assume it is $F_1(Z)$. Since $F(Z)$ is invariant under S_n, the action of S_n permutes these factors.

Lemma 4.7.7. *In the above situation, let $H' \subseteq S_n$ be the subgroup leaving $F_1(Z)$ invariant, i.e.,*

$$H' = \{\tau' \in S_n \mid \tau'(F_1(Z)) = F_1(Z)\}.$$

Then $G' = H'$, and hence G is isomorphic to H'.

Proof. We begin by making several observations.

Note that, for any $\tau' \in S_n$, $\tau'(Z - \theta)$ divides $\tau'(F_1(Z))$, so

$$H' = \{\tau' \in S_n \mid \tau'(Z - \theta) \text{ divides } F_1(Z)\}.$$

Note also that $G \times G' \subseteq G \times S_n$ acts on $\tilde{\mathbf{E}}$, and for any $\sigma \in G, \sigma\sigma'(\theta) = \theta$ (as $\sigma\sigma'$ simply permutes the terms of the sum $\alpha_1 Y_1 + \cdots + \alpha_n Y_n$), and so $\sigma(\theta) = (\sigma')^{-1}(\theta)$.

Consider

$$G_1(Z) = \prod_{\sigma \in G} (Z - \sigma(\theta)).$$

This is obviously invariant under G. Since the fix set of G operating on $\mathbf{E}$ is $\mathbf{F}$, the fix set of G operating on $\tilde{\mathbf{E}}$ is $\tilde{\mathbf{F}}$, by Corollary 3.4.9, so $G_1(Z) \in \tilde{\mathbf{F}}[Z]$. Since $\{\alpha_1, \ldots, \alpha_n\}$ is distinct, $\{\sigma(\theta) \mid \sigma \in G\}$ is distinct. Hence, by Lemma 2.7.12, $G_1(Z)$ is irreducible in $\tilde{\mathbf{F}}[Z]$. Since $G_1(Z)$ has $Z - \theta$ as a factor in $\tilde{\mathbf{F}}[Z]$, we must have that $G_1(Z) = F_1(Z)$. Also, by Lemma 3.5.4, $\tilde{\mathbf{E}} = \tilde{\mathbf{F}}(\theta)$. Hence the action of S_n on $\tilde{\mathbf{E}}$ is determined by its action on θ.

Suppose $\tau' \in H'$. Then $\tau'(Z - \theta)$ is also a factor of $F_1[Z] = G_1[Z]$, so it must be $Z - \sigma(\theta)$ for some $\sigma \in G$. But $\tau'(Z - \theta) = Z - \tau'(\theta) = Z - \sigma(\theta) = Z - (\sigma')^{-1}(\theta)$, so $\tau' = (\sigma')^{-1}$. But $\sigma' \in G'$ so $\tau' \in G'$.

On the other hand, if $\tau' \notin H'$, then $\tau'(Z - \theta) = Z - \tau'(\theta)$ is not a factor of $F_1(Z) = G_1(Z)$, so it is not equal to $Z - \sigma(\theta) = Z - (\sigma')^{-1}(\theta)$ for any $\sigma' \in G'$. Thus $\tau' \neq (\sigma')^{-1}$ for any $\sigma' \in G'$ and $\tau' \notin G'$. □

Remark 4.7.8. From the proof of Lemma 4.7.7, we see that

$$F_1(Z) = \prod_{\sigma' \in G'} (Z - \sigma'(\theta)),$$

and in fact

$$F_i(Z) = \prod_{\tau' \in \tau_i' G'} (Z - \tau'(\theta))$$

where the product is taken over a left coset of G'. In particular, since $\{\alpha_1, \ldots, \alpha_n\}$ are distinct, $\{\tau'(\theta) \mid \tau' \in S_n\}$ are also all distinct. We see from this that each $F_i(Z)$ has distinct roots and no two $F_i(Z)$ and $F_j(Z)$ have a root in common, for $i \neq j$. Furthermore, we also see from this that the action of S_n permutes $\{F_i(Z)\}$ transitively. ◇

Lemma 4.7.9. *Let p be a prime. Let $f(X) \in \mathbb{Z}[X]$ be a monic polynomial and let $\bar{f}(X)$ be its image in $\mathbf{F}_p[X]$ (identifying $\mathbf{F}_p$ with $\mathbb{Z}/p\mathbb{Z}$). Let $\mathbf{E}$ be a splitting field of $f(X)$ over $\mathbf{Q}$ and let $\bar{\mathbf{E}}$ be a splitting field of $\bar{f}(X)$ over $\mathbf{F}_p$. Suppose that all roots of $\bar{f}(X)$ in $\bar{\mathbf{E}}$ are simple. Then $\bar{K} = \mathrm{Gal}(\bar{\mathbf{E}}/\mathbf{F}_p)$ is isomorphic to a subgroup of $G = \mathrm{Gal}(\mathbf{E}/\mathbf{Q})$.*

Proof. Begin with the polynomial $f(X)$ and form $F(Z)$ as in Definition 4.7.6. Since $F(Z)$ is a polynomial in $Y_1, \ldots, Y_n, Z$, its factorization in the rational function field $\mathbf{Q}(Y_1, \ldots, Y_n)$ is actually a factorization in the polynomial ring $\mathbb{Z}[Y_1, \ldots, Y_n]$. This follows as the coefficients of $F(Z)$ are in fact polynomials in the coefficients of $f(X)$ by Lemma 3.1.14, and by the fact that Corollary 4.3.8 generalizes from $\mathbb{Z}$ to $\mathbb{Z}[Y_1, \ldots, Y_n]$ as $\mathbb{Z}[Y_1, \ldots, Y_n]$ is a unique factorization domain. Thus we have

$$F(Z) = F_1(Z) \cdots F_t(Z)$$

with each $F_i(Z)$ irreducible. Then reducing mod p gives a factorization (where the factors may not be irreducible)

$$\bar{F}(Z) = \bar{F}_1(Z) \cdots \bar{F}_t(Z).$$

Let $\{\bar{F}_{ij}(X)\}$ be the irreducible factors of $\bar{F}_i(X)$, for $i = 1, \ldots, t$. Let $\bar{K}'$ be the subgroup of S_n consisting of those permutations in S_n that preserve the irreducible factor $\bar{F}_{11}(X)$ of $\bar{F}_1(X)$. Now the action of S_n permutes $\{F_i(X)\}$ and hence $\{\bar{F}_{ij}(X)\}$. By Remark 4.7.8, applied to $\bar{\mathbf{E}}$, the $\bar{F}_i(X)$ have no common root and hence no common irreducible factor, so if $\bar{F}_{11}(X)$ is preserved, $\bar{F}_1(X)$ must be preserved. In other words, $\bar{K}' = \{\tau' \in S_n \mid \tau'(\bar{F}_{11}(X)) = \bar{F}_{11}(X)\} \subseteq \{\tau' \in S_n \mid \tau(F_1(X)) = F_1(X)\} = H'$.

But by the same logic as before, $\bar{K}'$ is isomorphic to $\bar{K} = \text{Gal}(\bar{\mathbf{E}}/\mathbf{F}_p)$. □

Proposition 4.7.10. *For every positive integer n there exists a polynomial $f_n(X) \in \mathbb{Z}[X]$ of degree n whose splitting field* $\mathbf{E}$ *has* Gal($\mathbf{E}/\mathbf{Q}$) *isomorphic to S_n.*

Proof. Choose a monic polynomial $f_2(X) \in \mathbb{Z}[X]$ of degree n whose mod 2 reduction $\bar{f}_2(X) \in \mathbf{F}_2[X]$ is irreducible, a monic polynomial $f_3(X) \in \mathbb{Z}[X]$ of degree n whose mod 3 reduction $\bar{f}_3(X) \in \mathbf{F}_3[X]$ is the product of an irreducible polynomial and a linear factor, and a monic polynomial $f_5(X) \in \mathbb{Z}[X]$ whose mod 5 reduction $\bar{f}_5(X) \in \mathbf{F}_5[X]$ is the product of an irreducible quadratic and other factors that are distinct irreducible polynomials of odd degree. Let

$$f(X) = -15 f_2(X) + 10 f_3(X) + 6 f_5(X).$$

Then $f(X)$ is a monic polynomial that reduces mod p to $\bar{f}_p(X)$ for $p = 2, 3, 5$. Since $\bar{f}_2(X)$ is irreducible, $f(X)$ is irreducible, and hence its roots are distinct.

Let G_p be the Galois group of $\bar{f}_p(X)$ over $\mathbf{F}_p$, $p = 2, 3, 5$. By Lemma 4.7.9, each G_p is isomorphic to a subgroup of $G = \text{Gal}(\mathbf{E}/\mathbf{Q})$.

Since $\bar{f}_2(X)$ is irreducible, G_2 operates transitively on its roots, so G is a transitive subgroup of S_n.

Since $\bar{f}_3(X)$ has an irreducible factor of degree $n-1$, and the Galois group of a finite extension of a finite field is cyclic, G_3 and hence G contains an $(n-1)$-cycle.

By the same logic, G_5 and hence G contains an element that is the product of a transposition and one or two cycles of odd order, so some (odd) power of it is a transposition in G.

Hence, by Lemma A.3.2, $G = S_n$. □

Remark 4.7.11. Our work actually gives an algorithm for computing Galois groups of polynomials $f(X) \in \mathbf{Q}[X]$: Form the polynomial $F(Z)$ as in Definition 4.7.6. Note that we do not need to know the roots of $f(X)$ to do so, since, as we have observed, its coefficients are functions of $Y_1, \ldots, Y_n$ and the coefficients of $f(X)$. Then factor $F(Z)$ into a product of irreducibles $F(Z) = F_1(Z) \cdots F_t(Z)$ in $\tilde{\mathbf{F}}[Z]$. Remark 4.1.13 gives an algorithm for doing so. As in Lemma 4.7.7, let H_i' be the stabilizer of $F_i(Z)$ in S_n. Note that, by Remark 4.7.8, the $\{H_i'\}$ are mutually conjugate subgroups of S_n so they are all isomorphic as abstract groups. (This is a general fact. If a group H acts transitively on a set $S = \{s_i\}$ with H_i the stabilizer of s_i, then, if $h \in H$ is any element with $h(s_i) = s_j$, we have that $H_j = hH_ih^{-1}$.) Thus we see from Lemma 4.7.7 that, for any i, the Galois group of $f(X)$ (i.e., the Galois group $\mathrm{Gal}(\mathbf{E}/\mathbf{Q})$, where $\mathbf{E}$ is a splitting field of $f(X)$) is isomorphic to H_i'. Thus we may simply pick any factor $F_i(X)$, and go through all of the (finitely many) elements of S_n, seeing which of them leave $F_i(X)$ invariant, to obtain H_i' and hence the Galois group.

It is evident that this algorithm is wholly impractical. We include it to show that the problem of computing Galois groups is algorithmically solvable (not to give a practical method for their computation). ◇

Corollary 4.7.12. *For any finite group G, there is a field $\mathbf{E}$ that is a finite extension of $\mathbf{Q}$, and a subfield $\mathbf{B}$ of $\mathbf{E}$ (necessarily also a finite extension of $\mathbf{Q}$), such that $\mathbf{E}$ is a Galois extension of $\mathbf{B}$ with* $\mathrm{Gal}(\mathbf{E}/\mathbf{B}) \cong G$.

Proof. This follows immediately from Proposition 4.7.10 and the proof of Lemma 3.1.2. □

4.8 The Discriminant

Let $f(X) \in \mathbf{F}[X]$ be a polynomial of degree n, with splitting field $\mathbf{E}$. As we have seen, the Galois group $\mathrm{Gal}(\mathbf{E}/\mathbf{F})$ is a subgroup of the symmetric group

S_n. In this section we develop a criterion for deciding whether G is a subgroup of A_n, the alternating group.

Definition 4.8.1. Let $f(X) \in \mathbf{F}[X]$ be a polynomial of degree n with roots $\alpha_1, \ldots, \alpha_n$ in some splitting field $\mathbf{E}$. Let

$$\delta = \prod_{i<j} (\alpha_i - \alpha_j)$$

and let

$$\Delta = \Delta(f(X)) = \delta^2.$$

Δ is the *discriminant* of $f(X)$. ◇

Remark 4.8.2. (1) Note that δ depends on the order of the roots $\alpha_1, \ldots, \alpha_n$ but $\Delta(f(X))$ does not, so $\Delta(f(X))$ is indeed an invariant of $f(X)$.

(2) If $f(X)$ is irreducible, then it has distinct roots, so in this case $\Delta(f(X)) \neq 0$.

(3) $\Delta(f(X))$ is a symmetric polynomial in the roots of $f(X)$, so, by Remark 3.1.5 and Lemma 3.1.14, it is a polynomial in the coefficients of $f(X)$. Thus $\Delta(f(X)) \in \mathbf{F}$ (and it is independent of the choice of splitting field $\mathbf{E}$).

(4) $\Delta(f(X))$ is visibly a square in $\mathbf{E}$, but may or may not be a square in $\mathbf{F}$. ◇

Lemma 4.8.3. *Let $f(X) \in \mathbf{F}[X]$ be an irreducible polynomial with splitting field $\mathbf{E}$ and with Galois group $G =$ Gal$(\mathbf{E}/\mathbf{F})$. Let $\Delta = \Delta(f(X))$ be the discriminant of $f(X)$.*

(1) If Δ is a square in $\mathbf{F}$, i.e., if $\delta \in \mathbf{F}$, then $G \subseteq A_n$.

(2) If Δ is not a square in $\mathbf{F}$, i.e., if $\delta \notin \mathbf{F}$, then $G \not\subseteq A_n$.

Proof. We observe that $\mathbf{E}$ is a Galois extension of $\mathbf{F}$ so $\mathrm{Fix}(G) = \mathbf{F}$.

Let $\sigma \in S_n$. Recall that $\sigma \in A_n$ if and only if $\mathrm{sign}(\sigma) = 1$. Also, $\mathrm{sign}(\sigma) = (-1)^i$, where i is the number of inversions in σ, i.e., i is the number of elements in the set $\mathcal{I} = \{(i, j) \mid i < j \text{ but } \sigma(j) < \sigma(i)\}$.

From this it follows immediately that $\sigma(\delta) = \mathrm{sign}(\sigma)\delta$. Thus, if $\delta \in \mathbf{F}$, then $\sigma(\delta) = \delta$ for every $\sigma \in G$, so $G \subseteq A_n$. On the other hand, if $\delta \notin \mathbf{F}$, then $\sigma_0(\delta) \neq \delta$ for some $\sigma_0 \in G$, and $\sigma_0 \notin A_n$, so $G \not\subseteq A_n$. □

Corollary 4.8.4. *In the situation of Lemma 4.8.3, suppose $\delta \notin \mathbf{F}$. Let $\mathbf{B} = \mathbf{F}(\delta)$. Also, let $H = G \cap A_n$. Then $\mathbf{E}$ is a Galois extension of $\mathbf{B}$ with Galois group* Gal$(\mathbf{E}/\mathbf{B}) = H$*, and $\mathbf{B}$ is a Galois extension of $\mathbf{F}$ with Galois group* Gal$(\mathbf{B}/\mathbf{F}) = G/H \cong \mathbb{Z}/2\mathbb{Z}$. □

Corollary 4.8.5. *Let $f(X) \in \mathbf{F}[X]$ be an irreducible cubic with splitting field* $\mathbf{E}$*, and let* $G = \mathrm{Gal}(\mathbf{E}/\mathbf{F})$*. If* $\Delta(f(X)) \notin \mathbf{F}$*, then* $G = S_3$*, while if* $\Delta(f(X)) \in \mathbf{F}$*, then* $G = A_3 \cong \mathbb{Z}/3\mathbb{Z}$*.*

Proof. By Theorem 2.8.19, G must be either S_3 or A_3, and then Lemma 4.8.3 decides between the two. □

In order to effectively use Lemma 4.8.3 we must be able to compute the discriminant. We do that now, in low degrees.

Proposition 4.8.6. *Let $f(X) \in \mathbf{F}[X]$, and let $\Delta = \Delta(f(X))$.*

(1) If $f(X) = X^2 + aX + b$, then $\Delta = a^2 - 4b$.

(2a) If $f(X) = X^3 + bX + c$, then $\Delta = -4b^3 - 27c^2$.

(2b) If $f(X) = X^3 + aX^2 + bX + c$, then $\Delta = -4a^3c + a^2b^2 + 18abc - 4b^3 - 27c^2$.

Proof. By Remark 4.8.2 (3), Δ is a polynomial in the elementary symmetric functions s_i of the roots of $f(X)$, or, equivalently, in the coefficients of $f(X)$.

(1) From Remark 3.1.5 we see that $a = -s_1 = -(\alpha_1 + \alpha_2)$ and that $b = s_2 = \alpha_1\alpha_2$. Directly from Definition 4.8.1 we see that $\Delta = (\alpha_1 - \alpha_2)^2 = \alpha_1^2 - 2\alpha_1\alpha_2 + \alpha_2^2 = \alpha_1^2 + 2\alpha_1\alpha_2 + \alpha_2^2 - 4\alpha_1\alpha_2 = a^2 - 4b$.

(2) From Remark 3.1.5 we see that $a = -s_1 = -(\alpha_1 + \alpha_2 + \alpha_3)$, that $b = s_2 = \alpha_1\alpha_2 + \alpha_2\alpha_3 + \alpha_1\alpha_3$, and that $c = -s_3 = -(\alpha_1\alpha_2\alpha_3)$. In principle, we can then simply find Δ by direct computation, but the computations get very messy, so we look for a better way.

(2a) We observe that Δ is a homogenous polynomial of degree 6 in s_1, s_2, and s_3, where s_i has degree i, or, equivalently, that Δ is a homogenous polynomial of degree 6 in a, b, and c, where a has degree 1, b has degree 2, and a has degree 3. Since $a = 0$ in this case, Δ must be of the form $\Delta = sb^3 + tc^2$ for some s and t. First, consider the polynomial $f(X) = X^3 - X$ with roots 1, 0, and -1. Direct computation shows $4 = \Delta = s(-1)^3 + t(0)^2$, so $s = -4$. Next consider $f(X) = X^3 - 1$ with roots 1, ω, and ω^2. Direct computation shows $-27 = \Delta = s(0)^2 + t(-1)^2$, so $t = -27$.

(2b) Note that adding the same value to each root leaves Δ unchanged, as it only depends on the difference between roots. If $f(X) = X^3 + aX^2 + bX + c$, let $X = Y - a/3$. Then $f(X) = g(Y) = Y^3 + b'Y + c'$ and so $\Delta = -4(b')^3 - 27(c')^2$. But, by elementary algebra, b' and c' can be expressed in terms of a, b and c, and doing so and substituting yields the result. □

Example 4.8.7. (1) Let $f(X) = X^3 + X + 1 \in \mathbf{Q}[X]$ and let $f(X)$ have splitting field $\mathbf{E}$. Then $f(X)$ is irreducible and $\Delta(f(X)) = -31$ so $\mathrm{Gal}(\mathbf{E}/\mathbf{Q}) = S_3$. On the other hand, if $\mathbf{B} = \mathbf{Q}[\sqrt{-31}]$, then $\mathrm{Gal}(\mathbf{E}/\mathbf{B}) = A_3$.

(2) Let $f(X) = X^3 - 3X + 1 \in \mathbf{Q}[X]$ and let $f(X)$ have splitting field $\mathbf{E}$. Then $f(X)$ is irreducible and $\Delta(f(X)) = 81$, so $\mathrm{Gal}(\mathbf{E}/\mathbf{Q}) = A_3 \cong \mathbb{Z}/3\mathbb{Z}$. Note we have already encountered this polynomial in Example 2.9.8. ◇

4.9 Practical Computation of Galois Groups

Our focus in this book has been the theoretical development of Galois theory, rather than its computational aspects. However, in the course of this development we have proved a number of results that are useful in (hand) computation of Galois groups of polynomials $f(X) \in \mathbf{F}[X]$. By this (common) language we mean the Galois groups $\mathrm{Gal}(\mathbf{E}/\mathbf{F})$, where $\mathbf{E}$ is a splitting field of $f(X)$. For the convenience of the reader, we shall collect most of these results in this section. We shall keep the original numbering to make it easy for the reader to refer back to the place they originally appeared.

It is not only the individual results we have derived that are useful, but also their interplay, and this better illustrated than summarized. Thus we conclude this section by working out a specific example that illustrates the application of many of these results together.

We begin by considering Galois groups of specific polynomials. Since we are in characteristic 0, the prime field $\mathbf{F}_0 = \mathbf{Q}$ and we may choose as primitive n^{th} root of unity $\zeta_n = \exp(2\pi i/n)$.

Recall we have the following definition:

Definition 3.7.5. Let $\alpha \in \mathbf{F}$. Then α is *n-powerless* in $\mathbf{F}$ if α is not an mth power in $\mathbf{F}$ for any m dividing n, $m > 1$.

In determining whether an element of $\mathbf{F}$ is n-powerless, the following results are useful:

Lemma 4.6.5. *(1) Let n be an odd integer, and let a be n-powerless in $\mathbf{Q}$. Then a is a n-powerless in $\mathbf{Q}(\zeta_N)$ for every N.*
(2) Let n be an even integer, and let a be n-powerless in $\mathbf{Q}$. If a is negative, assume also that $-a$ is not a square in $\mathbf{Q}$. Then one of the following two alternatives holds:
(a) a is n-powerless in $\mathbf{Q}(\zeta_N)$ for every N.
(b) $a = b^2$ for some $b \in \mathbf{Q}(\zeta_N)$, for some N, and b is $n/2$-powerless in $\mathbf{Q}(\zeta_{N'})$ for every multiple N' of N.

Corollary 4.5.5. *Let a be a square-free integer. If a is positive and odd, set $a' = a$. Otherwise set $a' = 4|a|$. Then $a = b^2$ for some $b \in \mathbf{Q}(\zeta_n)$ if and only if a' divides n.*

In terms of this definition, we have the following result:

Proposition 3.7.7. *Let* $\mathbf{F} \supseteq \mathbf{F}_0(\zeta_n)$. *Let* $\mathbf{E}$ *be an extension of* $\mathbf{F}$. *The following are equivalent:*
(1) $\mathbf{E}$ *is the splitting field of* $X^n - a \in \mathbf{F}[X]$, *for some* a *that is* n*-powerless in* $\mathbf{F}$.
(2) $\mathbf{E}$ *is a Galois extension of* $\mathbf{F}$ *with* $\mathrm{Gal}(\mathbf{E}/\mathbf{F}) \cong \mathbb{Z}/n\mathbb{Z}$.

This result has a generalization:

Theorem 3.7.11. *Let* $\mathbf{F} \supseteq \mathbf{F}_0(\zeta_n)$ *and let* $\mathbf{E}$ *be the splitting field of*

$$f(X) = (X^n - a_1) \cdots (X^n - a_t)$$

with each $a_i \neq 0$. *Then* $\mathrm{Gal}(\mathbf{E}/\mathbf{F})$ *is isomorphic to* $\langle a_1, \ldots, a_t \rangle \subseteq \mathbf{F}^*/(\mathbf{F}^*)^n$ *(i.e., to the subgroup of* $\mathbf{F}^*/(\mathbf{F}^*)^n$ *generated by* $a_1, \ldots, a_t$*).*

Now we turn to Galois groups of polynomials over $\mathbf{Q}$ itself.

We observe that the nth cyclotomic polynomial $\Phi_n \in \mathbf{Q}[X]$, and the polynomial $X^n - 1 \in \mathbf{Q}[X]$, both have splitting field $\mathbf{Q}(\zeta_n)$. Recall that $(\mathbb{Z}/n\mathbb{Z})^*$ denotes the multiplicative group of units in $(\mathbb{Z}/n\mathbb{Z})$, of order $\varphi(n)$. Then we have:

Corollary 4.2.7. $\mathrm{Gal}(\mathbf{Q}(\zeta_n)/\mathbf{Q}) \cong (\mathbb{Z}/n\mathbb{Z})^*$.

Now we have some results on extensions whose Galois groups are non-abelian.

Lemma 4.6.1. *Let* p *be an odd prime. Let* $a \in \mathbf{Q}$ *with* a *not a* p^{th} *power in* $\mathbf{Q}$, *and let* $\mathbf{E}$ *be the splitting field of* $X^p - a$. *Let* $G = \mathrm{Gal}(\mathbf{E}/\mathbf{Q})$. *Then*

$$G = \langle \sigma, \tau \mid \sigma^p = 1, \tau^{p-1} = 1, \tau\sigma\tau^{-1} = \sigma^r \rangle$$

for some primitive root r mod p.

Lemma 4.6.3. *Let* $a \in \mathbf{Q}$ *with* $a \neq \pm b^2$ *for any* $b \in \mathbf{Q}$, *and let* $\mathbf{E}$ *be the splitting field of* $X^4 - a$. *Let* $G = \mathrm{Gal}(\mathbf{E}/\mathbf{Q})$. *Then*

$$G = \langle \sigma, \tau \mid \sigma^4 = 1, \tau^2 = 1, \tau\sigma\tau^{-1} = \sigma^3 \rangle.$$

These two results have a common generalization:

Corollary 4.6.17. *Let* n *be an integer and let* $a_1, \ldots, a_t$ *be pairwise relatively prime* n*-powerless integers. If* n *is even, suppose that no* a_i *is the negative of a square and that* n *is relatively prime to each* a_i. *Let* $\mathbf{E}$ *be the splitting field of* $f(X) = \prod_{i=1}^t (X^n - a_i)$ *over* $\mathbf{Q}$. *Then* $\mathrm{Gal}(\mathbf{E}/\mathbf{Q})$ *is the semidirect product of a group isomorphic to* $(\mathbb{Z}/n\mathbb{Z})^t$ *and a normal subgroup isomorphic to* $(\mathbb{Z}/n\mathbb{Z})^*$.

Now we have two separate results that are useful in showing that Galois groups are large:

Theorem 4.7.4. *Let $f(X) \in \mathbf{Q}[X]$ be an irreducible polynomial of degree p, p a prime, with exactly $p-2$ real roots. Let $\mathbf{E}$ be the splitting field of $f(X)$. Then $\mathrm{Gal}(\mathbf{E}/\mathbf{Q}) = S_p$, the symmetric group on p elements.*

Lemma 4.7.9. *Let p be a prime. Let $f(X) \in \mathbb{Z}[X]$ be a monic polynomial and let $\bar{f}(X)$ be its image in $\mathbf{F}_p[X]$ (identifying $\mathbf{F}_p$ with $\mathbb{Z}/p\mathbb{Z}$). Suppose that all roots of both $f(X)$ and $\bar{f}(X)$ are simple. Let $\mathbf{E}$ be the splitting field of $f(X)$ over $\mathbf{Q}$ and let $\bar{\mathbf{E}}$ be the splitting field of $\bar{f}(X)$ over $\mathbf{F}_p$. Then $\bar{K} = \mathrm{Gal}(\bar{\mathbf{E}}/\mathbf{F}_p)$ is isomorphic to a subgroup of $G = \mathrm{Gal}(\mathbf{E}/\mathbf{Q})$.*

We now return to general fields $\mathbf{F}$.

For a polynomial $f(X) \in \mathbf{F}[X]$ we defined the discriminant $\Delta(f(X))$ in Definition 4.8.1. We then have:

Lemma 4.8.3. *Let $f(X) \in \mathbf{F}[X]$ be an irreducible polynomial with splitting field $\mathbf{E}$ and with Galois group $G = \mathrm{Gal}(\mathbf{E}/\mathbf{F})$. Let $\Delta = \Delta(f(X))$ be the discriminant of $f(X)$.*
(1) If Δ is a square in $\mathbf{F}$, i.e., if $\delta \in \mathbf{F}$, then $G \subseteq A_n$.
(2) If Δ is not a square in $\mathbf{F}$, i.e., if $\delta \notin \mathbf{F}$, then $G \not\subseteq A_n$.

In low degrees we have the following computation:

Proposition 4.8.6. *Let $f(X) \in \mathbf{F}[X]$, and let $\Delta = \Delta(f(X))$.*
(1) If $f(X) = X^2 + aX + b$, then $\Delta = a^2 - 4b$.
(2a) If $f(X) = X^3 + bX + c$, then $\Delta = -4b^3 - 27c^2$.
(2b) If $f(X) = X^3 + aX^2 + bX + c$, then $\Delta = -4a^3c + a^2b^2 + 18abc - 4b^3 - 27c^2$.

For irreducible cubics (and for quadratics, too, but there the result is trivial), the discriminant gives us the complete answer.

Corollary 4.8.5. *Let $f(X) \in \mathbf{F}[X]$ be an irreducible cubic with splitting field $\mathbf{E}$, and let $G = \mathrm{Gal}(\mathbf{E}/\mathbf{F})$. If $\Delta(f(X)) \notin \mathbf{F}$ then $G = S_3$ while if $\Delta(f(X)) \in \mathbf{F}$ then $G = A_3 \cong \mathbb{Z}/3\mathbb{Z}$.*

Now we turn to more general results which are useful in considering polynomials that are not irreducible.

Let $f(X) \in \mathbf{F}[X]$ with $f(X) = f_1(X)f_2(X)$. Let $\mathbf{B}_i$ be a splitting field of $f_i(X)$, $i = 1, 2$, with $\mathbf{B}_1$ and $\mathbf{B}_2$ both subfields of some field $\mathbf{A}$. Let $G_i = \mathrm{Gal}(\mathbf{B}_i/\mathbf{F})$, $i = 1, 2$, and let $G = \mathrm{Gal}(\mathbf{E}/\mathbf{F})$, where $\mathbf{E} = \mathbf{B}_1\mathbf{B}_2$ is a splitting field of $f(X)$. If $\mathbf{B}_1$ and $\mathbf{B}_2$ are disjoint extensions of $\mathbf{F}$, we can apply the following result.

Theorem 3.4.7. *Let* $\mathbf{B}_1$ *and* $\mathbf{B}_2$ *be disjoint Galois extensions of* $\mathbf{F}$. *Then* $\mathbf{E} = \mathbf{B}_1\mathbf{B}_2$ *is a Galois extension of* $\mathbf{F}$ *with* $\mathrm{Gal}(\mathbf{E}/\mathbf{F}) = \mathrm{Gal}(\mathbf{B}_1/\mathbf{F}) \times \mathrm{Gal}(\mathbf{B}_2/\mathbf{F})$.

In Section 3.4 we gave some criteria for determining when two extensions are disjoint. But, in any case, we have the following generalization of Theorem 3.4.7.

Corollary 3.4.8. *Let* $\mathbf{B}_1$ *and* $\mathbf{B}_2$ *be finite Galois extensions of* $\mathbf{F}$ *and let* $\mathbf{E} = \mathbf{B}_1\mathbf{B}_2$, $\mathbf{B}_0 = \mathbf{B}_1 \cap \mathbf{B}_2$. *Then* $\mathbf{E}$ *is a Galois extension of* $\mathbf{F}$ *and* $\mathrm{Gal}(\mathbf{E}/\mathbf{F}) = \{(\sigma_1, \sigma_2) \in \mathrm{Gal}(\mathbf{B}_1/\mathbf{F}) \times \mathrm{Gal}(\mathbf{B}_2/\mathbf{F}) \mid \sigma_1 \mid \mathbf{B}_0 = \sigma_2 \mid \mathbf{B}_0\}$.

Finally, we have a result that enables us to compare Galois groups over different fields.

Corollary 3.4.6 (Theorem on Natural Irrationalities). *Let* $f(X) \in \mathbf{F}[X]$ *be a separable polynomial and let* $\mathbf{B} \supseteq \mathbf{F}$. *Let* $\mathbf{E}$ *be a splitting field for* $f(X)$ *over* $\mathbf{F}$. *Then* $\mathbf{EB}$ *is a splitting field for* $f(X)$ *over* $\mathbf{B}$ *and* $\mathrm{Gal}(\mathbf{BE}/\mathbf{B})$ *is isomorphic to* $\mathrm{Gal}(\mathbf{E}/\mathbf{E} \cap \mathbf{B})$, *a subgroup of* $\mathrm{Gal}(\mathbf{E}/\mathbf{F})$, *with the isomorphism given by* $\sigma \mapsto \sigma \mid \mathbf{E}$.

We now present a single, rather elaborate, example, which shows how, with cleverness, hard work, and the use of a number of the results we have proved so far, we can determine the structure of a reasonably large Galois group. (We advise the reader to write out the diagram of intermediate fields and their inclusions in order to follow the argument here.)

Example 4.9.1. Let $\alpha = \sqrt[3]{1+\sqrt{3}}$ and let $\mathbf{E}$ be the splitting field of $m_\alpha(X) \in \mathbf{Q}[X]$. We determine $G = \mathrm{Gal}(\mathbf{E}/\mathbf{Q})$.

We observe that $\alpha^3 = 1 + \sqrt{3}$, so $\alpha^3 - 1 = \sqrt{3}$ and $(\alpha^3 - 1)^2 = 3$ and hence $\alpha^6 - 2\alpha^3 - 2 = 0$. Thus α is a root of the polynomial $X^6 - 2X^3 - 2 = 0$. This polynomial is irreducible by Eisenstein's criterion, so we see $m_\alpha(X) = X^6 - 2X^3 - 2$. We can readily find all the roots of this polynomial. Certainly $(\omega\alpha)^3 = 1 + \sqrt{3}$ and $(\omega^2\alpha)^3 = 1 + \sqrt{3}$, so $\omega\alpha$ and $\omega^2\alpha$ are also roots. But we can also see that if $\beta^3 - 1 = -\sqrt{3}$ then $(\beta^3 - 1)^2 = 3$ and hence $\beta^6 - 2\beta^3 - 2 = 0$, so β is also a root of $m_\alpha(X)$ then $\omega\beta$ and $\omega^2\beta$ are roots as well. Thus

$$\mathbf{E} = \mathbf{Q}(\alpha, \omega\alpha, \omega^2\alpha, \beta, \omega\beta, \omega^2\beta) = \mathbf{Q}(\omega, \alpha, \beta).$$

Clearly $\mathbf{E}$ contains the fields $\mathbf{Q}(\sqrt{3})$, $\mathbf{Q}(\omega) = \mathbf{Q}(\sqrt{-3})$, and $\mathbf{Q}(i)$ as well, as $i = (\sqrt{3})(\sqrt{-3})/3$. These are all extensions of $\mathbf{Q}$ of degree 2, and so we see that G contains at least 3 subgroups of index 2. Since $m_\alpha(X)$ is an irreducible polynomial of degree 6, $(\mathbf{Q}(\alpha)/\mathbf{Q}) = 6$, so $(\mathbf{Q}(\alpha)/\mathbf{Q}(\sqrt{3})) = 3$, and similarly $(\mathbf{Q}(\beta)/\mathbf{Q}(\sqrt{3})) = 3$. Since $\mathbf{Q}(\omega)$ and $\mathbf{Q}(\sqrt{3})$ are disjoint Galois extensions of $\mathbf{Q}$, both of degree 2, $\mathbf{Q}(\omega, \sqrt{3})$ is a Galois extension of $\mathbf{Q}$ of degree 4, and $\mathbf{Q}(\omega, \sqrt{3})$ is a Galois extension of $\mathbf{Q}(\sqrt{3})$ of degree 2. Since 2 and 3 are relatively prime, $\mathbf{Q}(\omega, \sqrt{3})$ and $\mathbf{Q}(\alpha)$ are disjoint extensions of $\mathbf{Q}(\sqrt{3})$, and so $\mathbf{Q}(\omega, \alpha) = \mathbf{Q}(\omega, \alpha, \sqrt{3})$ is an extension of $\mathbf{Q}(\sqrt{3})$ of degree 6. In fact, it is a Galois extension as it is the splitting field of the polynomial $X^3 - (1 + \sqrt{3})$. Similarly $\mathbf{Q}(\omega, \beta)$ is a Galois extension of $\mathbf{Q}(\sqrt{3})$ of degree 6.

We now make an important observation. We have just noted that $\mathbf{Q}(\omega, \sqrt{3})$ and $\mathbf{Q}(\alpha)$ are disjoint extensions of $\mathbf{Q}(\sqrt{3})$, so in particular $\omega \notin \mathbf{Q}(\alpha)$. Thus $\mathbf{Q}(\alpha)$ contains the root α of $m_\alpha(X)$ but does not contain the root $\omega\alpha$, so $\mathbf{Q}(\alpha)$ is not a normal extension of $\mathbf{Q}(\sqrt{3})$. Hence we see that $\mathrm{Gal}(\mathbf{Q}(\omega, \alpha)/(\mathbf{Q}(\sqrt{3}))$ is a nonabelian group of order 6, as is $\mathrm{Gal}(\mathbf{Q}(\omega, \beta)/(\mathbf{Q}(\sqrt{3}))$. Furthermore, since these two Galois groups are quotients of subgroups of G, we conclude that G is not abelian.

We claim $\beta \notin \mathbf{Q}(\omega, \alpha)$. For if $\beta \in \mathbf{Q}(\omega, \alpha)$, we would have $\mathbf{E} = \mathbf{Q}(\omega, \alpha)$ and then $(\mathbf{E}/\mathbf{Q}) = (\mathbf{Q}(\omega, \alpha)/\mathbf{Q}(\sqrt{3}))(\mathbf{Q}(\sqrt{3})/\mathbf{Q}) = (6)(2) = 12$, so G would be a nonabelian group of order 12 with at least 3 subgroups of order 6, and no such group exists. Similarly, $\alpha \notin \mathbf{Q}(\omega, \beta)$.

Now $6 = (\mathbf{Q}(\omega, \alpha)/\mathbf{Q}(\sqrt{3})) = (\mathbf{Q}(\omega, \alpha)/\mathbf{Q}(\omega, \sqrt{3}))(\mathbf{Q}(\omega, \sqrt{3})/\mathbf{Q}(\sqrt{3}))$, so we see that $(\mathbf{Q}(\omega, \alpha)/\mathbf{Q}(\omega, \sqrt{3})) = (\mathbf{Q}(\omega, \alpha)/\mathbf{Q}(\omega)) = 3$, and similarly we see that $(\mathbf{Q}(\omega, \beta)/\mathbf{Q}(\omega)) = 3$.

We claim that $\mathbf{Q}(\omega, \alpha)$ and $\mathbf{Q}(\omega, \beta)$ are disjoint extensions of $\mathbf{Q}(\omega, \sqrt{3})$. For we have the inclusions $\mathbf{Q}(\omega, \sqrt{3}) \subseteq \mathbf{Q}(\omega, \alpha) \cap \mathbf{Q}(\omega, \beta) \subseteq \mathbf{Q}(\omega, \alpha)$, so $(\mathbf{Q}(\omega, \alpha) \cap \mathbf{Q}(\omega, \beta)/\mathbf{Q}(\omega, \sqrt{3}))$ divides $(\mathbf{Q}(\omega, \alpha)/\mathbf{Q}(\omega, \sqrt{3})) = 3$, so it is either 1 or 3, i.e., $\mathbf{Q}(\omega, \alpha) \cap \mathbf{Q}(\omega, \beta) = \mathbf{Q}(\omega, \alpha)$ or $\mathbf{Q}(\omega, \sqrt{3})$. But $\mathbf{Q}(\omega, \alpha) \cap \mathbf{Q}(\omega, \beta) \neq \mathbf{Q}(\omega, \alpha)$ as $\alpha \notin \mathbf{Q}(\omega, \beta)$.

Since $\mathbf{Q}(\omega, \alpha)$ and $\mathbf{Q}(\omega, \beta)$ are disjoint Galois extensions of $\mathbf{Q}(\omega, \sqrt{3})$, we have that

$$(\mathbf{Q}(\omega, \alpha, \beta)/\mathbf{Q}(\omega, \sqrt{3})) = 9 \text{ and hence } (\mathbf{E}/\mathbf{Q}) = 36.$$

We also know that

$$\begin{aligned} 9 &= (\mathbf{Q}(\omega, \alpha, \beta)/\mathbf{Q}(\omega, \sqrt{3})) \\ &\leq (\mathbf{Q}(\alpha, \beta)/\mathbf{Q}(\sqrt{3})) \\ &\leq (\mathbf{Q}(\alpha)/\mathbf{Q}(\sqrt{3}))(\mathbf{Q}(\beta)/\mathbf{Q}(\sqrt{3})) = (3)(3) = 9, \end{aligned}$$

so we see that $(\mathbf{Q}(\alpha, \beta)/\mathbf{Q}(\sqrt{3})) = 9$. Hence $(\mathbf{Q}(\alpha, \beta)/\mathbf{Q}) = 18$ and $(\mathbf{E}/\mathbf{Q}(\alpha, \beta)) = 2$. Since

$$\begin{aligned} 36 = (\mathbf{E}/\mathbf{Q}) &= (\mathbf{Q}(\omega, \alpha, \beta)/\mathbf{Q}) \\ &= (\mathbf{Q}(\omega)/\mathbf{Q})(\mathbf{Q}(\alpha, \beta)/\mathbf{Q}) = (2)(18), \end{aligned}$$

we see that $\mathbf{Q}(\omega)$ and $\mathbf{Q}(\alpha, \beta)$ are disjoint extensions of $\mathbf{Q}$.

We now assemble all of this information to find the structure of G. First, since $(\mathbf{E}/\mathbf{Q}) = 36$, G is a group of order 36.

As we have observed, $\mathrm{Gal}(\mathbf{Q}(\omega, \alpha)/\mathbf{Q}(\sqrt{3}))$ is a nonabelian group of order 6. By the Theorem on Natural Irrationalities, $\mathrm{Gal}(\mathbf{E}/\mathbf{Q}(\beta))$ is isomorphic to a subgroup of this group. But $(\mathbf{E}/\mathbf{Q}(\beta)) = 6$ so these two groups are isomorphic. In other words, every automorphism of $\mathbf{Q}(\omega, \alpha)$ fixing $\mathbf{Q}(\sqrt{3})$ extends to an automorphism of $\mathbf{E}$ fixing $\mathbf{Q}(\beta)$, and similarly every automorphism of $\mathbf{Q}(\omega, \beta)$ fixing $\mathbf{Q}(\sqrt{3})$ extends to an automorphism of $\mathbf{E}$ fixing $\mathbf{Q}(\alpha)$. Similarly, since $(\mathbf{E}/\mathbf{Q}(\alpha, \beta)) = (\mathbf{Q}(\omega)/\mathbf{Q}) = 2$, we may again apply the Theorem on Natural Irrationalities to extend the nontrivial element of $\mathrm{Gal}(\mathbf{Q}(\omega)/\mathbf{Q})$ to an automorphism of $\mathbf{E}$ fixing $\mathbf{Q}(\alpha, \beta)$.

The elements we have found so far generate a subgroup H of G of order 18. Finally, although $\mathbf{Q}(\alpha, \beta)$ is not a Galois extension of $\mathbf{Q}$, $\mathbf{E}$ is a splitting field of $m_\alpha(X)$, an irreducible polynomial two of whose roots are α and β, so there is an automorphism of $\mathbf{E}$ taking α to β. This automorphism is not unique, but if we choose one such, we may multiply it by any element of H. (Note every element of H leaves each of the sets $\{\alpha, \omega\alpha, \omega^2\alpha\}$ and $\{\beta, \omega\beta, \omega^2\beta\}$ invariant while this new automorphism interchanges these two sets.)

Thus, assembling all of this information, we find that G is a group of order 36, generated by an element ρ of order 2, two elements σ_1 and σ_2 of order 3, and an element τ of order 2, whose actions on $\mathbf{E}$ are given by:

$$\begin{aligned} \rho(\omega) &= \omega^2, & \rho(\alpha) &= \alpha, & \rho(\beta) &= \beta \\ \sigma_1(\omega) &= \omega, & \sigma_1(\alpha) &= \omega\alpha, & \sigma_1(\beta) &= \beta \\ \sigma_2(\omega) &= \omega, & \sigma_2(\alpha) &= \alpha, & \sigma_2(\beta) &= \omega\beta \\ \tau(\omega) &= \omega, & \tau(\alpha) &= \beta, & \tau(\beta) &= \alpha. \end{aligned}$$

(Actually, H is generated by ρ, σ_1, and σ_2, and G is generated by ρ, σ_1, and τ, as $\sigma_2 = \rho\sigma_1\rho^{-1}$, but the more symmetric set of generators better reveals the structure of G.)

Finally, if we number the roots $\alpha, \omega\alpha, \omega^2\alpha, \beta, \omega\beta, \omega^2\beta$ of $m_\alpha(X)$ as $1, \ldots, 6$, the generators of G are given by the permutations

$$\rho = (23)(56), \quad \sigma_1 = (123), \quad \sigma_2 = (456), \quad \text{and } \tau = (14)(25)(36). \qquad \diamond$$

4.10 Exercises

Exercise 4.10.1. Show, without using Dirichlet's theorem about the existence of infinitely many primes in arithmetic progressions, and without using Lemma C.1.1, that:
(a) For any integer n, there exists a Galois extension $\mathbf{E}$ of $\mathbf{Q}$ of degree n.
(b) For any integer n, and any algebraic number field $\mathbf{F}$, there exists a Galois extension $\mathbf{E}$ of $\mathbf{F}$ of degree n.

Exercise 4.10.2. For a rational number $x = r/s$ in lowest terms, say that p divides x if p divides r. Show that Eisenstein's Criterion (Proposition 4.1.7) remains valid for polynomials in $\mathbf{Q}[X]$ with this definition of divisibility.

Exercise 4.10.3. Let $f(X) = a_nX^n + \cdots + a_0 \in \mathbb{Z}[X]$ and let $r \in \mathbf{Q}$ with $f(r) = 0$. Show that each of the n rational numbers

$$a_nr, \quad a_nr^2 + a_{n-1}r, \quad a_nr^3 + a_{n-1}r^2 + a_{n-2}r, \\ \ldots, \quad a_nr^n + a_{n-1}r^{n-1} + \cdots + a_1r$$

is in fact an integer. (This exercise is taken from the 2004 Putnam competition.)

Exercise 4.10.4. Use Exercise 2.10.10 to show that the polynomial $X^6 - 72$ is irreducible in $\mathbf{Q}[X]$.

Exercise 4.10.5. We might call an ordinary compass a 2-compass since, along with a straightedge, it allows us to geometrically find square roots of complex numbers. Suppose we also had a 3-compass, an instrument that allows us to geometrically find cube roots of complex numbers. For what values of $n < 1000$ could we construct a regular n-gon with these instruments?

Exercise 4.10.6. (a) Let n be an integer and let k be the product of the distinct prime factors of n. Show that $\Phi_n(X) = \Phi_k(X^{n/k})$.
(b) Let n be an integer not divisible by p. Show that

$$\Phi_{np}(X) = \Phi_n(X^p)/\Phi_n(X).$$

Exercise 4.10.7. (a) Find the n-the cyclotomic polynomial $\Phi_n(X)$ for $n = p^k$ a prime power.
(b) Find $\Phi_n(X)$ for $2 \le n \le 16$.
(c) Examining $\Phi_n(X)$ for small values of n might lead one to conjecture that the coefficients of $\Phi_n(X)$ are always 0, 1, or -1. Show, by hand computation, that $n = 105$ is a counterexample to this conjecture. (This is known to be the

smallest counterexample.) Note that $\Phi_{105}(X)$ is a polynomial of degree 48. With enough patience, this entire polynomial can be computed by hand. But only a partial, and much more manageable, computation is necessary to find a coefficient that is not 0, 1, or -1.

Exercise 4.10.8. Fix $n > 2$ and let $\mathbf{E} = \mathbf{Q}(\zeta_n)$ be the n-th cyclotomic field. Let $\mathbf{B} = \mathbf{E} \cap \mathbf{R}$.
(a) Show that $(\mathbf{E}/\mathbf{B}) = 2$ and that $\mathbf{B} = \mathbf{Q}(\zeta_n + \zeta_n^{-1})$.
(b) Show that $\mathbf{B}$ is the *unique* subfield of $\mathbf{E}$ with $(\mathbf{E}/\mathbf{B}) = 2$ if and only if $n = 4,\ p^k$, or $2p^k$ for p an odd prime.
(c) Find 3 subfields $\mathbf{B}$ of $\mathbf{E}$ with $(\mathbf{E}/\mathbf{B}) = 2$ in case $n = 8$.
(d) Show that there are exactly 3 fields $\mathbf{B}$ with $(\mathbf{E}/\mathbf{B}) = 2$ if $n = 2^k$ for any $k \geq 3$.

Exercise 4.10.9. Let $\mathbf{E} = \mathbf{Q}(\zeta_p)$ be the p-th cyclotomic field. Let $a, b \in \mathbf{Q}$ be arbitrary. Find $\mathrm{N}_{\mathbf{E}/\mathbf{Q}}(a + b\zeta_p)$ and $\mathrm{Tr}_{\mathbf{E}/\mathbf{Q}}(a + b\zeta_p)$.

Exercise 4.10.10. (a) Of course, $\zeta_1 = 1$, $\zeta_2 = -1$, $\zeta_3 = \omega = (-1 + i\sqrt{3})/2$, and $\zeta_4 = i$. Find $\zeta_5, \zeta_6, \zeta_8, \zeta_{10}, \zeta_{12}, \zeta_{15}$, and ζ_{16}.
(b) Find ζ_{17}. (This is an intricate computation. See Example 4.4.6 (3).)
(c) Find $\zeta_7, \zeta_9, \zeta_{13}$, and ζ_{14}. (This uses the solution of the cubic in Exercise 4.10.25.)

Exercise 4.10.11. Throughout this exercise, we assume $n > 2$.
(a) Show that $\Phi_n(x) > 0$ for all real numbers x.
(b) If $n = p^k$ for some k, show that $\Phi_n(1) = p$.
(c) If n is not a prime power, show that $\Phi_n(1) = 1$.
(d) If $n = 2^k$ for some k, show that $\Phi_n(-1) = 2$.
(e) If $n = 2p^k$ for some odd prime p, show that $\Phi_n(-1) = p$.
(f) If n is not a power of 2 and is not twice an odd prime power, show that $\Phi_n(-1) = 1$.
(g) Let $\mathbf{E} = \mathbf{Q}(\zeta_n)$. Show that the computations in (b) and (c) give the value of $\mathrm{N}_{\mathbf{E}/\mathbf{Q}}(1 - \zeta_n)$ and that the computations in (d), (e), and (f) give the value of $\mathrm{N}_{\mathbf{E}/\mathbf{Q}}(-1 - \zeta_n)$.

Exercise 4.10.12. (a) Let $\mathbf{F} = \mathbf{F}_p$ be the field with p elements and let let m be an integer relatively prime to p. Let $\bar{\Phi}_m(X) \in \mathbf{F}[X]$ be the mod p reduction of the cyclotomic polynomial $\Phi_m(X)$ and let $g(X) = X^m - 1 \in \mathbf{F}[X]$. Show that $\bar{\Phi}_m(X)$ and $g(X)$ have the same splitting field $\mathbf{E}$, and furthermore that $\mathbf{E} = \mathbf{F}_{p^r}$ where r is the smallest positive integer such that $p^r \equiv 1 \pmod{m}$.
(b) Let $\bar{\Phi}_m(X) = f_1(X) \cdots f_k(X) \in \mathbf{F}[X]$ be a factorization of $\bar{\Phi}_m(X)$ into a product of irreducible polynomials. Show that each $f_i(X)$ has degree r, and hence that $k = \varphi(m)/r$.

Exercise 4.10.13. Referring to Exercise 3.9.24 (b), let $k(Y, Z) = YZ$. Let m and n be arbitrary positive integers and let $d = \gcd(m, n)$ and $\ell = \operatorname{lcm}(m, n)$. Show that $Q_{YZ}(\Phi_m(X), \Phi_n(X)) = Q_{YZ}(\Phi_\ell(X), \Phi_d(X))$. In particular, if m and n are relatively prime, show that $Q_{YZ}(\Phi_m(X), \Phi_n(X)) = Q_{YZ}(\Phi_{mn}(X), \Phi_1(X)) = \Phi_{mn}(X)$.

Exercise 4.10.14. Factor each of the following polynomials into a product of irreducible polynomials in $\mathbf{Q}[X]$. (This includes the possibility that the given polynomial is irreducible.)
(a) $f(X) = X^3 - 7X^2 + 36$.
(b) $f(X) = X^3 + 5X^2 - 10X - 8$.
(c) $f(X) = X^3 - 4X^2 - 5X + 6$.
(d) $f(X) = X^4 + 2X^2 + 9$.
(e) $f(X) = X^4 + 7X^3 + 17X^2 + 18X + 6$.
(f) $f(X) = X^4 - 8X^2 + 15$.
(g) $f(X) = X^4 - 4X^2 + 16$.
(h) $f(X) = X^4 + X^3 + 7X^2 + 15X + 15$.
(i) $f(X) = X^5 + X^4 + 4X^3 + 4X^2 + 4X + 4$.
(j) $f(X) = X^6 - 2X^3 - 1$.
(k) $f(X) = X^7 - 7X^6 + 6X + 30$.

Exercise 4.10.15. For each α in Exercise 2.10.1, find $m_\alpha(X) \in \mathbf{Q}[X]$. (Consider the polynomial you found in solving that problem. If it is irreducible, it is $m_\alpha(X)$. Thus, you must decide whether it is irreducible. If it is, you are done. If not, you have more work to do.)

Exercise 4.10.16. For each α in Exercise 4.10.15, let $\mathbf{E}$ be a splitting field of $m_\alpha(X)$. Find $(\mathbf{E}/\mathbf{Q})$ and find $G = \operatorname{Gal}(\mathbf{E}/\mathbf{Q})$.

Exercise 4.10.17. In this exercise, let $\mathbf{E}$ be a splitting field of the polynomial $f(X) \in \mathbf{Q}[X]$ and let $G = \operatorname{Gal}(\mathbf{E}/\mathbf{Q})$.
(1) Find $(\mathbf{E}/\mathbf{Q})$ and find G.
(2) Find all subgroups H of G, and the fields $\mathbf{B}$ to which they belong.
(3) For each field $\mathbf{B}$ in (b), find a basis for $\mathbf{E}$ as a vector space over $\mathbf{B}$ and a basis for $\mathbf{B}$ as a vector space over $\mathbf{Q}$.
(4) Determine which subfields $\mathbf{B}$ in (b) are Galois extensions of $\mathbf{Q}$. For each of these fields $\mathbf{B}$, find a polynomial $g(X) \in \mathbf{Q}[X]$ with $\mathbf{B}$ a splitting field for $g(X)$, and find $\operatorname{Gal}(\mathbf{B}/\mathbf{Q})$.
(a) $f(X) = X^4 - 2$.
(b) $f(X) = X^5 - 2$.
(c) $f(X) = (X^3 - 2)(X^3 - 3)$.
(d) $f(X) = (X^2 - 3)(X^3 - 1)$.

(e) $f(X) = (X^2 + 3)(X^3 - 1)$.
(f) $f(X) = X^6 + 3$.
(g) $f(X) = X^6 + 5$.
(h) $f(X) = X^8 - 2$.
(i) $f(X) = X^8 - 3$.

Exercise 4.10.18. (a) Let $f(X) = X^3 + bX + c \in \mathbf{Q}[X]$ be irreducible. Suppose $b > 0$. Show that the Galois group of $f(X)$ is isomorphic to S_3.
(b) More generally, let $f(X) = X^3 + aX^2 + bX + c \in \mathbf{Q}[X]$ be irreducible. Suppose $b > a^2/3$. Show that the Galois group of $f(X)$ is isomorphic to S_3.

Exercise 4.10.19. (a) Let p be an odd prime and let

$$f(X) = X^2(X-2)(X-4)\cdots(X-2(p-2)) - 2.$$

Show that, for k sufficiently large, $f(X)$ is an irreducible polynomial with $p-2$ real roots.
(b) Let p be an odd prime and let

$$f(X) = (X^2+2)(X)(X-2)\cdots(X-2(p-3)) + 2/(2k+1).$$

Show that, for k sufficiently large, $f(X)$ is an irreducible polynomial with $p-2$ real roots. (Then, choosing $2k+1$ to be a pth power, i.e., $2k+1 = s^p$ for some integer s, we may make the substitution $Y = sX$ to obtain a monic irreducible polynomial with integer coefficients $g(Y) = (2k+1)f(X) = s^p f(X)$.)
(c) Let p be an odd prime and let

$$f(X) = (X^2+2k)(X)(X-2)\cdots(X-2(p-3)) + 2.$$

Show that, for k sufficiently large, $f(X)$ is an irreducible polynomial with $p-2$ real roots.

Hence, in any of these three cases, $f(X)$ has Galois group S_p. Compare Theorem 4.7.4 and Example 4.7.5.

Exercise 4.10.20. Let $f(X) \in \mathbf{Q}[X]$ be a reducible cubic. Show that $f(X)$ has splitting field $\mathbf{Q}(\delta)$, where δ is as in Definition 4.8.1.

Exercise 4.10.21. Let $\alpha = \sqrt[3]{7+5\sqrt{2}}$. Then certainly $(\alpha^3 - 7)^2 = 50$, so we see that α is a root of $f(X) = 0$, where $f(X) = X^6 - 14X^3 - 1 \in \mathbf{Q}[X]$. Observe that $\alpha = \beta^3$, where $\beta = 1+\sqrt{2}$. Use this observation to find all the roots of $f(X)$, the factorization of $f(X)$ as a product of irreducible polynomials in $\mathbf{Q}[X]$, the splitting field of $f(X)$, and the Galois group of $f(X)$.

Exercise 4.10.22. Let $f(X) = X^5 - 75X^4 + 15X^3 - 9X^2 - 16X - 56 \in \mathbf{Q}[X]$. Show that the Galois group of $f(X)$ is S_5.

Exercise 4.10.23. Let $f(X) = 4X^5 - 105X^4 + 840X^3 - 2160X^2 + 6000 \in \mathbf{Q}[X]$. Show that the Galois group of $f(X)$ is S_5.

Exercise 4.10.24. Let G be a finite group of order n and suppose G can be written as $G = H_1 \times \cdots \times H_k$ with H_i a group of prime power order, for each i. Such a group is called *nilpotent*. (This is not the definition of a nilpotent group but rather a characterization of finite nilpotent groups.)
(a) Show that G is solvable.
(b) Let $\mathbf{E}$ be an extension of $\mathbf{F}$ with $\mathrm{Gal}(\mathbf{E}/\mathbf{F}) \cong G$. Show that, for every divisor d of n, there is a field $\mathbf{B}$ with $\mathbf{F} \subseteq \mathbf{B} \subseteq \mathbf{E}$ and with $(\mathbf{B}/\mathbf{F}) = d$.
(c) Find a counterexample to (b) for G a solvable group.

Exercise 4.10.25. In this exercise we will show how to find the roots of a general cubic polynomial $f(X)$. Clearly it suffices to consider the case that $f(X)$ is monic, so let $f(X) = X^3 + aX^2 + bX + c$. By making the substitution $Y = X - a/3$, we may transform $f(X)$ to a cubic $g(Y) = Y^3 + b'Y + c'$, i.e., to one in which the quadratic term is absent. Thus it suffices to handle this case, so we assume $f(X) = X^3 + bX + c$. Recall from Section 4.8 that $f(X)$ has discriminant $\Delta = -4b^3 - 27c^2$. Let $\mathbf{E}$ be the splitting field of $f(X)$ and let $\mathbf{D} = \mathbf{E}(\omega)$. Then $(\mathbf{E}/\mathbf{Q}(\sqrt{\Delta})) = 1$ or 3 so we see from Section 3.7 that if $\mathbf{B} = \mathbf{Q}(\omega, \sqrt{\Delta})$, then either $\mathbf{D} = \mathbf{B}$ or that $\mathbf{D} = \mathbf{B}(\epsilon)$ with $\epsilon^3 = e \in \mathbf{B}$, i.e., that $\mathbf{D}$ is the splitting field of the polynomial $X^3 - e \in \mathbf{B}[X]$. We shall not try to find ϵ or e directly, but these considerations guide the form of our answer. Indeed, with these in mind, and the observation that e must be of the form $e = u + v\sqrt{\Delta}$ for some $u, v \in \mathbf{Q}(\omega)$, we look for roots of $f(X)$ of the form

$$\alpha = \sqrt[3]{u + v\sqrt{\Delta}} + \sqrt[3]{u - v\sqrt{\Delta}},$$
$$\beta = \omega\sqrt[3]{u + v\sqrt{\Delta}} + \omega^2\sqrt[3]{u - v\sqrt{\Delta}},$$
$$\gamma = \omega^2\sqrt[3]{u + v\sqrt{\Delta}} + \omega\sqrt[3]{u - v\sqrt{\Delta}}.$$

(Observe that α, β, and γ are all fixed by $\mathrm{Gal}(\mathbf{B}/\mathbf{Q})$, as we expect.) Now recall that the coefficients of $f(X)$ are, up to sign, the elementary symmetric functions in the roots of $f(X)$. Since $1+\omega+\omega^2 = 0$, we have that $\alpha+\beta+\gamma = 0$, which is consistent with the quadratic term of $f(X)$ being absent. Examining the linear and constant terms we obtain the equations

$$\alpha\beta + \beta\gamma + \alpha\gamma = b, \qquad \alpha\beta\gamma = -c.$$

Show that these equations yield

$$u = -c/2, \qquad v = \sqrt{-3}/18.$$

and verify by direct substitution that, with these values of u and v, α, β, and γ are roots of $f(X) = X^3 + aX + b$.

Exercise 4.10.26. Use the formula in Exercise 4.10.25 to find the roots of the following cubics:
(a) $f(X) = X^3 - 3X - 1$.
(b) $f(X) = X^3 - 21X - 37$.
(c) $f(X) = X^3 - 12X + 18$.
(d) $f(X) = X^3 - 27X + 30$.
(e) $f(X) = X^3 + 3X - 4$.
(f) $f(X) = X^3 - 13X - 12$.

Exercise 4.10.27. Observe that the polynomials in Exercise 4.10.26 parts (e) and (f) are not irreducible. Find their roots directly. Compare your answers here with your answers in Exercise 4.10.26. You may be surprised!

Exercise 4.10.28. (a) Let $\mathbf{F} \subseteq \mathbf{R}$ and let $\mathbf{E} = \mathbf{F}(\alpha)$ where $\alpha^n = a \in \mathbf{F}$ and $\alpha \in \mathbf{R}$. Let $\mathbf{B}$ be any Galois extension of $\mathbf{F}$ with $\mathbf{B} \subseteq \mathbf{E}$. Show that $(\mathbf{B}/\mathbf{F}) = 1$ or 2.
(b) Use part (a) to show that the solution of a cubic must involve nonreal radicals even when there are only real roots. (Examples of this are provided by parts (a) and (d) of Exercise 4.10.26.)

5

Further Topics in Field Theory

5.1 Separable and Inseparable Extensions

We now wish to further investigate questions related to separability and inseparability of algebraic extensions. Recall from Corollary 3.2.3 that every algebraic extension in characteristic 0 is separable, so in this case there is nothing more to be said. *Thus in this section we shall assume that* $\mathrm{char}(\mathbf{F}) = p > 0$.

Recall we have the Frobenius endomorphism $\Phi : \mathbf{F} \to \mathbf{F}$ given by $\Phi(x) = x^p$, and $\mathbf{F}^p = \mathrm{Im}(\Phi) = \{y \in \mathbf{F} \mid y = x^p \text{ for some } x \in \mathbf{F}\}$ is a subfield of $\mathbf{F}$. (See Definition 2.1.10 and Corollary 2.1.11.) If $\mathbf{F}$ is finite, then $\mathbf{F}^p = \mathbf{F}$, but in this case, too, every algebraic extension of $\mathbf{F}$ is separable (Theorem 3.2.6), so we will be interested in the case that $\mathbf{F}$ is infinite.

First, we shall deal with extensions that are separable, and then with extensions that are not.

We begin with a technical lemma.

Lemma 5.1.1. *Let* $\mathbf{E}$ *be an extension of* $\mathbf{F}$ *and let* $\alpha \in \mathbf{E}$, $\alpha \notin \mathbf{F}$. *Then* $\alpha^p \notin \mathbf{F}^p$.

Proof. Suppose $\alpha^p \in \mathbf{F}^p$, i.e., $\alpha^p = \beta^p$ for some $\beta \in \mathbf{F}$. Then $0 = \alpha^p - \beta^p = (\alpha - \beta)^p$ so $\alpha = \beta$. But $\alpha \notin \mathbf{F}$ and $\beta \in \mathbf{F}$, so this is impossible. □

We also remind the reader of the following result.

Lemma 5.1.2. *Let* $a \in \mathbf{F}$, $a \notin \mathbf{F}^p$. *Then* $f(X) = X^{p^t} - a \in \mathbf{F}[X]$ *is irreducible for every* $t \geq 1$.

Proof. Lemma 3.2.8. □

Lemma 5.1.3. *Let* $\mathbf{E}$ *be a finite extension of* $\mathbf{F}$. *If* $\mathbf{F}\mathbf{E}^p = \mathbf{E}$, *then* $\mathbf{E}$ *is a separable extension of* $\mathbf{F}$.

S.H. Weintraub, *Galois Theory*, DOI 10.1007/978-0-387-87575-0_5,

Proof. We claim that if $\{\alpha_i\}$ is a set of elements of $\mathbf{E}$ that is linearly independent over $\mathbf{F}$, then $\{\alpha_i^p\}$ is also linearly independent over $\mathbf{F}$. To see this, note that, extending $\{\alpha_i\}$ if necessary, we may assume that $\{\alpha_i\}_{i=1,\dots,d}$, $d = (\mathbf{E}/\mathbf{F})$, is a basis for $\mathbf{E}$ over $\mathbf{F}$. We will show that $\{\alpha_i^p\}_{i=1,\dots,d}$ spans $\mathbf{E}$ over $\mathbf{F}$. Hence, since $(\mathbf{E}/\mathbf{F}) = d$ is finite, $\{\alpha_i^p\}_{i=1,\dots,d}$ is a basis for $\mathbf{E}$ over $\mathbf{F}$ and hence this set is linearly independent.

Let $\gamma \in \mathbf{E}$. Since $\mathbf{E} = \mathbf{F}\mathbf{E}^p$, we may write

$$\gamma = b\alpha^p \qquad \text{for some } b \in \mathbf{F},\ \alpha \in \mathbf{E}.$$

Since $\{\alpha_i\}$ is a basis,

$$\alpha = c_1\alpha_1 + \cdots + c_d\alpha_d \qquad \text{with } c_1, \dots, c_d \in \mathbf{F},$$

and then

$$\gamma = b\alpha^p = b(c_1\alpha_1 \cdots + c_d\alpha_d)^p = (bc_1^p)\alpha_1^p + \cdots + (bc_d^p)\alpha_d^p,$$

proving the claim.

Now suppose $\alpha \in \mathbf{E}$ is not separable. Then, by Proposition 3.2.2, its minimum polynomial $m_\alpha(X)$ is of the form

$$m_\alpha(X) = \sum_{i=0}^{k} c_i X^{ip} \qquad \text{for some } k \text{ and } \{c_i\}, \text{ not all zero.}$$

Thus

$$0 = \sum_{i=0}^{k} c_i \alpha^{ip},$$

so $1 = \alpha^0, \alpha^p, \dots, \alpha^{kp}$ are linearly dependent and hence, by the contrapositive of our claim, $1 = \alpha^0, \alpha, \dots, \alpha^k$ are linearly dependent as well.

Thus

$$0 = \sum_{i=0}^{k} d_i \alpha^i \qquad \text{with not all } d_i \text{ equal to zero,}$$

so if

$$f(X) = \sum_{i=0}^{k} d_i X^i,$$

then $f(\alpha) = 0$. But $\deg f(X) < \deg m_\alpha(X)$, which is impossible. □

Lemma 5.1.4. *Let* $\alpha \in \mathbf{E}$. *Then* α *is separable over* $\mathbf{F}$ *if and only if* $\mathbf{F}(\alpha) = \mathbf{F}(\alpha^p)$.

Proof. Suppose that α is separable over $\mathbf{F}$. Clearly $\mathbf{F}(\alpha^p) \subseteq \mathbf{F}(\alpha)$, so we need only show the reverse inclusion. To show that, it suffices to show that $\alpha \in \mathbf{F}(\alpha^p)$.

Let $m_\alpha(X) \in \mathbf{F}[X]$ be the minimum polynomial of α over $\mathbf{F}$, and let $\tilde{m}_\alpha(X) \in \mathbf{F}(\alpha^p)[X]$ be the minimum polynomial of α over $\mathbf{F}(\alpha^p)$. Then $\tilde{m}_\alpha(X)$ divides $m_\alpha(X)$ in $\mathbf{F}(\alpha^p)[X]$.

Clearly α is a root of the polynomial $f(X) = X^p - \alpha^p \in \mathbf{F}(\alpha^p)[X]$. Suppose $\alpha \notin \mathbf{F}(\alpha^p)$. Then $\alpha^p \notin (\mathbf{F}(\alpha^p))^p$ by Lemma 5.1.1, so $f(X) = X^p - \alpha^p$ is irreducible in $\mathbf{F}(\alpha^p)[X]$ by Lemma 5.1.2. Since $f(X)$ is irreducible, $\tilde{m}_\alpha(X) = X^p - \alpha^p$. Since $\tilde{m}_\alpha(X)$ divides $m_\alpha(X)$ in $\mathbf{F}(\alpha^p)[X]$, $\tilde{m}_\alpha(X)$ divides $m_\alpha(X)$ in $\mathbf{F}(\alpha)[X]$. Hence $m_\alpha(X)$ is divisible by $X^p - \alpha^p = (X - \alpha)^p$ in $\mathbf{F}(\alpha)[X]$, contradicting the separability of α.

Conversely, if $\mathbf{F}(\alpha) = \mathbf{F}(\alpha^p)$, then

$$\mathbf{F}(\alpha) = \mathbf{F}(\alpha^p) = \mathbf{F}(\mathbf{F}(\alpha))^p$$

so, by Lemma 5.1.3, $\mathbf{F}(\alpha)$ is a separable extension of $\mathbf{F}$, i.e., α is a separable element of $\mathbf{E}$. □

Corollary 5.1.5. *If* $\mathbf{E}$ *is obtained from* $\mathbf{F}$ *by adjoining a root of a separable polynomial, then* $\mathbf{E}$ *is a separable extension of* $\mathbf{F}$.

Proof. Let $\mathbf{E} = \mathbf{F}(\alpha)$ where α is a root of a separable polynomial. Then $\mathbf{F}(\alpha) = \mathbf{F}(\alpha^p)$ by Lemma 5.1.4, and $\mathbf{F}(\alpha^p) = \mathbf{F}(\mathbf{F}(\alpha))^p$, so $\mathbf{F}(\alpha) = \mathbf{F}(\mathbf{F}(\alpha))^p$ and $\mathbf{F}(\alpha)$ is a separable extension of $\mathbf{F}$ by Lemma 5.1.3. □

Lemma 5.1.6. *Let* $\mathbf{B}$ *be a field intermediate between* $\mathbf{F}$ *and* $\mathbf{E}$. *If* $\mathbf{E}$ *is a separable extension of* $\mathbf{F}$, *then* $\mathbf{B}$ *is a separable extension of* $\mathbf{F}$ *and* $\mathbf{E}$ *is a separable extension of* $\mathbf{B}$.

Proof. As $\mathbf{B} \subseteq \mathbf{E}$, $\mathbf{B}$ is trivially a separable extension of $\mathbf{F}$.

Let $\alpha \in \mathbf{E}$. Since $\mathbf{E}$ is a separable extension of $\mathbf{F}$, α is a simple root of a polynomial $f(X) \in \mathbf{F}[X]$. But $\mathbf{F} \subseteq \mathbf{B}$ so $f(X) \in \mathbf{B}[X]$ and α is separable over $\mathbf{B}$. □

Lemma 5.1.3 has a converse (under weaker hypotheses, in fact).

Lemma 5.1.7. *Let* $\mathbf{E}$ *be an extension of* $\mathbf{F}$. *If* $\mathbf{E}$ *is a separable extension of* $\mathbf{F}$, *then* $\mathbf{F}\mathbf{E}^p = \mathbf{E}$.

Proof. Let $\mathbf{B} = \mathbf{F}\mathbf{E}^p$ so that $\mathbf{B}$ is intermediate between $\mathbf{F}$ and $\mathbf{E}$. By Lemma 5.1.6, $\mathbf{E}$ is a separable extension of $\mathbf{B}$.

Suppose $\mathbf{B} \subset \mathbf{E}$, and let $\alpha \in \mathbf{E}$, $\alpha \notin \mathbf{B}$. Then $a = \alpha^p \in \mathbf{B}$ so α is a root of $X^p - a \in \mathbf{B}[X]$. But, by Lemma 5.1.2, this polynomial is irreducible, so it is equal to $\tilde{m}_\alpha(X)$, the minimum polynomial of α over $\mathbf{B}$. Then this polynomial $\tilde{m}_\alpha(X) = X^p - a = X^p - \alpha^p = (X - \alpha)^p$ has repeated roots, contradicting the separability of $\mathbf{E}$ over $\mathbf{B}$. □

Theorem 5.1.8. *Let* $\mathbf{B}$ *be a field intermediate between* $\mathbf{F}$ *and* $\mathbf{E}$. *If* $\mathbf{E}$ *is separable extension of* $\mathbf{B}$ *and* $\mathbf{B}$ *is a separable extension of* $\mathbf{F}$, *then* $\mathbf{E}$ *is a separable extension of* $\mathbf{F}$.

Proof. First, assume both $(\mathbf{E}/\mathbf{B})$ and $(\mathbf{B}/\mathbf{F})$ are finite. Since $\mathbf{B}$ is a separable extension of $\mathbf{F}$, we have $\mathbf{B} = \mathbf{F}\mathbf{B}^p$, and since $\mathbf{E}$ is a separable extension of $\mathbf{B}$, we also have $\mathbf{E} = \mathbf{B}\mathbf{E}^p$, both by Lemma 5.1.7.

Then

$$\mathbf{E} = \mathbf{B}\mathbf{E}^p = \mathbf{F}\mathbf{B}^p(\mathbf{B}\mathbf{E}^p)^p \subseteq \mathbf{F}\mathbf{E}^p,$$

so $\mathbf{E} = \mathbf{F}\mathbf{E}^p$ and $(\mathbf{E}/\mathbf{F}) = (\mathbf{E}/\mathbf{B})(\mathbf{B}/\mathbf{F})$ is finite, and so $\mathbf{E}$ is a separable extension of $\mathbf{F}$ by Lemma 5.1.3.

Now for the general case. Let $\alpha \in \mathbf{E}$, and let $\tilde{m}_\alpha(X) \in \mathbf{B}[X]$ be the minimum polynomial of α over $\mathbf{B}$. Let $\mathbf{B}_0$ be the subfield of $\mathbf{B}$ obtained by adjoining the coefficients of $\tilde{m}_\alpha(X)$ to $\mathbf{F}$. Then $\mathbf{B}_0$ is a finite extension of $\mathbf{F}$ and $\mathbf{F} \subseteq \mathbf{B}_0 \subseteq \mathbf{B}$ with $\mathbf{B}$ a separable extension of $\mathbf{F}$, by hypothesis, so $\mathbf{B}_0$ is a separable extension of $\mathbf{F}$ by Lemma 5.1.6. Let $\mathbf{E}_0 = \mathbf{B}_0(\alpha)$. Then $\mathbf{E}_0$ is a finite extension of $\mathbf{B}_0$. Since $\mathbf{E}$ is a separable extension of $\mathbf{B}$, by hypothesis, $\tilde{m}_\alpha(X)$ is a separable polynomial, so, by Corollary 5.1.5, $\mathbf{E}_0$ is a separable extension of $\mathbf{B}_0$. Then, by the finite case, $\mathbf{E}_0$ is a separable extension of $\mathbf{F}$. Since α is arbitrary, $\mathbf{E}$ is separable. □

Theorem 5.1.9. *A finite extension* $\mathbf{E}$ *of* $\mathbf{F}$ *is separable if and only if* $\mathbf{E}$ *is obtained from* $\mathbf{F}$ *by adjoining root(s) of a separable polynomial* $f(X) \in \mathbf{F}[X]$.

Proof. $\mathbf{E}$ is obtained from $\mathbf{F}$ by successively adjoining finitely many elements. If one of them is not separable, then certainly $\mathbf{E}$ is not separable. If each of them is a root of a separable polynomial, then the theorem follows immediately from Corollary 5.1.5 and Theorem 5.1.8. □

Lemma 5.1.10. *Let* $\mathbf{E}$ *be an extension of* $\mathbf{F}$. *Then*

$$\mathbf{F}_s = \{\alpha \in \mathbf{E} \mid \alpha \textit{ is separable over } \mathbf{F}\}$$

is a field intermediate between $\mathbf{F}$ *and* $\mathbf{E}$.

Proof. Let $\alpha, \beta \in \mathbf{F}_s$. Then α is a simple root of $m_\alpha(X) \in \mathbf{F}[X] \subseteq \mathbf{F}(\beta)[X]$, so $\mathbf{F}(\beta)(\alpha) = \mathbf{F}(\alpha, \beta)$ is separable over $\mathbf{F}(\beta)$. Also, $\mathbf{F}(\beta)$ is separable over $\mathbf{F}$. Hence, by Theorem 5.1.8, $\mathbf{F}(\alpha, \beta)$ is separable over $\mathbf{F}$. Thus $\alpha + \beta \in \mathbf{F}_s$ and $\alpha\beta \in \mathbf{F}_s$. Also, if $\alpha \neq 0$ then $1/\alpha \in \mathbf{F}(\alpha)$ so $1/\alpha \in \mathbf{F}_s$. Thus $\mathbf{F}_s$ is a field. □

Definition 5.1.11. The field $\mathbf{F}_s$ of Lemma 5.1.10 is called the *separable closure* of $\mathbf{F}$ in $\mathbf{E}$. ◇

We now turn our attention to inseparable extensions.

Definition 5.1.12. Let $\mathbf{E}$ be an extension of $\mathbf{F}$. An element $\alpha \in \mathbf{E}$ is *purely inseparable* over $\mathbf{F}$ if $m_\alpha(X) = (X - \alpha)^m \in \mathbf{E}[X]$ for some m. The integer m is called the *degree of inseparability* of α. ◇

Remark 5.1.13. If $\alpha \in \mathbf{E}$ is both separable and purely inseparable over $\mathbf{F}$, then ($m = 1$ and) $\alpha \in \mathbf{F}$. ◇

Lemma 5.1.14. *Let $\alpha \in \mathbf{E}$ be purely inseparable over $\mathbf{F}$ with degree of inseparability m. Then $m = p^r$ for some r.*

Proof. Write $m = kp^r$ with $(k, p) = 1$. Then, in $\mathbf{E}[X]$,

$$\begin{aligned} m_\alpha(X) &= (X-\alpha)^{kp^r} = ((X-\alpha)^{p^r})^k = (X^{p^r} - \alpha^{p^r})^k \\ &= (X^{p^r})^k - k\alpha^{p^r}(X^{p^r})^{k-1} + \cdots \in \mathbf{F}[X] \end{aligned}$$

by the Binomial Theorem. Thus $k\alpha^{p^r} \in \mathbf{F}$ and so $\alpha^{p^r} \in \mathbf{F}$. Thus $f(X) = X^{p^r} - \alpha^{p^r} \in \mathbf{F}[X]$ and $f(X)$ divides $m_\alpha(X)$ in $\mathbf{E}[X]$ and hence in $\mathbf{F}[X]$, by Lemma 2.2.7. Since $m_\alpha(X)$ is irreducible in $\mathbf{F}[X]$, $m_\alpha(X) = f(X)$ and $k = 1$. □

Corollary 5.1.15. *If α is purely inseparable over $\mathbf{F}$, then $(\mathbf{F}(\alpha)/\mathbf{F}) = m = p^r$ for some r.*

Proof. $(\mathbf{F}(\alpha)/\mathbf{F}) = \deg m_\alpha(X)$. □

Definition 5.1.16. An extension $\mathbf{E}$ of $\mathbf{F}$ is *purely inseparable* if every $\alpha \in \mathbf{E}$ is inseparable over $\mathbf{F}$. ◇

Lemma 5.1.17. *Let $\mathbf{E}$ be an extension of $\mathbf{F}$. If $\alpha \in \mathbf{E}$ is purely inseparable, then $\mathbf{F}(\alpha)$ is a purely inseparable extension of $\mathbf{F}$.*

Proof. First observe that $m_\alpha(X) \in \mathbf{F}[X]$ factors as $m_\alpha(X) = (X - \alpha)^{p^r}$ in $\mathbf{E}[X]$, by Lemma 5.1.14, so $(X - \alpha)^{p^r} = X^{p^r} - \alpha^{p^r} \in \mathbf{F}[X]$ and $a = \alpha^{p^r} \in \mathbf{F}$.

Now let $\beta \in \mathbf{F}(\alpha)$. If $\beta \in \mathbf{F}$, there is nothing to prove (as then $m_\beta(X) = X - \beta$) so assume $\beta \notin \mathbf{F}$. Then $\beta = \sum_{i=0}^{m-1} b_i \alpha^i$ with $b_i \in \mathbf{F}$ (and $m = p^r$), so

$$\beta^{p^r} = \Big(\sum_{i=0}^{m-1} b_i \alpha^i\Big)^{p^r} = \sum_{i=0}^{m-1} b_i^{p^r} (\alpha^{p^r})^i = \sum_{i=0}^{m-1} b_i^{p^r} a^i \in \mathbf{F},$$

so $\mathbf{F} = \mathbf{F}(\beta^{p^r})$ and hence $\mathbf{F}(\beta^{p^r}) \subset \mathbf{F}(\beta)$. Then, by Lemma 5.1.4, β is inseparable over $\mathbf{F}$. (If β were separable over $\mathbf{F}$, we would have $\mathbf{F}(\beta) = \mathbf{F}(\beta^p) = \mathbf{F}((\beta^p)^p) = \cdots = \mathbf{F}(\beta^{p^r})$, a contradiction.) □

Lemma 5.1.18. *Let* $\mathbf{E}$ *be a purely inseparable extension of* $\mathbf{F}$. *Then every* $\alpha \in \mathbf{E}$ *is purely inseparable over* $\mathbf{F}$.

Proof. Let $\alpha \in \mathbf{E}$. If $\alpha \in \mathbf{F}$, there is nothing to prove, so let $\alpha \notin \mathbf{F}$. Consider its minimum polynomial $m_\alpha(X)$ and let p^e be the highest power of p dividing all the exponents of the nonzero terms in $m_\alpha(X)$. Then $m_\alpha(X) = f(X^{p^e})$ for some polynomial $f(X)$, and $f(X)$ is irreducible as $m_\alpha(X)$ is. Note that $f'(X) \neq 0$ as some term in $f(X)$ has positive degree not divisible by p, so, by Proposition 3.2.2, $f(X)$ is separable. Now α^{p^e} is a root of $f(X)$, so α^{p^e} is a separable element of $\mathbf{E}$. Hence $a = \alpha^{p^e} \in \mathbf{F}$, and α satisfies the polynomial $X^{p^e} - a$. Let f be the smallest integer with $a' = \alpha^{p^f} \in \mathbf{F}$. Then $a' \notin \mathbf{F}^p$, by Lemma 5.1.1, so, by Lemma 5.1.2, $g(X) = X^{p^f} - a'$ is irreducible, and $g(\alpha) = 0$. Hence $m_\alpha(X) = g(X) = X^{p^f} - a' = (X - \alpha)^{p^f}$ and α is purely inseparable (and $f = e$). □

Proposition 5.1.19. *Let* $\mathbf{E}$ *be a purely inseparable extension of* $\mathbf{F}$ *and let* $\mathbf{B}$ *be any field intermediate between* $\mathbf{F}$ *and* $\mathbf{E}$. *Then* $\mathbf{B}$ *is a purely inseparable extension of* $\mathbf{F}$ *and* $\mathbf{E}$ *is a purely inseparable extension of* $\mathbf{B}$.

Proof. Clearly $\mathbf{B}$ is a purely inseparable extension of $\mathbf{F}$.

Let $\alpha \in \mathbf{F}$. Then α has minimum polynomial $m_\alpha(X) \in \mathbf{F}[X]$ which factors as $(X - \alpha)^m \in \mathbf{E}[X]$. Let $\tilde{m}_\alpha(X) \in \mathbf{B}[X]$ be the minimum polynomial of α over $\mathbf{B}$. Then $\tilde{m}_\alpha(X)$ divides $m_\alpha(X)$ in $\mathbf{E}[X]$, so $\tilde{m}_\alpha(X) = (X - \alpha)^{m'}$ for some m', and α is purely inseparable over $\mathbf{B}$. Since α is arbitrary, $\mathbf{E}$ is purely inseparable over $\mathbf{B}$. □

Corollary 5.1.20. *Let* $\mathbf{E}$ *be a finite, purely inseparable extension of* $\mathbf{F}$. *Then* $(\mathbf{E}/\mathbf{F})$ *is a power of* p.

Proof. Let $\{\alpha_1, \ldots, \alpha_k\}$ be a basis for $\mathbf{E}$ over $\mathbf{F}$, so $\mathbf{E} = \mathbf{F}(\alpha_1, \ldots, \alpha_k)$. We prove the corollary by induction on k.

If $k = 1$, then $\mathbf{E} = \mathbf{F}(\alpha_1)$ and the result is immediate from Corollary 5.1.15.

Assume the result is true for $k-1$, and let $\mathbf{B} = \mathbf{F}(\alpha_1, \ldots, \alpha_{k-1})$. Since $\mathbf{E}$ is a purely inseparable extension of $\mathbf{F}$, $\mathbf{E}$ is a purely inseparable extension of $\mathbf{B}$ and $\mathbf{B}$ is a purely inseparable extension of $\mathbf{F}$, by Proposition 5.1.19. Since $(\mathbf{E}/\mathbf{F}) = (\mathbf{E}/\mathbf{B})(\mathbf{B}/\mathbf{F})$, the result follows. □

Lemma 5.1.21. *Let* $\mathbf{E}$ *be an algebraic extension of* $\mathbf{F}$ *and let* $\mathbf{F}_s$ *be the separable closure of* $\mathbf{F}$ *in* $\mathbf{E}$. *Then* $\mathbf{E}$ *is a purely inseparable extension of* $\mathbf{F}_s$.

Proof. Suppose $\alpha \in \mathbf{E}$ is separable over $\mathbf{F}_s$. Then $\mathbf{F}_s(\alpha)$ is a separable extension of $\mathbf{F}_s$ and $\mathbf{F}_s$ is a separable extension of $\mathbf{F}$ so, by Theorem 5.1.8, $\mathbf{F}_s(\alpha)$ is a separable extension of $\mathbf{F}$, and then $\alpha \in \mathbf{F}_s$ by the definition of $\mathbf{F}_s$. □

Corollary 5.1.22. *Let* $\mathbf{E}$ *be an algebraic extension of* $\mathbf{F}$. *Then there is a unique field* $\mathbf{B}$ *intermediate between* $\mathbf{F}$ *and* $\mathbf{E}$ *with* $\mathbf{B}$ *a separable extension of* $\mathbf{F}$ *and* $\mathbf{E}$ *a purely inseparable extension of* $\mathbf{B}$.

Proof. Take $\mathbf{B} = \mathbf{F}_s$. □

Note that Corollary 5.1.22 says that any algebraic extension of $\mathbf{F}$ can be obtained by first taking a separable extension, and then taking a purely inseparable extension.

Definition 5.1.23. Let $\mathbf{E}$ be an algebraic extension of $\mathbf{F}$ and let $\mathbf{F}_s$ be the separable closure of $\mathbf{F}$ in $\mathbf{E}$. Then $(\mathbf{F}_s/\mathbf{F})$ is the *separable degree* of $\mathbf{E}$ over $\mathbf{F}$ and $(\mathbf{E}/\mathbf{F}_s)$ is the *inseparable degree* of $\mathbf{E}$ over $\mathbf{F}$. ◇

Remark 5.1.24. (1) Of course, $(\mathbf{E}/\mathbf{F}) = (\mathbf{E}/\mathbf{F}_s)(\mathbf{F}_s/\mathbf{F})$.

(2) $\mathbf{E}$ is a separable extension of $\mathbf{F}$ if and only if $\mathbf{E} = \mathbf{F}_s$, or, equivalently, if and only if $(\mathbf{F}_s/\mathbf{F}) = (\mathbf{E}/\mathbf{F})$.

(3) $\mathbf{E}$ is a purely inseparable extension of $\mathbf{F}$ if and only if $\mathbf{F}_s = \mathbf{F}$, or, equivalently, if and only if $(\mathbf{F}_s/\mathbf{F}) = 1$.

(4) By Corollary 5.1.20, $(\mathbf{E}/\mathbf{F}_s)$ is a power of p.

(5) If $(\mathbf{E}/\mathbf{F})$ is relatively prime to p, then $\mathbf{E}$ is a separable extension of $\mathbf{F}$ (as in this case $(\mathbf{E}/\mathbf{F}_s)$, which is a power of p, divides $(\mathbf{E}/\mathbf{F})$, which is relatively prime to p, so we must have $(\mathbf{E}/\mathbf{F}_s) = 1$ and hence $\mathbf{E} = \mathbf{F}_s$). ◇

Remark 5.1.25. Observe that every purely inseparable extension of $\mathbf{F}$ is normal. (If $\alpha \in \mathbf{E}$ is purely inseparable, then $m_\alpha(X) = (X-\alpha)^m \in \mathbf{E}[X]$ is a product of linear factors.) ◇

Theorem 5.1.26. *Let* $\mathbf{B}$ *be a purely inseparable extension of* $\mathbf{F}$ *and let* $\mathbf{E}$ *be a purely inseparable extension of* $\mathbf{B}$. *Then* $\mathbf{E}$ *is a purely inseparable extension of* $\mathbf{F}$.

Proof. Let $\alpha \in \mathbf{E}$ and let $\tilde{m}_\alpha(X) \in \mathbf{B}[X]$ be the minimum polynomial of α over $\mathbf{B}$. Let $\mathbf{B}_0$ be the extension of $\mathbf{F}$ obtained by adjoining the coefficients of $\tilde{m}_\alpha(X)$ to $\mathbf{F}$. Then $\mathbf{F} \subseteq \mathbf{B}_0 \subseteq \mathbf{B}$ so $\mathbf{B}_0$ is purely inseparable over $\mathbf{F}$ by Proposition 5.1.19, and $\mathbf{B}_0$ is certainly a finite extension of $\mathbf{F}$. We now use Theorem 5.2.1 below to conclude that $\mathbf{B}_0$ is the splitting field of a polynomial $f(X) \in \mathbf{F}[X]$. Let $\{\beta_1, \ldots, \beta_k\}$ be the roots of $f(X)$, so $\mathbf{B}_0 = \mathbf{F}(\beta_1, \ldots, \beta_k)$.

Note also that $\tilde{m}_\alpha(X) = (X - \alpha)^m \in \mathbf{E}[X]$, so α is purely inseparable over $\mathbf{B}_0$.

Let $\mathbf{E}_0$ be a splitting field of $m_\alpha(X) \in \mathbf{F}[X]$ with $\mathbf{F}(\alpha) \subseteq \mathbf{E}_0$. Let $\alpha' \in \mathbf{E}_0$ be any root of $m_\alpha(X)$. Since $m_\alpha(X)$ is irreducible, by Lemma 2.6.1 there is a $\sigma : \mathbf{F}(\alpha) \to \mathbf{F}(\alpha')$ with $\sigma \mid \mathbf{F} = \mathrm{id}$ and $\sigma(\alpha) = \alpha'$. Since $\mathbf{B}_0$ is the splitting field of $f(X)$, by Lemma 2.6.3 σ extends to $\sigma : \mathbf{F}(\alpha)(\beta_1, \ldots, \beta_k) \to \mathbf{F}(\alpha', \beta_1, \ldots, \beta_k)$. Now $\mathbf{F}(\alpha)(\beta_1, \ldots, \beta_k) = \mathbf{F}(\beta_1, \ldots, \beta_k)(\alpha) = \mathbf{B}_0(\alpha)$. Similarly, $\mathbf{F}(\alpha', \beta_1, \ldots, \beta_k) = \mathbf{B}_0(\alpha')$. Now $\sigma(\beta_i) = \beta_i$ as σ must take a root of $m_{\beta_i}(X)$ to a root of $m_{\beta_i}(X)$, and β_i is the only root of $m_{\beta_i}(X)$. Hence $\sigma \mid \mathbf{B}_0 = \mathrm{id}$. Thus $\sigma : \mathbf{B}_0(\alpha) \to \mathbf{B}_0(\alpha')$ with $\sigma \mid \mathbf{B}_0 = \mathrm{id}$, $\sigma(\alpha) = \alpha'$. Then $0 = \sigma(0) = \sigma(\tilde{m}_\alpha(\alpha)) = \tilde{m}_\alpha(\sigma(\alpha)) = \tilde{m}_\alpha(\alpha')$. But α is purely inseparable over $\mathbf{B}_0$ so the only root of $\tilde{m}_\alpha(X)$ is α. Hence $\alpha' = \sigma(\alpha) = \alpha$, and so α is the only root of $m_\alpha(X)$ and hence α is purely inseparable over $\mathbf{F}$. Since α is arbitrary, $\mathbf{E}$ is purely inseparable over $\mathbf{F}$. □

We now extend the second part of Definition 5.1.12.

Definition 5.1.27. Let $\mathbf{E}$ be an algebraic extension of $\mathbf{F}$ and let $\alpha \in \mathbf{E}$. The *degree of inseparability* $i(\alpha)$ is the multiplicity of α as a root of $m_\alpha(X)$. ◇

Lemma 5.1.28. *(1) $i(\alpha)$ is independent of* $\mathbf{E}$.

(2) If α and α' are two roots of the irreducible polynomial $f(X) \in \mathbf{F}[X]$ in $\mathbf{E}$, *then $i(\alpha) = i(\alpha')$.*

Proof. (1) Let $\mathbf{E}' \supseteq \mathbf{E}$, $i = i(\alpha)$ in $\mathbf{E}$, $i' = i(\alpha)$ in $\mathbf{E}'$. In $\mathbf{E}[X]$, $m_\alpha(X) = (X - \alpha)^i g(X)$ with $g(X)$ relatively prime to $X - \alpha$, and then $g(X)$ is relatively prime to $X - \alpha$ in $\mathbf{E}'[X]$ as well (Lemma 2.2.7), so $i' = i$.

(2) By Lemma 2.6.1, there is an isomorphism $\sigma : \mathbf{F}(\alpha) \to \mathbf{F}(\alpha')$ with $\sigma \mid \mathbf{F} = \mathrm{id}$ and $\sigma(\alpha) = \alpha'$. Then $\sigma(f(X)) = f(X)$, and also if we write $f(X) = (X - \alpha)^i g(X) \in \mathbf{F}(\alpha)[X]$ with $g(\alpha) \neq 0$, then we see that $f(X) = (X - \alpha')^i g(X) \in \mathbf{F}(\alpha')[X]$ with $g(\alpha') \neq 0$. □

Lemma 5.1.29. *Let* $\mathbf{E}$ *be an algebraic extension of* $\mathbf{F}$ *and let* $\alpha \in \mathbf{E}$. *Then $i(\alpha)$ is a power of p. More precisely, $i(\alpha)$ is the highest power of p dividing the exponent of every nonzero term in $m_\alpha(X)$.*

Proof. Let p^r be the highest power of p such that $f(X) = g(X^{p^r})$, $g(X) \in \mathbf{F}[X]$. Since $m_\alpha(X)$ is irreducible, so is $g(X)$, and $g'(X) \neq 0$, so $g(X)$ is separable (Proposition 3.2.2) and hence α^{p^r} is a simple root of $g(X)$. Hence $g(X) = (X - \alpha^{p^r})h(X)$ with $h(\alpha^{p^r}) \neq 0$, i.e., with $X - \alpha^{p^r}$ not dividing $h(X)$. Thus, one the one hand, $X^{p^r} - \alpha^{p^r} = (X - \alpha)^{p^r}$ divides $f(X)$ and, on the other hand, this is the highest power of $X - \alpha$ dividing $f(X)$. □

Corollary 5.1.30. *Let* $\mathbf{E}$ *be an algebraic extension of* $\mathbf{F}$ *and let* $\alpha \in \mathbf{E}$. *If* $m_\alpha(X)$ *splits in* $\mathbf{E}[X]$, *then*

$$m_\alpha(X) = (X - \alpha_1)^{p^r} \cdots (X - \alpha_k)^{p^r}$$

with $\alpha_i = \alpha$ *for some* i, $p^r = i(\alpha)$, *and* $kp^r = \deg m_\alpha(X)$. □

5.2 Normal Extensions

In this section we investigate normal extensions.

Theorem 5.2.1. $\mathbf{E}$ *is a finite normal extension of* $\mathbf{F}$ *if and only if* $\mathbf{E}$ *is the splitting field of a polynomial* $f(X) \in \mathbf{F}[X]$.

Proof. If $\mathbf{E}$ is normal and $\{\epsilon_1, \ldots, \epsilon_n\}$ is a basis for $\mathbf{E}$ over $\mathbf{F}$, then $m_{\epsilon_i}(X) \in \mathbf{F}[X]$ splits into a product of linear factors for each i, so $f(X) = m_{\epsilon_1}(X) \cdots m_{\epsilon_n}(X) \in \mathbf{F}[X]$ splits into a product of linear factors, and then $\mathbf{E}$ is the splitting field of $f(X)$, as any field in which $f(X)$ splits must contain $\{\epsilon_1, \ldots, \epsilon_n\}$.

Suppose now that $\mathbf{E}$ is the splitting field of $f(X) \in \mathbf{F}[X]$ and let $\alpha_1, \ldots, \alpha_k$ be the roots of $f(X)$ in $\mathbf{E}$, so $\mathbf{E} = \mathbf{F}(\alpha_1, \ldots, \alpha_k)$. Let $g(X) \in \mathbf{F}[X]$ have a root $\beta \in \mathbf{E}$. We need to show $g(X)$ splits into a product of linear factors in $\mathbf{E}[X]$. Clearly it suffices to consider the case that $g(X)$ is irreducible.

Let $\tilde{\mathbf{E}} \supseteq \mathbf{E}$ be a splitting field of the irreducible polynomial $g(X)$ and let $\tilde{\beta} \in \tilde{\mathbf{E}}$ be a root of $g(X)$.

Then, by Lemma 2.6.1, there is a unique isomorphism $\sigma : \mathbf{F}(\beta) \to \mathbf{F}(\tilde{\beta})$ with $\sigma(\beta) = \tilde{\beta}$ and $\sigma \mid \mathbf{F} = \mathrm{id}$.

Now $\mathbf{E}$ is a splitting field for $f(X)$ over $\mathbf{F}(\beta)$ and $\mathbf{E}(\tilde{\beta}) = \mathbf{F}(\alpha_1, \ldots, \alpha_k, \tilde{\beta})$ is a splitting field for $f(X)$ over $\mathbf{F}(\tilde{\beta})$ so by Lemma 2.6.3 there is an isomorphism $\tau : \mathbf{E} = \mathbf{E}(\beta) \to \mathbf{E}(\tilde{\beta})$ extending σ, and hence with $\tau(\beta) = \tilde{\beta}$.

Now τ extends σ, which is the identity on $\mathbf{F}$, so τ is itself the identity on $\mathbf{F}$. Thus τ permutes $\{\alpha_1, \ldots, \alpha_k\}$, as it must take any root of $f(X)$ to a root of $f(X)$. Since $\beta \in \mathbf{E} = \mathbf{F}(\alpha_1, \ldots, \alpha_k)$, $\beta = h(\alpha_1, \ldots, \alpha_k)$ for some polynomial $h(X_1, \ldots, X_k) \in \mathbf{F}[X_1, \ldots, X_k]$. Then $\tilde{\beta} = \tau(\beta) = \tau(h(\alpha_1, \ldots, \alpha_k)) = h(\tau(\alpha_1), \ldots, \tau(\alpha_k))$ so $\tilde{\beta} \in \mathbf{E}$. □

Corollary 5.2.2. *Let* $\mathbf{E}$ *be a normal extension of* $\mathbf{F}$ *and let* $\sigma : \mathbf{B} \to \mathbf{B}'$ *be any isomorphism of fields* $\mathbf{B}$ *and* $\mathbf{B}'$ *intermediate between* $\mathbf{E}$ *and* $\mathbf{F}$ *such that* $\sigma \mid \mathbf{F} = \mathrm{id}$. *Then* σ *extends to an automorphism* $\tilde{\sigma} : \mathbf{E} \to \mathbf{E}$.

Proof. First, suppose that $\mathbf{E}$ is a finite extension of $\mathbf{F}$. Then $\mathbf{E}$ is the splitting field of a polynomial $f(X) \in \mathbf{F}[X]$. We may regard $\mathbf{E}$ as the splitting field of $f(X) \in \mathbf{B}[X]$ and note that $\sigma(f(X)) = f(X)$. Then the corollary follows immediately from Lemma 2.6.3.

The general case then follows by a Zorn's Lemma argument. (See the proof of Lemma 5.4.1 below.) □

Definition 5.2.3. Let $\mathbf{E}$ be an algebraic extension of $\mathbf{F}$. An extension $\mathbf{D}$ of $\mathbf{E}$ is a *normal closure* of $\mathbf{E}$ if $\mathbf{D}$ is a normal extension of $\mathbf{F}$ but no field intermediate between $\mathbf{D}$ and $\mathbf{E}$ is a normal extension of $\mathbf{F}$. ◇

Lemma 5.2.4. *Every algebraic extension* $\mathbf{E}$ *of* $\mathbf{F}$ *has a normal closure* $\mathbf{D}$, *and any two normal closures of* $\mathbf{E}$ *are isomorphic.*

Proof. First, suppose $\mathbf{E}$ is a finite extension of $\mathbf{F}$. Let $\mathbf{E} = \mathbf{F}(\alpha_1, \ldots, \alpha_k)$ and let $f(X) = m_{\alpha_1}(X) \cdots m_{\alpha_k}(X)$. Let $\mathbf{D} \supseteq \mathbf{E}$ be a splitting field of $f(X)$. Then $\mathbf{D}$ is a normal extension of $\mathbf{F}$ by Theorem 5.2.1, and $f(X)$ does not split in any proper subfield of $\mathbf{D}$ containing $\mathbf{E}$, so $\mathbf{D}$ is a normal closure of $\mathbf{E}$. Furthermore, $\mathbf{D}$ is unique up to isomorphism as any two splitting fields of $f(X)$ are isomorphic (Lemma 2.6.4).

The general case then follows by a Zorn's Lemma argument. □

Corollary 5.2.5. *Let* $\mathbf{E}$ *be a separable extension of* $\mathbf{F}$. *Then the normal closure* $\mathbf{D}$ *of* $\mathbf{E}$ *is a Galois extension of* $\mathbf{F}$.

Proof. First, suppose $\mathbf{E}$ is a finite extension of $\mathbf{F}$. Then, following the proof of Lemma 5.2.4, $\mathbf{D}$ is a splitting field of the separable polynomial $f(X)$ and so, by Theorem 2.7.14, $\mathbf{D}$ is a Galois extension of $\mathbf{F}$.

The general case then follows from Corollary 5.4.3 below. □

This corollary justifies the following definition.

Definition 5.2.6. Let $\mathbf{E}$ be a separable extension of $\mathbf{F}$. A normal closure $\mathbf{D}$ of $\mathbf{E}$ is called a *Galois closure* of $\mathbf{E}$. ◇

Remark 5.2.7. (1) Note that if $\mathbf{E}$ is normal over $\mathbf{F}$ and $\mathbf{B}$ is an intermediate field between $\mathbf{F}$ and $\mathbf{E}$, then $\mathbf{E}$ is normal over $\mathbf{B}$, but $\mathbf{B}$ need not be normal over $\mathbf{F}$. $\mathbf{E}$ is normal over $\mathbf{B}$ as for any $\alpha \in \mathbf{E}$, $m_\alpha(X)$ splits into linear factors in $\mathbf{E}$, and hence $\tilde{m}_\alpha(X)$, the minimum polynomial of α over $\mathbf{B}$, which divides $m_\alpha(X)$, also splits into linear factors in $\mathbf{E}$. On the other hand, if $\mathbf{E} = \mathbf{Q}(\omega, \sqrt[3]{2})$ is the splitting field of $X^3 - 2 \in \mathbf{Q}[X]$, and $\mathbf{B} = \mathbf{Q}(\sqrt[3]{2})$, $\mathbf{F} = \mathbf{Q}$, then $\mathbf{E}$ is normal over $\mathbf{B}$ but $\mathbf{B}$ is not normal over $\mathbf{F}$.

(2) If $\mathbf{B}$ is a normal extension of $\mathbf{F}$ and $\mathbf{E}$ is a normal extension of $\mathbf{B}$, then $\mathbf{E}$ need not be a normal extension of $\mathbf{F}$. Taking $\mathbf{F} = \mathbf{Q}$, $\mathbf{B} = \mathbf{Q}(\sqrt{2})$, and $\mathbf{E} = \mathbf{Q}(\sqrt[4]{2})$ provides an example of this. ◇

Now we turn to inseparable extensions. We first recall from Remark 5.1.25 that a purely inseparable extension is automatically normal.

Lemma 5.2.8. *Let* $\mathbf{E}$ *be a normal extension of* $\mathbf{F}$. *Then* $\mathbf{E}$ *is separable if and only if* $\mathrm{Fix}(\mathrm{Gal}(\mathbf{E}/\mathbf{F})) = \mathbf{F}$, *and* $\mathbf{E}$ *is purely inseparable if and only if* $\mathrm{Fix}(\mathrm{Gal}(\mathbf{E}/\mathbf{F})) = \mathbf{E}$, *or, equivalently, if and only if* $\mathrm{Gal}(\mathbf{E}/\mathbf{F}) = \{\mathrm{id}\}$.

Proof. The first claim is immediate from the fact that $\mathbf{E}$ is a Galois extension of $\mathbf{F}$ if and only if $\mathbf{E}$ is normal and separable (Theorem 2.7.14 in case $(\mathbf{E}/\mathbf{F})$ is finite and Theorem 5.4.2 below in case $(\mathbf{E}/\mathbf{F})$ is infinite).

For the second claim, first suppose that $(\mathbf{E}/\mathbf{F})$ is finite. Suppose next that $\mathbf{E}$ is a purely inseparable extension of $\mathbf{F}$. Let $\alpha \in \mathbf{E}$. Then $m_\alpha(X) = (X - \alpha)^m$ in $\mathbf{E}[X]$ for some m. Now for any $\sigma \in \mathrm{Gal}(\mathbf{E}/\mathbf{F})$, $\sigma(\alpha)$ is also a root of $m_\alpha(X)$. But the only root of $m_\alpha(X)$ is α, so $\sigma(\alpha) = \alpha$. Since α is arbitrary, $\sigma = \mathrm{id}$.

The general case follows by a Zorn's lemma argument.

Conversely, let $\mathbf{E}$ be normal and finite over $\mathbf{F}$, and suppose that $\mathbf{E}$ is not purely inseparable over $\mathbf{F}$. Then there is an $\alpha \in \mathbf{E}$ such that $m_\alpha(X)$ has (at least) two distinct roots α and α'. By Lemma 2.6.1 there is a $\sigma : \mathbf{F}(\alpha) \to \mathbf{F}(\alpha')$ with $\sigma(\alpha) = \alpha'$ and, by Corollary 5.2.2, σ extends to $\tilde{\sigma} : \mathbf{E} \to \mathbf{E}$ with $\tilde{\sigma}(\alpha) = \alpha'$ and $\tilde{\sigma} \mid \mathbf{F} = \mathrm{id}$, so $\tilde{\sigma} \in \mathrm{Gal}(\mathbf{E}/\mathbf{F})$ and $\tilde{\sigma} \neq \mathrm{id}$. □

We have seen in Section 5.1 that any algebraic extension may be obtained by first taking a separable extension and then taking a purely inseparable extension. In the case of a normal extension, the order may be reversed.

Proposition 5.2.9. *Let* $\mathbf{E}$ *be a normal extension of* $\mathbf{F}$. *If* $\mathbf{F}_i = \mathrm{Fix}(\mathrm{Gal}(\mathbf{E}/\mathbf{F}))$ *then* $\mathbf{F}_i$ *is a purely inseparable extension of* $\mathbf{F}$ *and* $\mathbf{E}$ *is a separable extension of* $\mathbf{F}_i$. *Furthermore,* $\mathbf{E}$ *is a Galois extension of* $\mathbf{F}_i$ *and* $\mathrm{Gal}(\mathbf{E}/\mathbf{F}_i) = \mathrm{Gal}(\mathbf{E}/\mathbf{F})$.

Proof. By definition, if G is a group of automorphisms of $\mathbf{E}$, then $\mathbf{E}$ is a Galois extension of $\mathrm{Fix}(G)$. Apply that here with $G = \mathrm{Gal}(\mathbf{E}/\mathbf{F})$. Then $\mathbf{E}$ is a Galois extension of $\mathbf{F}_i$ and hence $\mathbf{E}$ is a separable extension of $\mathbf{F}$ (Theorem 2.7.14 in case $(\mathbf{E}/\mathbf{F}_i)$ is finite and Theorem 5.4.2 below in case $(\mathbf{E}/\mathbf{F}_i)$ is infinite).

Now we consider the extension $\mathbf{F}_i$ of $\mathbf{F}$. If $\mathbf{F}_i$ is not purely inseparable then, as in the proof of Lemma 5.2.8, there is an element $\alpha \in \mathbf{F}_i$ and an isomorphism

$\sigma : \mathbf{F}(\alpha) \to \mathbf{F}(\alpha')$ with $\sigma \mid \mathbf{F} = \text{id}$ and $\sigma(\alpha) = \alpha' \neq \alpha$. Then, by Corollary 5.2.2, σ extends to $\tilde{\sigma} : \mathbf{E} \to \mathbf{E}$ with $\tilde{\sigma} \mid \mathbf{F} = \text{id}$, i.e., to $\tilde{\sigma} \in \text{Gal}(\mathbf{E}/\mathbf{F})$ with $\tilde{\sigma}(\alpha) \neq \alpha$, contradicting the definition of $\mathbf{F}_i$. □

In fact, in this situation we may say more. Recall that $\mathbf{F}_s$ was defined in Definition 5.1.11 and characterized in Corollary 5.1.22.

Theorem 5.2.10. *Let* $\mathbf{E}$ *be a normal extension of* $\mathbf{F}$. *Then* $\mathbf{F}_s$ *and* $\mathbf{F}_i$ *are disjoint extensions of* $\mathbf{F}$ *and* $\mathbf{E} = \mathbf{F}_s\mathbf{F}_i$. *Furthermore,* $\mathbf{F}_s$ *is a Galois extension of* $\mathbf{F}$ *and* $\text{Gal}(\mathbf{F}_s/\mathbf{F}) = \text{Gal}(\mathbf{E}/\mathbf{F}_i)$. *Also,* $(\mathbf{E}/\mathbf{F}_i) = (\mathbf{F}_s/\mathbf{F})$ *and* $(\mathbf{E}/\mathbf{F}_s) = (\mathbf{F}_i/\mathbf{F})$.

Proof. Let $\alpha \in \mathbf{F}_s \cap \mathbf{F}_i$. Then α is both separable and purely inseparable over $\mathbf{F}$, so, by Remark 5.1.13, $\alpha \in \mathbf{F}$. Thus $\mathbf{F}_s$ and $\mathbf{F}_i$ are disjoint extensions of $\mathbf{F}$.

Let $\alpha \in \mathbf{F}_s$. Then α has separable minimum polynomial $m_\alpha(X)$. Since $\mathbf{E}$ is a normal extension of $\mathbf{F}$, $m_\alpha(X)$ splits into a product of linear factors $(X - \alpha_1) \cdots (X - \alpha_t)$ in $\mathbf{E}[X]$. But each α_i is a root of a separable polynomial, namely $m_\alpha(X)$, so $\alpha_i \in \mathbf{F}_s$ for each i, and hence $m_\alpha(X)$ splits into a product of linear factors $(X - \alpha_1) \cdots (X - \alpha_t)$ in $\mathbf{F}_s[X]$. Thus $\mathbf{F}_s$ is a normal and separable extension of $\mathbf{F}$, so is a Galois extension of $\mathbf{F}$ (by Theorem 2.7.14 in case $\mathbf{E}$ is a finite extension of $\mathbf{F}$, and by Theorem 5.4.2 below in general).

Suppose that $\mathbf{E}$ is a finite extension of $\mathbf{F}$. By Theorem 3.4.6, $\text{Gal}(\mathbf{E}/\mathbf{F}_i)$ is isomorphic to a subgroup of $\text{Gal}(\mathbf{F}_s/\mathbf{F})$. But let $\sigma \in \text{Gal}(\mathbf{F}_s/\mathbf{F})$. Now $\mathbf{E}$ is a purely inseparable extension of $\mathbf{F}_s$, so it is a normal extension of $\mathbf{F}$ (Remark 5.1.15) and hence it is the splitting field of some polynomial $f(X) \in \mathbf{F}[X]$ by Theorem 5.2.1. Hence, by Lemma 2.6.3, σ extends to an automorphism of $\mathbf{E}$, i.e., to an element of $\text{Gal}(\mathbf{E}/\mathbf{F}) = \text{Gal}(\mathbf{E}/\mathbf{F}_i)$, so these two groups are isomorphic. But then also

$$(\mathbf{E}/\mathbf{F}) = (\mathbf{E}/\mathbf{F}_i)(\mathbf{F}_i/\mathbf{F}) = (\mathbf{E}/\mathbf{F}_s)(\mathbf{F}_s/\mathbf{F})$$

giving the claimed equalities between the degrees of the extensions. Furthermore, since $\mathbf{F}_s$ is a Galois extension of $\mathbf{F}$, we have from Theorem 2.4.5 that

$$(\mathbf{F}_s\mathbf{F}_i/\mathbf{F}) = (\mathbf{F}_s/\mathbf{F})(\mathbf{F}_i/\mathbf{F}) = (\mathbf{E}/\mathbf{F}),$$

so $\mathbf{F}_s\mathbf{F}_i = \mathbf{E}$.

Now for the general case. Let $\alpha \in \mathbf{E}$ and let $\mathbf{E}_0 \subseteq \mathbf{E}$ be the splitting field of $m_\alpha(X)$. Then $\mathbf{E}_0$ is a finite extension of $\mathbf{F}$, so we may apply the above argument to $\mathbf{E}_0$, $\mathbf{F}_s \cap \mathbf{E}_0$, and $\mathbf{F}_i \cap \mathbf{E}_0$. Then we may apply Zorn's Lemma to obtain the conclusion of the theorem for $\mathbf{E}$. (The final point to see is that we have equalities among the degrees even if they are not all finite, when we cannot simply use the argument above. But $(\mathbf{E}/\mathbf{F}_s) \leq (\mathbf{F}_i/\mathbf{F})$ by Lemma

2.3.4. If we did not have equality, then there would be some nontrivial linear combination of basis elements of **E** over $\mathbf{F}_s$ that would be zero, and this linear combination would involve only finitely many basis elements and hence would have coefficients in some finite extension of **F**, contradicting equality of the degrees in the finite case.) □

5.3 The Algebraic Closure

In this section we show that every field **F** has an algebraic closure $\bar{\mathbf{F}}$ that is unique up to isomorphism.

Definition 5.3.1. A field **E** is *algebraically closed* if the only algebraic extension of **E** is **E** itself. ◇

Lemma 5.3.2. *The following are equivalent:*

(1) **E** *is algebraically closed.*

(2) Every nonconstant polynomial $f(X) \in \mathbf{E}[X]$ *is a product of linear factors in* $\mathbf{E}[X]$.

(3) Every nonconstant polynomial $f(X) \in \mathbf{E}[X]$ *has a root in* **E**.

Proof. (1) implies (2): Let **E** be algebraically closed. Let $f(X) \in \mathbf{E}[X]$ be irreducible. Then $f(X)$ has a root α in some algebraic extension $\mathbf{E}(\alpha)$ of **E**. But $\mathbf{E}(\alpha) = \mathbf{E}$ by assumption, so $X - \alpha$ divides $f(X)$ in **E** and hence $f(X) = X - \alpha$. Now let $g(X) \in \mathbf{E}[X]$ be arbitrary. Then $g(X)$ splits into a product of irreducible and hence linear factors in **E**, yielding (2).

(2) implies (1): Let α be an element of some extension of **E** with α algebraic over **E**. Then $m_\alpha(X) = \mathbf{E}[X]$ splits into a product of linear factors in **E**, one of which must be $X - \alpha$. Hence $\alpha \in \mathbf{E}$, yielding (1).

(2) implies (3): Let $f(X) \in \mathbf{E}[X]$. Then $f(X) = (X - \alpha_1) \cdots (X - \alpha_k)$ for some $\alpha_1, \ldots, \alpha_k \in \mathbf{E}$, so $f(\alpha_1) = 0$, yielding (3).

(3) implies (2): By induction on $k = \deg f(X)$. For $k = 1$ this is trivial. Assume it is true for $k-1$ and let $f(X)$ have degree k. Then $f(X)$ has a root α_1 so $f(X) = (X - \alpha_1)g(X)$ with $\deg g(X) = k - 1$. By the inductive hypothesis $g(X) = (X - \alpha_2) \cdots (X - \alpha_k)$ for some $\alpha_2, \ldots, \alpha_k$, and then $f(X) = (X - \alpha_1)(X - \alpha_2) \cdots (X - \alpha_k)$, yielding (2). □

Definition 5.3.3. Let **F** be a field. An *algebraic closure* $\bar{\mathbf{F}}$ of **F** is an algebraic extension of **F** that is algebraically closed. ◇

Theorem 5.3.4. *Let* **F** *be an arbitrary field. Then* **F** *has an algebraic closure* $\bar{\mathbf{F}}$ *and* $\bar{\mathbf{F}}$ *is unique up to isomorphism.*

Proof. Both existence and uniqueness are Zorn's Lemma arguments.

The idea of the proof is clear. We wish to take the set of all algebraic extensions of $\mathbf{F}$, order this set by inclusion, use Zorn's Lemma to conclude this set has a maximal element, and show this maximal element is an algebraic closure of $\mathbf{F}$. But, in carefully considering this argument, set-theoretic difficulties appear, as, a priori, it is not clear what inclusion means. (We can say, for example, that $\mathbf{Q}(\sqrt{2}) \subset \mathbf{Q}(\sqrt{2}, \sqrt{3})$ as we can regard these both as subfields of $\mathbf{C}$. But this is because we already know that $\mathbf{C}$ exists. In the abstract, what we mean by this inclusion is the following: Let $\mathbf{F}_1$ be the field obtained from $\mathbf{Q}$ by adjoining an element α with $\alpha^2 = 2$ and let $\mathbf{F}_2$ be the field obtained from $\mathbf{Q}$ by adjoining elements β_1 and β_2 with $\beta_1^2 = 2$ and $\beta_2^2 = 3$. Then there is a map $\sigma : \mathbf{F}_1 \to \mathbf{F}_2$ given by $\sigma(\alpha) = \beta_1$ and using this map we may *identify* $\mathbf{F}_1$ with a subfield of $\mathbf{F}_2$. But this is not quite the same thing as saying that $\mathbf{F}_1$ *is* a subset of $\mathbf{F}_2$.) The way out of this dilemma is to assume that all fields we need to consider are contained in some universal set. (Note we cannot assume they are all contained in some universal field as that would be presuming the existence of the algebraic closure, which is what we are trying to prove.)

Thus we proceed as follows in order to show the existence of $\bar{\mathbf{F}}$.

Given a set $\mathcal{U}$, and a field $\mathbf{E}$, we shall write $\mathbf{E} \subseteq \mathcal{U}$ if $\mathbf{E}$ is a subset of $\mathcal{U}$. Then the field operations (addition, multiplication, inversion) are defined on this subset of $\mathcal{U}$. For two such fields $\mathbf{E}_1$ and $\mathbf{E}_2$, we shall write $\mathbf{E}_1 \subseteq \mathbf{E}_2$ if $\mathbf{E}_1 \subseteq \mathbf{E}_2$ as subsets of $\mathcal{U}$ and the restrictions of the field operations on $\mathbf{E}_2$ to $\mathbf{E}_1$ agree with the field operations on $\mathbf{E}_1$.

We now set $\mathcal{U}_0 = 2^{\mathbf{F}}$, the set of subsets of $\mathbf{F}$. If $\mathbf{F}$ is infinite we let $\mathcal{U} = \mathcal{U}_0$. If $\mathbf{F}$ is finite we let $\mathcal{U} = 2^{\mathcal{U}_0}$. Regard $\mathbf{F} \subset \mathcal{U}$ by identifying a with $\{a\}$, $a \in \mathbf{F}$, if $\mathbf{F}$ is infinite, or with $\{\{a\}\}$, if $\mathbf{F}$ is finite.

As is well known, for any set A, the cardinality of 2^A is greater than the cardinality of A, so the cardinality of $\mathbf{F}$ is less than the cardinality of $\mathcal{U}$. Furthermore, by construction, $\mathcal{U}$ is uncountable.

Let

$$\mathcal{S} = \{\text{algebraic extensions } \mathbf{E} \text{ of } \mathbf{F} \text{ with } \mathbf{E} \subseteq \mathcal{U}\}$$

and order $\mathcal{S}$ by inclusion. Then every chain has a maximal element: If $\mathbf{E}_1 \subseteq \mathbf{E}_2 \subseteq \cdots$, then $\mathbf{E} = \cup \mathbf{E}_i$ is algebraic. (Any $\alpha \in \mathbf{E}$ is an element of some $\mathbf{E}_i$ and hence is algebraic over $\mathbf{F}$.) Hence, by Zorn's Lemma, $\mathcal{S}$ has a maximal element $\bar{\mathbf{F}}$.

Claim: $\bar{\mathbf{F}}$ is algebraically closed.

Proof of claim: Let $\bar{\mathbf{F}} \subseteq \tilde{\mathbf{F}}$ with $\tilde{\mathbf{F}}$ an algebraic extension of $\bar{\mathbf{F}}$. Then, by Theorem 2.4.12, $\tilde{\mathbf{F}}$ is an algebraic extension of $\mathbf{F}$. To complete the existence part of the proof we must show that it is possible to embed $\tilde{\mathbf{F}}$ in $\mathcal{U}$. Assuming that, by the maximality of $\bar{\mathbf{F}}$ we must have that this image is just $\bar{\mathbf{F}}$, and then $\tilde{\mathbf{F}} = \bar{\mathbf{F}}$, so we see that $\bar{\mathbf{F}}$ is algebraically closed.

Claim: Let $\mathbf{E}$ be an algebraic extension of $\mathbf{F}$. If $\mathbf{F}$ is finite, $\mathbf{E}$ is finite or countably infinite. If $\mathbf{F}$ is infinite, $\mathbf{E}$ has the same cardinality as $\mathbf{F}$.

Assuming this claim, $\bar{\mathbf{F}} \subset \mathcal{U}$ is a set of smaller cardinality than that of $\mathcal{U}$ and $\bar{\mathbf{F}} \subset \tilde{\mathbf{F}}$ with $\tilde{\mathbf{F}}$ also a set of smaller cardinality than $\mathcal{U}$, so we may choose an arbitrary embedding of $\tilde{\mathbf{F}} - \bar{\mathbf{F}}$ in $\mathcal{U}$ as a set and then define the field operations on this set to agree with those on $\mathbf{F}$.

Proof of claim: We consider the set $\mathcal{P}$ of all monic polynomials in $\mathbf{F}[X]$. Then $\mathcal{P} = \bigcup_{n=1}^{\infty} \mathcal{P}_n$, where $\mathcal{P}_n$ is the set of all monic polynomials of degree n. Now there is a bijection between $\mathbf{F}^n$ and $\mathcal{P}_n$ given by $(a_0, \ldots, a_{n-1}) \longleftrightarrow a_0 + a_1 X + \cdots + a_{n-1} X^{n-1} + X^n$. If $\mathbf{F}$ is finite, so is $\mathbf{F}^n$ for each n, and hence $\mathcal{P}$ is countable union of finite sets and so is countable. If $\mathbf{F}$ is infinite, then $\mathbf{F}^n$ has the same cardinality as $\mathbf{F}$ for each n, and then $\mathcal{P}$ is a countable union of such sets, so also has the same cardinality as $\mathbf{F}$.

Now if $\mathbf{E}$ is an algebraic extension of $\mathbf{F}$, we have a map $\rho \colon \mathbf{E} \to \mathcal{P}$ given by $\rho(\alpha) = m_\alpha(X)$. Then ρ is a finite-one mapping as if $\rho(\alpha) = f(X)$, then $f(\alpha) = 0$, and any polynomial only has finitely many roots. But, in general, if $\rho \colon A \to B$ is a finite-to-one mapping of the set A onto the infinite set B, then the cardinality of A is equal to the cardinality of B.

This completes the proof of the existence of $\bar{\mathbf{F}}$. Now we must show that $\bar{\mathbf{F}}$ is unique up to isomorphism.

Thus let $\bar{\mathbf{F}}_1$ and $\bar{\mathbf{F}}_2$ be two algebraic closures of $\mathbf{F}$. Let

$$\mathcal{S} = \{(\mathbf{E}, \sigma) \mid \mathbf{E} \subseteq \bar{\mathbf{F}}_1, \ \sigma \colon \mathbf{E} \to \bar{\mathbf{F}}_2\}$$

where σ is a map of fields.

Order $\mathcal{S}$ by $(\mathbf{E}_1, \sigma_1) \leq (\mathbf{E}_2, \sigma_2)$ if $\mathbf{E}_1 \subseteq \mathbf{E}_2$ and $\sigma_2 \mid \mathbf{E}_1 = \sigma_1$. Again by Zorn's Lemma every chain has a maximal element so $\mathcal{S}$ has a maximal element $(\tilde{\mathbf{F}}_1, \tilde{\sigma})$.

Claim: $\tilde{\mathbf{F}}_1 = \bar{\mathbf{F}}_1$.

Proof of claim: Suppose not, and let $\alpha \in \bar{\mathbf{F}}_1$, $\alpha_1 \notin \tilde{\mathbf{F}}_1$. Let $\tilde{\sigma}(\tilde{\mathbf{F}}_1) = \tilde{\mathbf{F}}_2 \subseteq \bar{\mathbf{F}}_2$. Since α_1 is algebraic over $\mathbf{F}$, it is algebraic over $\tilde{\mathbf{F}}_1$, so let $\tilde{m}_{\alpha_1}(X) \in \tilde{\mathbf{F}}_1(X)$ be its minimum polynomial over $\tilde{\mathbf{F}}_1$. Since $\bar{\mathbf{F}}_2$ is algebraically closed, there is a root α_2 of $\sigma(\tilde{m}_{\alpha_1}(X))$ in $\bar{\mathbf{F}}_2$. Then, by Lemma 2.6.1, $\tilde{\sigma}$ extends to a map $\tilde{\tau} \colon \tilde{\mathbf{F}}_1(\alpha_1) \to \tilde{\mathbf{F}}_2(\alpha_2) \subseteq \bar{\mathbf{F}}_2$ (with $\tilde{\tau}(\alpha_1) = \alpha_2$), contradicting the maximality of $\tilde{\mathbf{F}}_1$. Thus $\tilde{\mathbf{F}}_1 = \bar{\mathbf{F}}_1$ and we have a map $\bar{\sigma} \colon \bar{\mathbf{F}}_1 \to \bar{\mathbf{F}}_2$.

As $\bar{\sigma}$ is a map of fields, it is injective. To complete the proof we must show it is surjective. Let $\beta_2 \in \bar{\mathbf{F}}_2$. Then β_2 is algebraic over $\mathbf{F}$, so it is algebraic over

$\sigma(\bar{\mathbf{F}}_1)$. But $\sigma(\bar{\mathbf{F}}_1)$ is algebraically closed so $\beta_2 \in \sigma(\bar{\mathbf{F}}_1)$, i.e., $\beta_2 = \sigma(\beta_1)$ for some $\beta_1 \in \bar{\mathbf{F}}_1$, as required. □

We now wish to examine some concrete algebraically closed fields. But first we shall prove a lemma that enables us to weaken the hypotheses in Lemma 5.3.2.

Lemma 5.3.5. *Let* $\mathbf{E}$ *be an algebraic extension of* $\mathbf{F}$*. Then* $\mathbf{E}$ *is algebraically closed if and only if every polynomial* $f(X) \in \mathbf{F}[X]$ *is a product of linear factors in* $\mathbf{E}[X]$*. If* $\mathbf{E}$ *is a normal extension of* $\mathbf{F}$*, then* $\mathbf{E}$ *is algebraically closed if and only if every polynomial* $f(X) \in \mathbf{F}[X]$ *has a root in* $\mathbf{E}$*.*

Proof. First, let $\mathbf{E}$ be an algebraic extension of $\mathbf{F}$. Then the "only if" part is clear. For the "if" part, let α be algebraic over $\mathbf{E}$. Then $\mathbf{E}(\alpha)$ is an algebraic extension of $\mathbf{E}$ (Corollary 2.4.9) and $\mathbf{E}$ is algebraic over $\mathbf{F}$ (by hypothesis), so $\mathbf{E}(\alpha)$ is algebraic over $\mathbf{F}$ (Theorem 2.4.12). Hence α is a root of $m_\alpha(X) \in \mathbf{F}[X]$. But $m_\alpha(X) = (X - \alpha_1)\cdots(X - \alpha_n)$ in $\mathbf{E}[X]$, so $\alpha = \alpha_i \in \mathbf{E}$ for some i. Now if $\mathbf{E}$ is a normal extension of $\mathbf{F}$, then by definition, any polynomial in $\mathbf{F}[X]$ with a root in $\mathbf{E}$ splits into a product of linear factors in $\mathbf{E}[X]$, so we are back in the previous case. □

Theorem 5.3.6. *The field of rational numbers* $\mathbf{Q}$ *has an algebraic closure* $\bar{\mathbf{Q}}$ *that is a subfield of the field of complex numbers* $\mathbf{C}$*.*

Proof. Let $f(X) \in \mathbf{Q}[X]$ be an arbitrary irreducible polynomial and let $\mathbf{E}$ be a splitting field of $f(X)$. Let $G = \mathrm{Gal}(\mathbf{E}/\mathbf{Q})$, and let H be the 2-Sylow subgroup of G. Set $\mathbf{B} = \mathrm{Fix}(H)$. Then $\mathbf{E}$ is a Galois extension of $\mathbf{B}$ with Galois group H. Also, $d = [G : H] = (\mathbf{B}/\mathbf{Q})$ is odd. By Corollary 3.5.3, there is an element α of $\mathbf{B}$ with $\mathbf{B} = \mathbf{Q}(\alpha)$. Then $m_\alpha(X) \in \mathbf{Q}[X]$ is a monic irreducible polynomial of odd degree d.

By elementary calculus, $m_\alpha(X)$ has a real root x_0. (Proof: $\lim_{x\to-\infty} m_\alpha(x) = -\infty$, so $m_\alpha(x_-) < 0$ for some $x_- \in \mathbf{R}$, and $\lim_{x\to\infty} m_\alpha(x) = \infty$, so $m_\alpha(x_+) > 0$ for some $x_+ \in \mathbf{R}$. Then, by the Intermediate Value Theorem, $m_\alpha(x_0) = 0$ for some $x_0 \in \mathbf{R}$.)

Let $\mathbf{B}' = \mathbf{B}(x_0) \subseteq \mathbf{R}$. Then $f(X)$ has a splitting field $\mathbf{E}' \supseteq \mathbf{B}'$. Let $H' = \mathrm{Gal}(\mathbf{E}'/\mathbf{B}')$. By Corollary 3.4.6 (the Theorem on Natural Irrationalities), H' is isomorphic to a subgroup of H, so H' is also a 2-group. Thus, by Corollary A.2.3, there is a sequence $H' = H'_0 \supset H'_1 \supset \cdots \supset H'_n = \{1\}$ with $[H_{i-1} : H_i] = 2$, and, setting $\mathbf{E}'_i = \mathrm{Fix}(H'_i)$, $\mathbf{B}' = \mathbf{E}'_0 \subset \mathbf{E}'_1 \subset \ldots \mathbf{E}'_n = \mathbf{E}'$ with $(\mathbf{E}'_i/\mathbf{E}'_{i-1}) = 2$. In other words, for each i, $\mathbf{E}'_i$ is a quadratic extension of $\mathbf{E}'_{i-1}$, i.e., $\mathbf{E}'_i = \mathbf{E}'_{i-1}(\sqrt{\alpha_i})$ for some $\alpha_i \in \mathbf{E}'_{i-1}$. But if $\mathbf{F} \subseteq \mathbf{C}$ and $\mathbf{D} = \mathbf{F}(\sqrt{z})$ for some $z \in \mathbf{F}$, then $\mathbf{D}$ is isomorphic to a subfield of $\mathbf{C}$. (If $z = re^{i\theta}$, $\mathbf{D} =$

$\mathbf{F}(\sqrt{r}e^{i\theta/2})$.) Thus, proceeding inductively, we see that $f(X)$ has a splitting field that is a subfield of $\mathbf{C}$.

Now let $\{f_i(X)\}$ be the set of all monic irreducible polynomials in $\mathbf{Q}[X]$ and let $\mathbf{E}_i \subseteq \mathbf{C}$ be the splitting field of $f_i(X)$. Let $\bar{\mathbf{Q}}$ be the composite of $\{\mathbf{E}_i\}$. Then $\bar{\mathbf{Q}}$ is algebraic over $\mathbf{F}$ and every $f(X) \in \mathbf{Q}[X]$ splits in $\bar{\mathbf{Q}}$, so, by Lemma 5.3.5, $\bar{\mathbf{Q}}$ is an algebraic closure of $\mathbf{Q}$. □

Actually, a similar argument enables us to prove a stronger result.

Theorem 5.3.7 (Fundamental Theorem of Algebra). *The field of complex numbers* $\mathbf{C}$ *is algebraically closed.*

Proof. First, observe that it suffices to prove that every polynomial $f(X) \in \mathbf{R}[X]$ has a root in $\mathbf{C}$. On the one hand, this follows directly from Lemma 5.3.5, but on the other hand, we have the following direct argument in this case: Let $g(X) \in \mathbf{C}[X]$ be an arbitrary polynomial, and set $f(X) = g(X)\bar{g}(X)$. Then $f(X) \in \mathbf{R}[X]$. If $f(X)$ has a root γ, then $g(\gamma) = 0$, in which case γ is a root of $g(X)$, or $\bar{g}(\gamma) = 0$, in which case $g(\bar{\gamma}) = 0$ and $\bar{\gamma}$ is a root of $g(X)$.

Thus let $f(X) \in \mathbf{R}[X]$ be an arbitrary irreducible polynomial and let $\mathbf{E}$ be a splitting field of $f(X)$. Let $G = \mathrm{Gal}(\mathbf{E}/\mathbf{R})$, and let H be the 2-Sylow subgroup of G. Set $\mathbf{B} = \mathrm{Fix}(H)$. Then $\mathbf{E}$ is a Galois extension of $\mathbf{B}$ with Galois group H. Also, $d = [G : H] = (\mathbf{B}/\mathbf{Q})$ is odd. By Corollary 3.5.3, there is an element α of $\mathbf{B}$ with $\mathbf{B} = \mathbf{R}(\alpha)$. Then $m_\alpha(X) \in \mathbf{R}[X]$ is an irreducible polynomial of odd degree d. But, by the same elementary calculus argument as in the proof of Theorem 5.3.6, $m_\alpha(X)$ has a real root x_0. Thus $(X - x_0)$ divides $m_\alpha(X)$. Since $m_\alpha(X)$ is irreducible, we must have $m_\alpha(X) = (X - x_0)$, and so $d = 1$ and $\mathbf{B} = \mathbf{R}$.

Then $\mathbf{E}$ is a splitting field of $f(X)$ with $(\mathbf{E}/\mathbf{R}) = 2^n$ for some n, and, arguing as in the proof of Theorem 5.3.6, we see that $\mathbf{E}$ is isomorphic to a subfield of $\mathbf{C}$, so $f(X)$ has a root in $\mathbf{C}$, and we are done. □

Remark 5.3.8. (1) $\mathbf{Q}$ has an algebraic closure $\bar{\mathbf{Q}} \subseteq \mathbf{C}$ but $\bar{\mathbf{Q}} \neq \mathbf{C}$ as $\bar{\mathbf{Q}}$ is countable, being an algebraic extension of $\mathbf{Q}$, but $\mathbf{C}$ is uncountable.

(2) Theorem 5.3.7 implies Theorem 5.3.6, for, once we know $\mathbf{C}$ is algebraically closed, it is easy to see that $\{z \in \mathbf{C} \mid z \text{ is algebraic over } \mathbf{Q}\}$ is an algebraic closure of $\mathbf{Q}$. This field is known as the *field of algebraic numbers*.

(3) The Fundamental Theorem of Algebra cannot have a purely algebraic proof, as the definitions of $\mathbf{R}$ and $\mathbf{C}$ are not purely algebraic. We have given a proof that uses only a minimum of analysis (the argument from elementary calculus given in the proof of Theorem 5.3.6). But this important theorem has many proofs. Here is one well-known proof that uses results from complex analysis:

Let $f(X) \in \mathbf{C}[X]$ be a nonconstant polynomial, $f(X) = \sum_{i=0}^{n} a_i X^i$. Then for any $z \in \mathbf{C}$, $|f(z)| \geq |a_n||z|^n - \sum_{i=0}^{n-1} |a_i||z|^i$ from which it easily follows that there is an A such that $|z| \geq A$ implies $|f(z)| \geq 1$. Now suppose $f(X)$ does not have a root in $\mathbf{C}$. Consider $g(X) = 1/f(X)$. Then $|g(z)| \leq 1$ for $|z| \geq A$. On the other hand, $\{z \mid |z| \leq A\}$ is a compact set so there is a C with $|g(z)| \leq C$ for all Z with $|z| \leq A$. Thus $g(z)$ is a bounded analytic function and hence, by Liouville's theorem, a constant. But that implies $f(z)$ is a constant, which is absurd. ◇

5.4 Infinite Galois Extensions

In this section we will consider infinite algebraic extensions. Our main goals will be to show that Theorem 2.7.14 and Theorem 2.8.8 (the Fundamental Theorem of Galois Theory) have appropriate generalizations to this case.

First, we must generalize Lemma 2.6.3.

Lemma 5.4.1. *Let* $\mathbf{E}$ *be a normal extension of* $\mathbf{F}$ *and let* $\sigma : \mathbf{B} \to \mathbf{B}'$ *be any isomorphism with* $\sigma \mid \mathbf{F} = \mathrm{id}$, *where* $\mathbf{B}$ *and* $\mathbf{B}'$ *are any fields intermediate between* $\mathbf{F}$ *and* $\mathbf{E}$. *Then* σ *extends to an automorphism of* $\mathbf{E}$, *i.e., to an element of* $\mathrm{Gal}(\mathbf{E}/\mathbf{F})$.

Proof. This is an application of Zorn's Lemma. Let

$$\mathcal{S} = \{(\mathbf{D}, \tau) \mid \mathbf{B} \subseteq \mathbf{D},\ \tau : \mathbf{D} \to \mathbf{D},\ \tau \mid \mathbf{B} = \sigma\}.$$

Order $\mathcal{S}$ by $(\mathbf{D}_1, \tau_1) \leq (\mathbf{D}_2, \tau_2)$ if $\mathbf{D}_1 \subseteq \mathbf{D}_2$ and $\tau_2 \mid \mathbf{D}_1 = \tau_1$. Then every chain $\{(\mathbf{D}_i, \tau_i)\}$ in $\mathcal{S}$ has a maximal element $(\mathbf{D}, \tau)$ given by

$$\mathbf{D} = \bigcup \mathbf{D}_i, \quad \tau(d) = \tau_i(d) \text{ if } d \in \mathbf{D}_i.$$

Hence by Zorn's Lemma $\mathcal{S}$ has a maximal element $(\mathbf{E}_0, \tau_0)$. We claim $\mathbf{E}_0 = \mathbf{E}$. Suppose not and let $\alpha \in \mathbf{E}$, $\alpha \notin \mathbf{E}_0$. Consider its minimum polynomial $m_\alpha(X)$ over $\mathbf{F}$. Since $m_\alpha(X) \in \mathbf{F}[X]$, $\tau_0(m_\alpha(X)) = m_\alpha(X)$.

Now consider $m_\alpha(X)$ as an element of $\mathbf{E}_0[X]$, and let $\mathbf{E}_1 \subseteq \mathbf{E}$ be the splitting field of $m_\alpha(X)$ over $\mathbf{E}_0$. Note $\mathbf{E}_0 \subset \mathbf{E}_1$, as $\alpha \notin \mathbf{E}_0$ but $\alpha \in \mathbf{E}_1$. Then, by Lemma 2.6.3, there is an automorphism τ_1 of $\mathbf{E}_1$ extending τ_0, contradicting the maximality of $(\mathbf{E}_0, \tau_0)$. □

Now we generalize Theorem 2.7.14.

Theorem 5.4.2. *Let* **E** *be an algebraic extension of* **F**. *The following are equivalent:*

(1) **E** *is a Galois extension of* **F**.

(2) **E** *is a normal and separable extension of* **F**.

(3) Every field **B** *intermediate between* **F** *and* **E** *with* **B** *a finite extension of* **F** *is contained in a field* **D** *intermediate between* **E** *and* **F** *that is the splitting field of a separable polynomial in* $\mathbf{F}[X]$, *or, equivalently, that is a finite Galois extension of* **F**.

Proof. Although this not the logically most economical proof, we shall first show that (1) and (2) are equivalent and then that (3) is equivalent to both of these.

First, we show that (1) implies (2). Let $\alpha \in \mathbf{E}$. Then α is algebraic over **F**, so has minimum polynomial $m_\alpha(X)$. Let $\{\alpha_1, \ldots, \alpha_k\}$ be the roots of $m_\alpha(X)$ in **E**, with $\alpha_1 = \alpha$. For any $\sigma \in \mathrm{Gal}(\mathbf{E}/\mathbf{F})$, $m_\alpha(\sigma(\alpha)) = \sigma(m_\alpha(\alpha)) = \sigma(0) = 0$ so $\sigma(\alpha) \in \{\alpha_1, \ldots, \alpha_k\}$, and thus the action of $\mathrm{Gal}(\mathbf{E}/\mathbf{F})$ is to permute $\{\alpha_1, \ldots, \alpha_k\}$. Consider $f(X) = \prod_{i=1}^{k}(X - \alpha_i)$. This polynomial is invariant under $\mathrm{Gal}(\mathbf{E}/\mathbf{F})$ so $f(X) \in \mathbf{F}[X]$, as **E** is a Galois extension of **F**. If $g(X) \in \mathbf{F}[X]$ is any polynomial with $g(\alpha) = 0$, then by the same logic $g(\alpha_i) = 0$ for each i, so $f(X)$ divides $g(X)$ in $\mathbf{E}[X]$ and hence in $\mathbf{F}[X]$. Thus $f(X)$ is irreducible and so $f(X) = m_\alpha(X)$. Hence $m_\alpha(X)$ is a separable polynomial that splits in **E**.

Next, we show that (2) implies (1). Let $\alpha \in \mathbf{E}$, and consider $m_\alpha(X) \in \mathbf{F}[X]$. Then $m_\alpha(X) = \prod_{i=1}^{k}(X - \alpha_i)$ in $\mathbf{E}[X]$ for some $\{\alpha_1, \ldots, \alpha_k\}$ with $\alpha_1 = \alpha$. Thus if we let $\mathbf{B} = \mathbf{F}(\alpha_1, \ldots, \alpha_k)$, we have that $m_\alpha(X) = \prod_{i=1}^{k}(X - \alpha_i)$ in $\mathbf{B}[X]$. Thus we see that **B** is a finite extension of **F** and **B** is the splitting field of a separable polynomial, so **B** is a Galois extension of **F**. Hence $\mathrm{Fix}(\mathrm{Gal}(\mathbf{B}/\mathbf{F})) = \mathbf{F}$. Now let $\alpha \notin \mathbf{F}$. Then $\alpha \notin \mathrm{Fix}(\mathrm{Gal}(\mathbf{B}/\mathbf{F}))$, and so there is an automorphism σ of **B** with $\sigma|\mathbf{F} = \mathrm{id}$ and with $\sigma(\alpha) \neq \alpha$. By Lemma 5.4.1, σ extends to an element $\tau \in \mathrm{Gal}(\mathbf{E}/\mathbf{F})$, and $\tau(\alpha) = \sigma(\alpha) \neq \alpha$, so $\alpha \notin \mathrm{Fix}(\mathrm{Gal}(\mathbf{E}/\mathbf{F}))$. Hence, as $\alpha \notin \mathbf{F}$ is arbitrary, we have that $\mathrm{Fix}(\mathrm{Gal}(\mathbf{E}/\mathbf{F})) = \mathbf{F}$, i.e., **E** is a Galois extension of **F**.

Next, we show that (2) implies (3). Since **B** is finite, $\mathbf{B} = \mathbf{F}(\alpha_1, \ldots, \alpha_k)$ for some finite set $\{\alpha_1, \ldots, \alpha_k\}$. Then $m_{\alpha_i}(X)$, $i = 1, \ldots, k$, are all separable, and all split in **E**. Let **D** be the splitting field of the polynomial $m_{\alpha_1}(X) \cdots m_{\alpha_k}(X)$. Then $\mathbf{B} \subseteq \mathbf{D} \subseteq \mathbf{E}$ and **D** is a finite extension of **F** that is the splitting field of a separable polynomial.

Finally, we show that (3) implies (1). Let $\alpha \in \mathbf{E}$, $\alpha \notin \mathbf{F}$. Let $\mathbf{B} = \mathbf{F}(\alpha)$. Then **B** is a finite extension of **F**, so $\mathbf{B} \subseteq \mathbf{D}$ where **D** is the splitting field of a separable polynomial $f(X) \in \mathbf{F}[X]$. (In fact, we may take $f(X) = m_\alpha(X)$.)

Thus $\mathbf{D}$ is a Galois extension of $\mathbf{F}$, so there is an automorphism $\sigma \in \mathrm{Gal}(\mathbf{D}/\mathbf{F})$ with $\sigma(\alpha) \neq \alpha$. By Lemma 5.4.1, σ extends to an element $\tau \in \mathrm{Gal}(\mathbf{E}/\mathbf{F})$, and $\tau(\alpha) = \sigma(\alpha) \neq \alpha$, so $\alpha \notin \mathrm{Fix}(\mathrm{Gal}(\mathbf{E}/\mathbf{F}))$. Hence, as $\alpha \notin \mathbf{F}$ is arbitrary, we have that $\mathrm{Fix}(\mathrm{Gal}(\mathbf{E}/\mathbf{F})) = \mathbf{F}$, i.e., $\mathbf{E}$ is a Galois extension of $\mathbf{F}$. □

Corollary 5.4.3. *Let* $\mathbf{E}$ *be an algebraic extension of* $\mathbf{F}$. *The following are equivalent:*

(1) $\mathbf{E}$ *is a Galois extension of* $\mathbf{F}$.

(2) $\mathbf{E}$ *is the splitting field of a set of separable polynomials in* $\mathbf{F}[X]$.

Proof. Suppose that $\mathbf{E}$ is a Galois extension of $\mathbf{F}$. Then, by Theorem 5.4.2, $\mathbf{E}$ is normal and separable. Thus for every $\alpha \in \mathbf{E}$, $m_\alpha(X) \in \mathbf{F}[X]$ is a separable polynomial that splits in $\mathbf{E}[X]$. Then $\mathbf{E}$ is the splitting field of the set of separable polynomials $\{m_\alpha(X) \mid \alpha \in \mathbf{E}\}$.

On the other hand, let $\mathbf{E}$ be the splitting field of $\{f_i(X)\}_{i\in I}$, a set of separable polynomials in $\mathbf{F}[X]$. First, observe that any $\epsilon \in \mathbf{E}$ is contained in the splitting field of a finite subset of $\{f_i(X)\}_{i\in I}$.

Now let $\mathbf{B}$ be any field intermediate between $\mathbf{F}$ and $\mathbf{E}$ that is a finite extension of $\mathbf{F}$. Then $\mathbf{B} = \mathbf{F}(\epsilon_1, \ldots, \epsilon_n)$ for some finite set of elements $\{\epsilon_1, \ldots, \epsilon_n\}$ of $\mathbf{E}$, so, by the above observation, $\mathbf{B}$ is contained in the splitting field $\mathbf{D} \subseteq \mathbf{E}$ of a finite subset $\{f_1(X), \ldots, f_k(X)\}$ of $\{f_i(X)\}_{i\in I}$. Then $\mathbf{D}$ is the splitting field of the separable polynomial $f(X) = f_1(X)\cdots f_k(X) \in \mathbf{F}[X]$, so, by Theorem 5.4.2, $\mathbf{E}$ is a Galois extension of $\mathbf{F}$. □

Corollary 5.4.4. *Let* $\mathbf{E}$ *be a Galois extension of* $\mathbf{F}$ *and let* $\mathbf{B}$ *be any field intermediate between* $\mathbf{F}$ *and* $\mathbf{E}$. *Then* $\mathbf{E}$ *is a Galois extension of* $\mathbf{B}$.

Proof. This follows from Theorem 5.4.2. We restate condition (3) of this theorem in order to avoid confusion of notation: Every field $\mathbf{A}$ intermediate between $\mathbf{F}$ and $\mathbf{E}$ with $\mathbf{A}$ a finite extension of $\mathbf{F}$ is contained in a field $\mathbf{D}$ intermediate between $\mathbf{E}$ and $\mathbf{F}$ that is the splitting field of a separable polynomial. We want to show this remains true when $\mathbf{F}$ is replaced by $\mathbf{B}$.

Let $\mathbf{A}$ be a finite extension of $\mathbf{B}$, so that $\mathbf{A} = \mathbf{B}(\alpha_1, \ldots, \alpha_n)$. Let $\mathbf{A}' = \mathbf{F}(\alpha_1, \ldots, \alpha_n)$. We claim that $\mathbf{A}'$ is a finite extension of $\mathbf{F}$. (Of course, if $\mathbf{B}$ is a finite extension of $\mathbf{F}$, then $\mathbf{A}$ is a finite extension of $\mathbf{F}$ and $\mathbf{A}' \subseteq \mathbf{A}$, so this is clear, but we need to prove this also in case $\mathbf{B}$ is not a finite extension of $\mathbf{F}$.) Since $\mathbf{A}' = \mathbf{F}(\alpha_1)\cdots\mathbf{F}(\alpha_n)$, in order to prove this it suffices to prove that $\mathbf{F}(\alpha_i)$ is a finite extension of $\mathbf{F}$, for each i. Thus, fix i and consider $\mathbf{F}(\alpha_i)$. Let $\tilde{m}_{\alpha_i}(X) \in \mathbf{B}[X]$ be the minimum polynomial of α_i regarded as an element of $\mathbf{B}$. Let $\mathbf{B}_i$ be the field obtained by adjoining the coefficients of $\tilde{m}_{\alpha_i}(X)$ to $\mathbf{F}$. Then $\mathbf{B}_i(\alpha)$ is a finite extension of $\mathbf{B}_i$ and $\mathbf{B}_i$ is a finite extension of $\mathbf{F}$, so $\mathbf{B}_i(\alpha_i)$

is a finite extension of $\mathbf{F}$, and $\mathbf{F}(\alpha_i) \subseteq \mathbf{B}_i(\alpha_i)$, so $\mathbf{F}(\alpha_i)$ is a finite extension of $\mathbf{F}$, as claimed.

Then, arguing as in the proof of Corollary 5.4.3, $\mathbf{A}' \subseteq \mathbf{D}' \subseteq \mathbf{E}$, where $\mathbf{D}'$ is an extension of $\mathbf{F}$ that is the splitting field of a separable polynomial $f(X) \in \mathbf{F}[X]$. But $\mathbf{A} = \mathbf{B}\mathbf{A}' \subseteq \mathbf{B}\mathbf{D}'$, and $\mathbf{B}\mathbf{D}'$ is an extension of $\mathbf{B}$ that is the splitting field of the separable polynomial $f(X) \in \mathbf{B}[X]$, as required. □

In order to generalize Theorem 2.8.8, the *FTGT*, we must introduce a topology on the Galois group.

Definition 5.4.5. Let $\mathbf{E}$ be an algebraic extension of $\mathbf{F}$, and let $G = \mathrm{Gal}(\mathbf{E}/\mathbf{F})$. The *Krull topology* on $\mathbf{E}$ is defined as follows: The identity element of G has a neighborhood basis $\{U_{\mathbf{B}} = \mathrm{Gal}(\mathbf{E}/\mathbf{B}) \mid \mathbf{B} \text{ is a finite extension of } \mathbf{F}\}$. An element $\sigma \in G$ has neighborhood basis $\{\sigma U_{\mathbf{B}}\}$. ◇

Remark 5.4.6. We explicitly observe that $\tau \in U_{\mathbf{B}}$ if and only if $\tau \mid \mathbf{B} = \mathrm{id}$, and $\tau \in \sigma U_{\mathbf{B}}$ if and only if $\tau \mid \mathbf{B} = \sigma \mid \mathbf{B}$. ◇

Remark 5.4.7. If $\mathbf{E}$ is a finite extension of $\mathbf{F}$, then we may let $\mathbf{B} = \mathbf{E}$ in the above definition and we see that every element σ of G has a neighborhood consisting of σ alone. In other words, in this case G has the discrete topology. All of the results in this section remain true in this case, but the topology on G yields no additional information. ◇

Definition 5.4.5 is not complete until we show these sets define a topology on G. In fact we show a bit more.

Proposition 5.4.8. *The sets $\{\sigma U_{\mathbf{B}}\}$ define a topology on G. Under this topology, G is a topological group.*

Proof. Let $U_i = U_{\mathbf{B}_i}$, $i = 1, 2$, and let $U_3 = U_1 \cap U_2$. Let $\mathbf{B}_3 = \mathbf{B}_1\mathbf{B}_2$. Then $\mathbf{B}_3$ is a finite extension of $\mathbf{F}$, and, by Proposition 2.8.14, $\mathbf{B}_3 = \mathrm{Fix}(U_1 \cap U_2) = \mathrm{Fix}(U_3)$. Furthermore, it is easy to check that if $\sigma \notin U_3$, then $\sigma \mid \mathbf{B}_3$ is not the identity. Hence $U_3 = \mathrm{Gal}(\mathbf{E}/\mathbf{B}_3)$, i.e., $U_3 = U_{\mathbf{B}_3}$. This shows that $\{U_{\mathbf{B}}\}$ form a neighborhood basis of the identity, and hence that $\{\sigma U_{\mathbf{B}}\}$ form a basis for a topology on G.

Note that if $\sigma \in G$, then $\sigma U_{\mathbf{B}}\sigma^{-1} = U_{\sigma(\mathbf{B})}$ and $\sigma(\mathbf{B})$ is a finite extension of $\mathbf{F}$. Then $U_{\mathbf{B}}\sigma = \sigma(\sigma^{-1}U_{\mathbf{B}}\sigma) = \sigma U_{\sigma^{-1}(\mathbf{B})}$ so

$$\{\sigma U_{\mathbf{B}} \mid \sigma \in G,\ \mathbf{B}/\mathbf{F} \text{ finite}\} = \{U_{\mathbf{B}}\sigma \mid \sigma \in G,\ \mathbf{B}/\mathbf{F},\ \text{finite}\}$$

so, in defining the topology, we may take left cosets, right cosets, or both left and right cosets.

To show G is a topological group we must show multiplication and inversion are continuous. If $m = G \times G \to G$ is multiplication, and $\sigma\tau U$ is a neighborhood of $\sigma\tau$, then $m^{-1}(\sigma\tau U) \supset (\sigma(\tau U \tau^{-1})) \times (\tau U)$, a neighborhood of (σ, τ) in $G \times G$, and if $i\colon G \to G$ is inversion, and $\sigma^{-1}U$ is a neighborhood of σ^{-1}, then $i^{-1}(\sigma^{-1}U) = U^{-1}\sigma^{-1} = U\sigma^{-1}$, a neighborhood of σ^{-1} in G. □

Lemma 5.4.9. *Let* **E** *be a Galois extension of* **F** *and let* **B** *be any field intermediate between* **F** *and* **E**. *Then the Krull topology in* Gal(**E**/**B**) *is the induced topology on* Gal(**E**/**B**) *as a subspace of* Gal(**E**/**F**) *with the Krull topology.*

Proof. Recall that, by Corollary 5.4.4, **E** is a Galois extension of **B**. Let $\tilde{U}$ be a neighborhood of the identity in $\mathrm{Gal}(\mathbf{E}/\mathbf{B})$, so $\tilde{U} = \mathrm{Gal}(\mathbf{E}/\mathbf{A})$ for some finite extension **A** of **B**. Then $\mathbf{A} = \mathbf{B}(\alpha)$ for some $\alpha \in \mathbf{E}$, by Corollary 3.5.3. Let $\mathbf{D} = \mathbf{F}(\alpha)$, so $\mathbf{A} = \mathbf{BD}$, and let $U = \mathrm{Gal}(\mathbf{E}/\mathbf{D})$. Then

$$\tilde{U} = \mathrm{Gal}(\mathbf{E}/\mathbf{A}) = \mathrm{Gal}(\mathbf{E}/\mathbf{BD}) = \mathrm{Gal}(\mathbf{E}/\mathbf{B}) \cap \mathrm{Gal}(\mathbf{E}/\mathbf{D}) = \mathrm{Gal}(\mathbf{E}/\mathbf{B}) \cap U,$$

and, similarly, if U is a neighborhood of the identity in $\mathrm{Gal}(\mathbf{E}/\mathbf{F})$, then $U = \mathrm{Gal}(\mathbf{E}/\mathbf{D})$ with **D** a finite extension of **F**, so

$$\tilde{U} = \mathrm{Gal}(\mathbf{E}/\mathbf{B}) \cap U = \mathrm{Gal}(\mathbf{E}/\mathbf{B}) \cap \mathrm{Gal}(\mathbf{E}/\mathbf{D}) = \mathrm{Gal}(\mathbf{E}/\mathbf{BD})$$

is a neighborhood of the identity in $\mathrm{Gal}(\mathbf{E}/\mathbf{B})$, as **BD** is a finite extension of **B**. □

Now we generalize the Fundamental Theorem of Galois Theory (Theorem 2.8.8) to infinite Galois extensions.

Theorem 5.4.10 (Fundamental Theorem of Infinite Galois Theory). *Let* **E** *be a Galois extension of* **F** *and let* $G = \mathrm{Gal}(\mathbf{E}/\mathbf{F})$ *have the Krull topology.*

(1) There is a 1-1 correspondence between intermediate fields $\mathbf{E} \supseteq \mathbf{B} \supseteq \mathbf{F}$ *and closed subgroups* $\{1\} \subseteq G_{\mathbf{B}} \subseteq G$ *given by*

$$\mathbf{B} = \mathrm{Fix}(G_{\mathbf{B}}).$$

(2) **B** *is a normal extension of* **F** *if and only if* $G_{\mathbf{B}}$ *is a normal subgroup of* G. *This is the case if and only if* **B** *is a Galois extension of* **F**. *In this case there is an isomorphism of topological groups*

$$\mathrm{Gal}(\mathbf{B}/\mathbf{F}) \cong G/G_{\mathbf{B}}.$$

Proof. (1) For each closed subgroup H of G, let

$$\mathbf{B}_H = \mathrm{Fix}(H).$$

This gives a map

$$\Gamma\colon \{\text{closed subgroups of } G\} \to \{\text{fields intermediate between } \mathbf{F} \text{ and } \mathbf{E}\}.$$

We show Γ is a 1–1 correspondence.

In this case it is easiest to work with the inverse correspondence. Thus, for each field $\mathbf{B}$ intermediate between $\mathbf{F}$ and $\mathbf{E}$, we let

$$G_{\mathbf{B}} = \Delta(\mathbf{B}) = \mathrm{Gal}(\mathbf{E}/\mathbf{B}).$$

First, we claim that $G_{\mathbf{B}}$ is a closed subgroup of G.

Let $\sigma \in \bar{G}_{\mathbf{B}}$, the closure of $G_{\mathbf{B}}$. Let $\beta \in \mathbf{B}$ be arbitrary. Then $\mathbf{F}(\beta)$ is a finite extension of $\mathbf{F}$, so σ has a neighborhood σ_N where $N = \mathrm{Gal}(\mathbf{E}/\mathbf{F}(\beta))$. Since every neighborhood of σ has a nonempty intersection with $G_{\mathbf{B}}$, we may choose $\tau \in G_{\mathbf{B}} \cap \sigma N$. Thus $\tau = \sigma\nu$ for some $\nu \in N$, and so, in particular, $\tau(\beta) = \sigma\nu(\beta)$. Since $\tau \in G_{\mathbf{B}}$, $\tau(\beta) = \beta$, and, by the definition of N, $\nu(\beta) = \beta$. Then $\sigma(\beta) = \sigma(\nu(\beta)) = \sigma\nu(\beta) = \tau(\beta) = \beta$. Since $\beta \in \mathbf{B}$ is arbitrary, we see that σ fixes B. Then, since $\sigma \in \bar{G}_{\mathbf{B}}$ is arbitrary, we see that $\bar{G}_{\mathbf{B}}$ fixes B. Now by definition $G_{\mathbf{B}}$ consists of all automorphisms of $\mathbf{E}$ fixing $\mathbf{B}$. Hence $G_{\mathbf{B}} = \bar{G}_{\mathbf{B}}$ and so $G_{\mathbf{B}}$ is closed.

Thus we have a map

$$\Delta\colon \{\text{fields } \mathbf{B} \text{ intermediate between } \mathbf{F} \text{ and } \mathbf{E}\} \to \{\text{closed subgroups of } G\}$$

and this map is 1–1 since if $G_{\mathbf{B}_1} = \Delta(\mathbf{B}_1) = \Delta(\mathbf{B}_2) = G_{\mathbf{B}_2}$, then, by Corollary 5.4.4, $\mathbf{E}$ is a Galois extension of $\mathbf{B}_1$ and of $\mathbf{B}_2$, so $\mathbf{B}_1 = \mathrm{Fix}(G_{\mathbf{B}_1}) = \mathrm{Fix}(G_{\mathbf{B}_2}) = \mathbf{B}_2$.

It remains to show Δ is onto. Thus, let H be a closed subgroup of G and let $\mathbf{B} = \mathrm{Fix}(H)$. Then $K = \mathrm{Gal}(\mathbf{E}/\mathbf{B}) \supseteq H$ and we need only show they are equal. Let $\sigma \in K$ be arbitrary and consider an arbitrary basic open neighborhood σU of σ. We claim $\sigma U \cap H \neq \emptyset$. Assuming this claim, we have that σ is in the closure $\bar{H}$ of H. Hence, as $\sigma \in K$ is arbitrary, $K \subseteq \bar{H}$. But H is closed, so $\bar{H} = H$ and hence $K = H$.

To see the claim, let $U = \mathrm{Gal}(\mathbf{E}/\mathbf{A})$ with $\mathbf{A}$ a finite extension of $\mathbf{F}$. By Corollary 5.4.4, $\mathbf{E}$ is a Galois extension of $\mathbf{B}$, and, by Lemma 2.3.4, $\mathbf{AB}$ is a finite extension of $\mathbf{B}$, so, by Theorem 5.4.2, $\mathbf{AB}$ is contained in a field $\mathbf{D}$ that is the splitting field of a separable polynomial $f(X) \in \mathbf{B}[X]$. Thus every element of $\mathrm{Gal}(\mathbf{E}/\mathbf{B})$ restricts to an element of $\mathrm{Gal}(\mathbf{E}/\mathbf{D})$, as it must take any root of $f(X)$ to a root of $f(X)$, and conversely any element of Gal($\mathbf{D}/\mathbf{B}$) extends to an element (not, in general, a unique element, but that is irrelevant) of Gal($\mathbf{E}/\mathbf{B}$), by Lemma 5.4.1. In other words, restriction to $\mathbf{D}$ give a well-defined epimorphism $R: K = \mathrm{Gal}(\mathbf{E}/\mathbf{B}) \to \mathrm{Gal}(\mathbf{D}/\mathbf{B})$. Let $\sigma_0 = R(\sigma) \in \mathrm{Gal}(\mathbf{D}/\mathbf{B})$.

Now $\mathbf{B} \subseteq \mathbf{AB} \subseteq \mathbf{D}$ and $\mathbf{D}$ is a finite Galois extension of $\mathbf{B}$. Let $H_0 = R(H) \subseteq \mathrm{Gal}(\mathbf{D}/\mathbf{B})$. Then $\mathrm{Fix}(H_0) = \mathrm{Fix}(H) = \mathbf{B} = \mathrm{Fix}(\mathrm{Gal}(\mathbf{D}/\mathbf{B}))$, so by the Fundamental Theorem of (finite) Galois Theory, $H_0 = \mathrm{Gal}(\mathbf{D}/\mathbf{B})$. Thus there is an element $\tau \in H$ with $R(\tau) = \sigma_0 = R(\sigma)$, i.e., with $\tau \mid \mathbf{B} = \sigma \mid \mathbf{B}$, and we see, by Remark 5.4.6, that $\tau \in \sigma U_{\mathbf{B}}$, so $\sigma U \cap H \neq \emptyset$, as claimed.

(2) We have proved much of this in proving (1). Just as in the finite case, for any $\sigma \in \mathrm{Gal}(\mathbf{E}/\mathbf{F})$, $\sigma(\mathbf{B}) = \mathrm{Fix}(\sigma G_{\mathbf{B}} \sigma^{-1})$, so if $G_{\mathbf{B}}$ is a normal subgroup of G, then $\sigma(\mathbf{B}) = \mathbf{B}$ for every $\sigma \in G$, while if $G_{\mathbf{B}}$ is not a normal subgroup of G, then $\sigma(\mathbf{B}) \neq \mathbf{B}$ for some $\sigma \in G$, and conversely. If $\mathbf{B}$ is a normal extension of $\mathbf{F}$, then for every $\beta \in \mathbf{B}$, $m_\beta(X) \in \mathbf{F}[X]$ splits in $\mathbf{B}$. Now for any $\sigma \in \mathrm{Gal}(\mathbf{E}/\mathbf{F})$, $\beta' = \sigma(\beta)$ is a root of $\sigma(m_\beta(X)) = m_\beta(X)$, so $\beta' \in \mathbf{B}$, and so $\sigma(\mathbf{B}) = \mathbf{B}$. On the other hand, if $\mathbf{B}$ is not a normal extension of $\mathbf{F}$, then there is some $\beta \in \mathbf{B}$ such that $m_\beta(X)$ has a root $\beta' \in \mathbf{E}$, $\beta' \notin \mathbf{B}$, and then there is an element σ of Gal($\mathbf{E}/\mathbf{F}$) with $\sigma(\beta) = \beta'$, by Lemma 5.4.1, and hence $\sigma(\mathbf{B}) \neq \mathbf{B}$.

Now if $\mathbf{B}$ is normal, then, just as before, R: $\mathrm{Gal}(\mathbf{E}/\mathbf{F}) \to \mathrm{Gal}(\mathbf{B}/\mathbf{F})$, given by $R(\sigma) = \sigma \mid \mathbf{B}$, is a well-defined epimorphism, whose kernel is clearly $G_{\mathbf{B}} = \mathrm{Gal}(\mathbf{E}/\mathbf{B})$. It remains only to show that the Krull topology on Gal($\mathbf{B}/\mathbf{F}$) is the quotient topology on $G/G_{\mathbf{B}}$.

Let U_0 be a basic neighborhood of the identity in Gal($\mathbf{B}/\mathbf{F}$). Then $U_0 = \{\sigma \in \mathrm{Gal}(\mathbf{B}/\mathbf{F}) \mid \sigma \mid \mathbf{A} = \mathrm{id}\}$ for some finite extension $\mathbf{A}$ of $\mathbf{F}$ (where of course $\mathbf{A} \subseteq \mathbf{B}$) and U_0 is the image under the quotient map R of the basic neighborhood of the identity $U_{\mathbf{A}} \subseteq \mathrm{Gal}(\mathbf{E}/\mathbf{F})$. On the other hand, if $U_{\mathbf{A}}$ is a basic neighborhood of the identity in Gal($\mathbf{E}/\mathbf{F}$), its image under the quotient map R is $U_0 = \{\sigma \in \mathrm{Gal}(\mathbf{B}/\mathbf{F}) \mid \sigma \mid \mathbf{B} \cap \mathbf{A} = \mathrm{id}\}$, and $\mathbf{B} \cap \mathbf{A} \subseteq \mathbf{A}$ is a finite extension of $\mathbf{F}$, so U_0 is a basic neighborhood of the identity in Gal($\mathbf{B}/\mathbf{F}$). □

Proposition 5.4.11. *Let* $\mathbf{E}$ *be a Galois extension of* $\mathbf{F}$ *and let* $\mathbf{B}$ *be an intermediate field between* $\mathbf{F}$ *and* $\mathbf{E}$. *Let* $G = \mathrm{Gal}(\mathbf{E}/\mathbf{F})$. *Then:*

(1) $(\mathbf{E}/\mathbf{B})$ *is finite if and only if* $|G_{\mathbf{B}}|$ *is finite. In this case,* $(\mathbf{E}/\mathbf{B}) = |G_{\mathbf{B}}|$.

(2) $(\mathbf{B}/\mathbf{F})$ *is finite if and only if* $[G : G_{\mathbf{B}}]$ *is finite. In this case,* $(\mathbf{B}/\mathbf{F}) = [G : G_{\mathbf{B}}]$.

(3) $G_{\mathbf{B}}$ *is always a closed subgroup of* G. $G_{\mathbf{B}}$ *is an open subgroup of* G *if and only if* $[G : G_{\mathbf{B}}]$ *is finite.*

Proof. (1) This follows immediately from Theorem 2.8.5.

(2)Assume $\mathbf{B}/\mathbf{F}$ is finite and let $\mathbf{B} \subseteq \mathbf{D}$ with $\mathbf{D}/\mathbf{F}$ finite Galois (Theorem 5.4.2). Then, by the *FTGT*, $(\mathbf{B}/\mathbf{F}) = [\mathrm{Gal}(\mathbf{D}/\mathbf{F}) : \mathrm{Gal}(\mathbf{D}/\mathbf{B})]$. But $\mathrm{Gal}(\mathbf{E}/\mathbf{F}) \to \mathrm{Gal}(\mathbf{D}/\mathbf{F})$ is an epimorphism so

$$[G : G_{\mathbf{B}}] = [\mathrm{Gal}(\mathbf{E}/\mathbf{F}) : \mathrm{Gal}(\mathbf{E}/\mathbf{B})] = [\mathrm{Gal}(\mathbf{D}/\mathbf{F}) : \mathrm{Gal}(\mathbf{D}/\mathbf{B})] = (\mathbf{B}/\mathbf{F}).$$

Assume $\mathbf{B}/\mathbf{F}$ is infinite and let $\{\beta_i\}$ be an infinite set of elements of $\mathbf{B}$ with $\mathbf{F} \subset \mathbf{F}(\beta_1) \subset \mathbf{F}(\beta_1, \beta_2) \subset \cdots \subset \mathbf{B}$. By Lemma 5.4.1, for each $i = 1, 2, \ldots$ there is an element $\sigma_i \in \mathrm{Gal}(\mathbf{E}/\mathbf{F})$ with $\sigma_i(\beta_j) = \beta_j$ for $j < i$ but $\sigma_i(\beta_i) \neq \beta_i$, so we see that $\sigma_i \neq \sigma_j$ for $i \neq j$. Since two elements of $\mathrm{Gal}(\mathbf{E}/\mathbf{F})$ are in the same left coset of $G_{\mathbf{B}}$ if and only if they agree on $\mathbf{B}$, we see that $[G : G_{\mathbf{B}}]$ is infinite.

(3) The fact that $G_{\mathbf{B}}$ is closed is part of Theorem 5.4.10 (1).

Suppose $[G : G_{\mathbf{B}}]$ is finite and let $\{\sigma_1, \ldots, \sigma_k\}$ be a set of left coset representatives of $G_{\mathbf{B}}$ with $\sigma_1 = 1$. Since $G_{\mathbf{B}}$ is closed, the finite union $\sigma_2 G_{\mathbf{B}} \cup \cdots \sigma_k G_{\mathbf{B}}$ is closed and hence its complement $\sigma_1 G_{\mathbf{B}} = G_{\mathbf{B}}$ is open.

Suppose $G_{\mathbf{B}}$ is open. Then it contains a basic open set $U_{\mathbf{A}} = \mathrm{Gal}(\mathbf{E}/\mathbf{A})$, with $\mathbf{A}$ a finite extension of $\mathbf{F}$. Then $\mathbf{B} = \mathrm{Fix}(G_{\mathbf{B}}) \subseteq \mathrm{Fix}(U_{\mathbf{A}}) = \mathbf{A}$, so $(\mathbf{B}/\mathbf{F})$ is finite and hence $[G : G_{\mathbf{B}}]$ is finite. □

We now further investigate the Krull topology on G.

Lemma 5.4.12. *Each basic open set $\sigma U_{\mathbf{B}}$ in G is also closed.*

Proof. Let $\tau \notin \sigma U_{\mathbf{B}}$. Then $\emptyset = \tau_{\mathbf{B}} \cap \sigma U_{\mathbf{B}}$ (as $U_{\mathbf{B}}$ is a subgroup of G), i.e., $\tau U_{\mathbf{B}} \subseteq G - \sigma U_{\mathbf{B}}$. In other words, the complement of $\sigma U_{\mathbf{B}}$ contains a neighborhood of each of its points, so is an open set, and hence $\sigma U_{\mathbf{B}}$ is a closed set. □

We now prove a lemma we will need in describing the topology of $\mathrm{Gal}(\mathbf{E}/\mathbf{F})$, but one which is interesting in its own right.

Lemma 5.4.13. *Let* $\mathbf{E}$ *be an algebraic extension of* $\mathbf{F}$ *and let* $\sigma : \mathbf{E} \to \mathbf{E}$ *be a map of fields with* $\sigma \mid \mathbf{F} = \mathrm{id}$. *Then* σ *is an automorphism of* $\mathbf{E}$ *(and hence* $\sigma \in \mathrm{Gal}(\mathbf{E}/\mathbf{F})$*).*

Proof. Since $\sigma(1) = 1$, σ is an injection, so we need only show σ is a surjection. Let $\alpha \in \mathbf{E}$ be arbitrary. Set

$$S = \{\text{roots of } m_\alpha(X) \text{ in } \mathbf{E}\} = \{\beta \in \mathbf{E} \mid m_\alpha(\beta) = 0\}.$$

Then for any $\beta \in S$, $m_\alpha(\sigma(\beta)) = \sigma((m_\alpha(\beta)) = \sigma(0) = 0$ so $\sigma(\beta) \in S$. Thus $\sigma \mid S$ is an injective map from S to itself. Since S is finite (as any polynomial has only finitely many roots), $\sigma \mid S$ is surjective as well, so there is a $\beta_0 \in S$ with $\sigma(\beta_0) = \alpha$, as required. □

Remark 5.4.14. This lemma is false for arbitrary extensions. For example, let $\mathbf{E} = \mathbf{F}(X)$, the field of rational functions over $\mathbf{F}$ in the variable X. Then $\sigma(X) = X^2$ induces $\sigma\colon \mathbf{E} \to \mathbf{E}$, an injective, but not surjective, map of fields. ◇

Theorem 5.4.15. *Let* $\mathbf{E}$ *be a Galois extension of* $\mathbf{F}$ *and let* $G = \mathrm{Gal}(\mathbf{E}/\mathbf{F})$ *have the Krull topology. Then* G *is Hausdorff, totally disconnected, and compact.*

Proof. We shall prove this in several steps.

(1) G is Hausdorff: Since G is a topological group, to show G is Hausdorff it suffices to show that

$$\bigcap U_{\mathbf{B}} = \{\mathrm{id}\}$$

where the intersection is taken over all finite extensions $\mathbf{B}$ of $\mathbf{F}$.

Let $\sigma \in G$, $\sigma \neq \mathrm{id}$. Then for some $\alpha \in \mathbf{E}$, $\sigma(\alpha) \neq \alpha$. Let $\mathbf{B} = \mathbf{F}(\alpha)$. Then $\sigma \notin U_{\mathbf{B}}$, as required.

(2) G is totally disconnected: Let $\sigma, \tau \in G$, $\sigma \neq \tau$. Then $\tau \notin \sigma U_{\mathbf{B}}$ for some $U_{\mathbf{B}}$, and then $G = \sigma U_{\mathbf{B}} \cup (G - \sigma U_{\mathbf{B}})$ is a union of two relatively open sets with σ (resp. τ) in the first (resp. second). ($\sigma U_{\mathbf{B}}$ is open by definition and $G - \sigma U_{\mathbf{B}}$ is open as in the proof of Lemma 5.4.12.) Thus σ and τ are in different components of G, so the only components of G are single points.

(3) G is compact: This will require considerably more work.

For each $\alpha \in \mathbf{E}$, let $R_\alpha = \{\beta \in \mathbf{E} \mid m_\alpha(\beta) = 0\}$. Let

$$\mathcal{G} = \prod_{\alpha\in\mathbf{E}} R_\alpha.$$

By identifying an element $\prod_{\alpha\in\mathbf{E}}(\beta_\alpha)_\alpha$ with $f\colon \mathbf{E} \to \mathbf{E}$ by $f(\alpha) = \beta_\alpha$ for each $\alpha \in \mathbf{E}$, we may regard $\mathcal{G}$ as a set of functions from $\mathbf{E}$ to itself. (This is identifying a function with its graph.) Note that $R_\alpha = \{\alpha\}$ for $\alpha \in \mathbf{F}$, so every $f \in \mathcal{G}$ satisfies $f(\alpha) = \alpha$ for every $\alpha \in \mathbf{F}$. Note also that R_α is finite for each α. We give $\mathbf{E}$ the discrete topology, and each R_α the induced topology as a subset of $\mathbf{E}$, which is also the discrete topology. Then R_α is a finite discrete space, so is compact. We give $\mathcal{G}$ the product topology. Then, by Tychonoff's Theorem, $\mathcal{G}$ is the product of compact sets and hence is compact. (Recall that if $X = \prod X_i$ is a product of spaces, the product topology has a basis of open sets $\prod Y_i$ where each Y_i is open and $Y_i = X_i$ for all but finitely many i. If each X_i is discrete, then every subset of X_i is open, so this condition reduces to $Y_i = X_i$ for all but finitely many i.)

Since every $\sigma \in \mathrm{Gal}(\mathbf{E}/\mathbf{F})$ satisfies $m_\alpha(\sigma(\alpha)) = 0$ for every $\alpha \in \mathbf{E}$, we may regard G as a subset of $\mathcal{G}$.

Claim: The induced topology on G as a subset of $\mathcal{G}$ is the Krull topology on G.

Proof of claim: Let $\sigma U_{\mathbf{B}}$ be a basic open set in G. Let $\{\beta_1, \ldots, \beta_k\}$ be a basis for $\mathbf{B}$ over $\mathbf{F}$. Then for $\tau \in G$, $\tau \in \sigma U_{\mathbf{B}}$ if and only if $\tau(\beta_i) = \sigma(\beta_i)$, $i = 1, \ldots, k$. Thus $\sigma U_{\mathbf{B}} = G \cap V$ where $V = \prod_{\alpha \in \mathbf{E}} S_\alpha$ with $S_\alpha = R_\alpha$ for $\alpha \neq \beta_i$ and $S_{\beta_i} = \{\sigma(\beta_i)\}$, and so V is an open set in $\mathcal{G}$.

Conversely, let V be a basic open set in $\mathcal{G}$. Then $V = \prod_{\alpha \in \mathbf{E}} S_\alpha$ with $S_\alpha = R_\alpha$ for all but finitely many α. Since each R_α is finite, we see that V is a finite union of open sets of the form $V = \prod_{\alpha \in \mathbf{E}} S_\alpha$ with $S_\alpha = R_\alpha$ for all but finitely many α, say for $\alpha \neq \alpha_1, \ldots, \alpha_k$, and $S_{\alpha_i} = \{\alpha_i'\}$ is a single point for each $i = 1, \ldots, k$. Let $\mathbf{B} = \mathbf{F}(\alpha_1, \ldots, \alpha_k)$, a finite extension of $\mathbf{F}$. If $\sigma \in G$ satisfies $\sigma(\alpha_i) = \alpha_i'$, $i = 1, \ldots, k$, then $G \cap V = \sigma U_{\mathbf{B}}$ is open, while if there is no $\sigma \in G$ satisfying $\sigma(\alpha_i) = \alpha_i'$, $i = 1, \ldots, k$, then $G \cap V$ is the empty set, which is also open.

Claim: $\mathcal{G}$ is a Hausdorff space.

Proof of claim: Let $f_1, f_2 \in \mathcal{G}$ with $f_1 \neq f_2$. Then $f_1(\alpha_0) \neq f_2(\alpha_0)$ for some $\alpha_0 \in \mathbf{E}$ (regarding $\mathcal{G}$ as a space of functions). Let $A_1 = \prod_{\alpha \in \mathbf{E}} S_\alpha$ with $S_\alpha = R_\alpha$ for $\alpha \neq \alpha_0$ and $S_{\alpha_0} = \{f_1(\alpha_0)\}$, and let $A_2 = \prod_{\alpha \in \mathbf{E}} S_\alpha$ with $S_\alpha = R_\alpha$ for $\alpha \neq \alpha_0$ and $S_{\alpha_0} = \{f_2(\alpha_0)\}$. Then $f_1 \in A_1$, $f_2 \in A_2$, and A_1 and A_2 are disjoint open sets.

(Since G is a subspace of a Hausdorff space, it is Hausdorff, but for simplicity we chose to prove that directly in step (1). Observe that the proofs that G and $\mathcal{G}$ are Hausdorff are very similar. In the same way, we may adapt the proof that G is totally disconnected to prove that $\mathcal{G}$ is totally disconnected. However, it is G and not $\mathcal{G}$ that we are interested in.)

Claim: G is a closed subset of $\mathcal{G}$.

Proof of claim: Let $f \in \mathcal{G}$, $f \notin G$. We need to show f has an open neighborhood V that is disjoint from G. Note every $f \in \mathcal{G}$ satisfies $f \mid \mathbf{F} = \mathrm{id}$, so if $f \notin G$, f is not an automorphism of $\mathbf{E}$. If $f : \mathbf{E} \to \mathbf{E}$ is an injective map of fields, then, by Lemma 5.4.13, f is an automorphism of $\mathbf{E}$, contradicting our assumption. Thus there are two possibilities:

(1) f is not an injection; or

(2) f is an injection but not a map of fields.

In case (1) let $\alpha_1, \alpha_2 \in \mathbf{E}$ with $f(\alpha_1) = f(\alpha_2)$, $\alpha_1 \neq \alpha_2$. Let $V = \prod_{\alpha \in \mathbf{E}} S_\alpha$ with $S_{\alpha_1} = \{f(\alpha_1)\}$, $S_{\alpha_2} = \{f(\alpha_2)\}$, and $S_\alpha = R_\alpha$ for $\alpha \neq \alpha_1, \alpha_2$. Then $f \in V$, V is open, and $G \cap V$ is the empty set (as any $g \in G$ is an injection, so $g(\alpha_1) \neq g(\alpha_2)$).

In case (2) we must have one of the following: $f(\alpha_1 + \alpha_2) \neq f(\alpha_1) + f(\alpha_2)$ for some $\alpha_1, \alpha_2 \in \mathbf{E}$, $f(\alpha_1\alpha_2) \neq f(\alpha_1)f(\alpha_2)$ for some $\alpha_1, \alpha_2 \in \mathbf{E}$, or $f(\alpha_1^{-1}) \neq f(\alpha_1)^{-1}$ for some $0 \neq \alpha_1 \in \mathbf{E}$. For simplicity assume it is the first of these. Let $V = \prod_{\alpha \in \mathbf{E}} S_\alpha$ with $S_{\alpha_1} = \{f(\alpha_1)\}$, $S_{\alpha_2} = \{f(\alpha_2)\}$, $S_{\alpha_1+\alpha_2} = \{f(\alpha_1 + \alpha_2)\}$, and $S_\alpha = R_\alpha$ for $\alpha \neq \alpha_1, \alpha_2, \alpha_1 + \alpha_2$. Then $f \in V$, V is open, and $G \cap V$ is the empty set (as any $g \in G$ is a map of fields, so $g(\alpha_1 + \alpha_2) = g(\alpha_1) + g(\alpha_2)$.

Claim: G is compact in the Krull topology.

Proof of claim: We merely need to assemble our previous work. The Krull topology on G is the topology G inherits as a subspace of $\mathcal{G}$. We have shown that $\mathcal{G}$ is a compact Hausdorff space and that G is a closed subspace of $\mathcal{G}$. But it is a basic topological fact that a closed subspace of a compact Hausdorff space is compact. □

Remark 5.4.16. While we do not need this for our purposes, it is worthwhile to put things in a larger context. An important topology on function spaces is the *compact-open* topology. If $\mathcal{F} = \{f : X \to Y\}$ is the set of continuous functions from X to Y, the compact-open topology on $\mathcal{F}$ has as subbases the sets $Z_{A,B} = \{f \in \mathcal{F} \mid f(A) \subseteq B\}$ where $A \subseteq X$ is compact and $B \subseteq Y$ is open. In our case, $\mathbf{E}$ is discrete so every $f : \mathbf{E} \to \mathbf{E}$ is continuous, $A \subseteq \mathbf{E}$ is compact if and only if it is finite, and every $B \subseteq \mathbf{E}$ is open. Now $\mathcal{G} \subseteq \mathcal{F}$ and it is easy to see that the topology we have defined on $\mathcal{G}$ is the restriction of the compact-open topology on $\mathcal{F}$ to $\mathcal{G}$, i.e., that this topology is the compact-open topology on $\mathcal{G}$, and then we further see that the Krull topology on G agrees with the compact-open topology on G. ◇

Let $\{\mathbf{B}_i\}$ be a set of fields intermediate between $\mathbf{E}$ and $\mathbf{F}$. Generalizing Definition 3.4.1, we say that $\{\mathbf{B}_i\}$ are *disjoint* extensions of $\mathbf{F}$ if $\mathbf{B}_i \cap \mathbf{D}_i = \mathbf{F}$ for each i, where $\mathbf{D}_i$ is the composite of the fields $\{\mathbf{B}_j\}$ for $j \neq i$. Then we have the following generalization of Theorem 3.4.7.

Theorem 5.4.17. *Let* $\mathbf{E}$ *be an extension of* $\mathbf{F}$ *and let* $\{\mathbf{B}_i \subseteq \mathbf{E}\}$ *be disjoint Galois extensions of* $\mathbf{F}$ *with* $G_i = \mathrm{Gal}(\mathbf{B}_i/\mathbf{F})$. *Let* $\mathbf{B}$ *be the composite of* $\{\mathbf{B}_i\}$ *and let* $G = \mathrm{Gal}(\mathbf{B}/\mathbf{F})$. *Then* $\mathbf{B}$ *is a Galois extension of* $\mathbf{F}$ *and* $G = \prod_i G_i$ *with the product topology.*

Proof. A Zorn's Lemma argument allows us to generalize Theorem 3.4.7 and conclude that $\mathbf{B}$ is a Galois extension of $\mathbf{F}$ and $G = \prod_i G_i$. We show G has the product topology.

Let U be a basic open neighborhood of the identity in the product topology. Then, by the definition of the product topology, $U = \prod_{i \in I} U_i$ where $U_i \subseteq G_i$ is a basic open set and $U_i = G_i$ except for finitely many i. Then, for each $i \in I$, $U_i = \mathrm{Gal}(\mathbf{B}/\mathbf{A}_i)$ where $\mathbf{A}_i$ is a finite extension of $\mathbf{F}$, and $\mathbf{A}_i = \mathbf{F}$ except for finitely many i. Let $\mathbf{A}$ be the composite of the $\{\mathbf{A}_i\}$. Then $\mathbf{A}$ is a finite extension of $\mathbf{F}$, and $U = U_{\mathbf{A}}$ is a basic open neighborhood of the identity in the Krull topology on G.

Conversely, let U be a basic open neighborhood of the identity in the Krull topology on G. Then, by definition, $U = U_{\mathbf{A}}$ for some finite extension $\mathbf{A}$ of $\mathbf{F}$. Let $\mathbf{A}_i = \mathbf{A} \cap \mathbf{B}_i$ and note that, since $\{\mathbf{B}_i\}$ are disjoint extensions of $\mathbf{F}$, $\mathbf{A}_i = \mathbf{F}$ for all but finitely many i. Then, setting $U_i = U_{\mathbf{A}_i}$, we have that each U_i is open, $U_i = G_i$ for all but finitely many i, and $U = \prod_{i \in I} U_i$ is a basic open neighborhood of the identity in the product topology. □

Example 5.4.18. Here are two cases of infinite Galois extensions $\mathbf{E}$ of $\mathbf{F}$ and two subgroups $H_1 \neq H_2$ of $G = \mathrm{Gal}(\mathbf{E}/\mathbf{F})$ with $\mathrm{Fix}(H_1) = \mathrm{Fix}(H_2)$. (Of course, $\bar{H}_1 = \bar{H}_2$.)

(1) Let $J = \{-1, 2, 3, 5, \ldots\}$ consist of -1 and the primes. For each $j \in J$, let $\mathbf{B}_j = \mathbf{Q}(\sqrt{j}) \subset \mathbf{C}$, and let $\mathbf{B}$ be the composite of the $\{\mathbf{B}_j\}$. Note $\{\mathbf{B}_j\}$ are disjoint extensions of $\mathbf{Q}$. Let $G_j = \mathrm{Gal}(\mathbf{B}_j/\mathbf{Q})$. Note G_j is isomorphic to $\mathbb{Z}/2\mathbb{Z}$ for each j, so, by Theorem 5.4.17, $G = \mathrm{Gal}(\mathbf{B}/\mathbf{Q}) = \prod_{j \in J}(\mathbb{Z}/2\mathbb{Z})$, and G is uncountable. Hence G has uncountably many subgroups of index 2. But for any subgroup H of index 2, $(\mathrm{Fix}(H)/\mathbf{Q}) = 2$ by Proposition 5.4.11. However, $\mathbf{Q}$ has only countably many extensions of degree 2 (as each is obtained by adjoining a root of a quadratic equation to $\mathbf{Q}$, and there are only countably many quadratic equations with rational coefficients). Hence there are two (and indeed, uncountably many) distinct subgroups H_1 and H_2 of G of index 2 with $\mathrm{Fix}(H_1) = \mathrm{Fix}(H_2)$.

(2) Fix a prime p and an algebraic closure $\bar{\mathbf{F}}_p$ of $\mathbf{F}_p$. Then $\bar{\mathbf{F}}_p$ is the composite of the fields $\mathbf{B}_n = \mathbf{F}_{p^n}$ contained in $\bar{\mathbf{F}}_p$. Let $G = \mathrm{Gal}(\bar{\mathbf{F}}_p/\mathbf{F}_p)$ and let $\Phi \in G$ be the Frobenius map, $\Phi(\alpha) = \alpha^p$. Let H be the subgroup of G generated by Φ. Then $\mathrm{Fix}(H) = \mathbf{F}_p = \mathrm{Fix}(G)$. We claim $H \neq G$. To see this, let $\mathbf{A}$ be the composite of the fields $\mathbf{B}_n$ where n is a power of 2, and let $\mathbf{D}$ be the composite of the fields $\mathbf{B}_n$ for n odd. Then, by Lemma 3.4.2, $\mathbf{A}$ and $\mathbf{D}$ are disjoint extensions of $\mathbf{F}_p$, and $\mathbf{AD} = \bar{\mathbf{F}}_p$. Hence $G = \mathrm{Gal}(\mathbf{A}/\mathbf{F}_p) \times \mathrm{Gal}(\mathbf{D}/\mathbf{F}_p)$. Let $\Phi \mid \mathbf{A} = \Phi_{\mathbf{A}}$ and $\Phi \mid \mathbf{D} = \Phi_{\mathbf{D}}$. Then neither $\Phi_{\mathbf{A}}$ nor $\Phi_{\mathbf{D}}$ is the identity map. Now H is a cyclic group, so $H = \{\Phi^k\} = \{\Phi_{\mathbf{A}}^k \Phi_{\mathbf{D}}^k\}$. But G also has the subgroups $\{\Phi_{\mathbf{A}}^k\}$ and $\{\Phi_{\mathbf{D}}^k\}$ (and uncountably many others), so $H \neq G$. ◇

5.5 Exercises

Exercise 5.5.1. Let $\mathbf{B}_1$ and $\mathbf{B}_2$ be finite extensions of $\mathbf{F}$ satisfying property "P". Show that $\mathbf{B}_1\mathbf{B}_2$ and $\mathbf{B}_1 \cap \mathbf{B}_2$ also satisfy property "P". Here "P" is any one of the following: finite separable, finite purely inseparable, finite normal, separable, purely inseparable, normal, Galois. ("P" finite Galois is handled in Corollary 3.4.8 and in Exercise 3.9.13.)

Exercise 5.5.2. (a) Let $\mathbf{A}_0$ be the field obtained by adjoining all the complex roots of every irreducible polynomial $f(X) \in \mathbf{Q}[X]$ of odd degree. For $i > 0$, let $\mathbf{A}_i$ be the field obtained from $\mathbf{A}_{i-1}$ by adjoining all complex square roots of all elements of $\mathbf{A}_{i-1}$. Show that $\mathbf{A}_0 \subset \mathbf{A}_1 \subset \mathbf{A}_2 \subset \cdots$ and that

$$\bar{\mathbf{Q}} = \bigcup_{i=0}^{\infty} \mathbf{A}_i.$$

(b) Show that the conclusion of part (a) remains true if $\mathbf{A}_0$ is the field obtained by adjoining a single real root of every irreducible polynomial $f(X) \in \mathbf{Q}[X]$ of odd degree.

Exercise 5.5.3. Let $\mathbf{Q}^{\text{solv}} = \{z \in \mathbf{C} \mid m_z(X) \text{ is solvable by radicals}\}$. Let $\mathbf{B}_0 = \mathbf{Q}$. For $i > 0$, let $\mathbf{B}_i$ be the field obtained from $\mathbf{B}_{i-1}$ by adjoining all complex n^{th}-roots of all elements of $\mathbf{B}_{i-1}$, for all positive integers n. Show that $\mathbf{B}_0 \subset \mathbf{B}_1 \subset \mathbf{B}_2 \subset \cdots$ and that

$$\mathbf{Q}^{\text{solv}} = \bigcup_{i=0}^{\infty} \mathbf{B}_i.$$

Exercise 5.5.4. Recall that $\bar{\mathbf{Q}}$ denotes the field of algebraic numbers. Show that the only nontrivial element of finite order in $\text{Gal}(\bar{\mathbf{Q}}/\mathbf{Q})$ is the automorphism of $\bar{\mathbf{Q}}$ given by complex conjugation (an element of order 2).

Exercise 5.5.5. Let $\mathbf{F} = \mathbf{C}$ and let $\mathbf{E} = \mathbf{C}(X)$. Note that $\mathbf{E}$ is *not* an algebraic extension of $\mathbf{F}$. Let $G = \{\sigma : \mathbf{E} \to \mathbf{E} \mid \sigma \mid \mathbf{F} = \text{id}\}$. Show that $\text{Fix}(G) = \mathbf{F}$. (This shows that in Section 5.4, it was necessary to require that $\mathbf{E}$ be an *algebraic* extension of $\mathbf{F}$. Otherwise we would have nonalgebraic "Galois" extensions and the theory developed there would not hold.)

Exercise 5.5.6. In the notation of Example 5.4.18 (1), show that the closed subgroups of G of index 2 are in 1–1 correspondence with the nonempty finite subsets of J.

Exercise 5.5.7. Explicitly exhibit uncountably many subgroups of the group G of Example 5.4.18 (2).

6

Transcendental Extensions

The preceding chapters have dealt (almost) exclusively with algebraic extensions of fields. In this chapter we present an introduction to the theory of transcendental extensions.

6.1 General Results

Definition 6.1.1. (1) Let $\mathbf{E}$ be an extension of $\mathbf{F}$. Then $\alpha \in \mathbf{E}$ is *transcendental* over $\mathbf{F}$ if α is not a root of any polynomial $f(X) \in \mathbf{F}[X]$.

(2) $\mathbf{E}$ is a *transcendental extension* of $\mathbf{F}$ (or $\mathbf{E}/\mathbf{F}$ is *transcendental*) if some $\alpha \in \mathbf{E}$ is transcendental over $\mathbf{F}$. ◇

It is convenient to adopt the following nonstandard terminology.

Definition 6.1.2. Let $\mathbf{E}$ be an extension of $\mathbf{F}$. Then $\mathbf{E}$ is a *completely transcendental extension* of $\mathbf{F}$ (or $\mathbf{E}/\mathbf{F}$ is *completely transcendental*) if every $\alpha \in \mathbf{E}$, $\alpha \notin \mathbf{F}$, is transcendental over $\mathbf{F}$. ◇

Lemma 6.1.3. *Let* $\mathbf{E}$ *be any extension of* $\mathbf{F}$. *Then there is a unique field* $\mathbf{B}$ *with* $\mathbf{F} \subseteq \mathbf{B} \subseteq \mathbf{E}$ *and with* $\mathbf{B}$ *algebraic over* $\mathbf{F}$ *and* $\mathbf{E}$ *completely transcendental over* $\mathbf{B}$.

Proof. $\mathbf{B} = \{\alpha \in \mathbf{E} \mid \alpha \text{ is algebraic over } \mathbf{F}\}$. □

Definition 6.1.4. Let $\mathbf{E}$ be an extension of $\mathbf{F}$. Then $\mathbf{E}$ is a *purely transcendental extension* of $\mathbf{F}$ (or $\mathbf{E}/\mathbf{F}$ is *purely transcendental*) if $\mathbf{E}$ is isomorphic to the field of rational functions $\mathbf{F}(X_1, X_2, \ldots)$ in the (finite or infinite) set of variables $\{X_1, X_2, \ldots\}$. ◇

S.H. Weintraub, *Galois Theory*, DOI 10.1007/978-0-387-87575-0_6,

Lemma 6.1.5. *Let* **E** *be a purely transcendental extension of* **F**. *Then* **E** *is a completely transcendental extension of* **F**.

Proof. We may assume that $\mathbf{E} = \mathbf{F}(X_1, X_2, \dots)$. Let $\alpha \in \mathbf{E}$. Then $\alpha = p(X_1, \dots, X_k)/q(X_1, \dots, X_k)$ for some polynomials $p(X_1, \dots, X_k)$ and $q(X_1, \dots, X_k)$. (Note that any rational function can only involve finitely many variables.) Suppose that α is algebraic over **F**. Then α is a root of some irreducible polynomial $\sum_{i=0}^{n} a_i X^i$ in $\mathbf{F}[X]$. Substituting, and clearing denominators, we have $a_0 q(X_1, \dots, X_k)^n + \sum_{i=1}^{n-1} a_i p(X_1, \dots, X_k)^i q(X_1, \dots, X_k)^{n-i} + a_n p(X_1, \dots, X_k)^n = 0$. Now $\mathbf{F}[X_1, X_2, \dots, X_k]$ is a unique factorization domain, and $q(X_1, \dots, X_k)$ divides every term in this expression except possibly the last one, so $q(X_1, \dots, X_k)$ must divide $p(X_1, \dots, X_k)^n$ and hence also $p(X_1, \dots, X_k)$; similarly $p(X_1, \dots, X_k)$ must divide $q(X_1, \dots, X_k)$. Hence $\alpha = p(X_1, \dots, X_k)/q(X_1, \dots, X_k) \in \mathbf{F}$. □

Remark 6.1.6. Of course, if **F** is algebraically closed then every extension of **F** is completely transcendental. We will give examples of extensions of **C** that are not purely transcendental in Section 6.3. ◇

Definition 6.1.7. Let **E** be an extension of **F**. A subset $S = \{s_1, s_2, \dots\}$ of **E** is *algebraically independent* over **F** if $f(s_1, s_2, \dots) \neq 0$ for any nonzero polynomial $f(X_1, X_2, \dots)$ in $\mathbf{F}[X_1, X_2, \dots]$. ◇

Lemma 6.1.8. *Let* **E** *be an extension of* **F** *and let* $S = \{s_1, s_2, \dots\} \subset \mathbf{E}$ *be algebraically independent over* **F**. *Let* $\mathbf{B} = \mathbf{F}(S) = \mathbf{F}(s_1, s_2, \dots)$. *Then* **B** *is isomorphic to* $\mathbf{F}(X_1, X_2, \dots)$. *In particular,* **B** *is purely transcendental.*

Proof. Let $\varphi_0 : \mathbf{F}[X_1, X_2, \dots] \to \mathbf{F}[s_1, s_2, \dots]$ be defined by $\varphi_0(X_i) = s_i$ for each i. φ_0 is obviously surjective. It is injective as S is algebraically independent over **F**, so it is an isomorphism. But then, again as S is algebraically independent over **F**, φ_0 extends to an isomorphism $\varphi : \mathbf{F}(X_1, X_2, \dots) \to \mathbf{F}(s_1, s_2, \dots)$. □

We adopt the convention that if S is the empty set, then S is algebraically independent and $\mathbf{F}(S) = \mathbf{F}$.

We now develop the theory of transcendence bases. The reader should note the very strong analogy between the notion, and properties of, a transcendence basis of a field extension and a basis of a vector space. We begin with a technical lemma, which is of interest in itself.

Lemma 6.1.9. *Let* **E** *be an extension of* **F** *and let* S *be a subset of* **E**. *Let* s_1 *be an arbitrary element of* S. *Then* S *is algebraically independent if and only if* $S' = S - \{s_1\}$ *is algebraically independent and* s *is transcendental over* $\mathbf{F}(S')$.

Proof. We prove the contrapositive of the statement of the lemma.

Clearly, if S' is not algebraically independent or if s_1 is algebraic over $\mathbf{F}(S')$, then S is not algebraically independent, so it remains to prove the reverse implication.

Let $S' = \{s_2, s_3, \dots\}$ and let $\mathbf{B} = \mathbf{F}(S')$. Suppose that S is not algebraically independent. Then there is some polynomial $f(X_1, X_2, \dots)$ with coefficients in $\mathbf{F}$ with $f(s_1, s_2, \dots) = 0$. If the variable X_1 does not appear in this polynomial then $f(s_1, s_2, \dots) = g(s_2, \dots)$ and S' is not algebraically independent. If the variable X_1 appears in this polynomial then we gather terms in like powers of X_1 together, and substitute s_i for X_i for $i > 1$. Then, writing $h(X_1) = f(X_1, s_2, s_3, \dots) = \sum h_k(s_2, s_3, \dots)X_1^k \in \mathbf{B}[X_1]$, we have that $h(s_1) = 0$, so s_1 is algebraic over S'. □

Definition 6.1.10. Let $\mathbf{E}$ be an extension of $\mathbf{F}$ and let $S = \{s_1, s_2, \dots\}$ be a subset of $\mathbf{E}$. S is a *transcendence basis* of $\mathbf{E}$ over $\mathbf{F}$ if S is algebraically independent over $\mathbf{F}$ and $\mathbf{E}$ is an algebraic extension of $\mathbf{F}(S)$. ◇

Remark 6.1.11. At first glance, it may seem more logical to require that $\mathbf{E} = \mathbf{F}(S)$ in the definition of a transcendence basis. But if we were to make that definition, by Lemma 6.1.8 only purely transcendental extensions would have transcendence bases, and that would greatly restrict the utility of the theory. So we make the more general definition. ◇

Theorem 6.1.12. *Let* $\mathbf{E}$ *be an extension of* $\mathbf{F}$ *and let* $R \subseteq T$ *be subsets of* $\mathbf{E}$ *such that* R *is algebraically independent over* $\mathbf{F}$ *and* $\mathbf{E}$ *is an algebraic extension of* $\mathbf{F}(T)$. *Then there is a transcendence basis* S *of* $\mathbf{E}$ *over* $\mathbf{F}$ *with* $R \subseteq S \subseteq T$.

Proof. The proof is a Zorn's Lemma argument. Let $\mathcal{S}$ be the set of subsets of $\mathbf{E}$ defined by

$$\mathcal{S} = \{S \mid R \subseteq S \subseteq T \text{ and } S \text{ is algebraically independent over } \mathbf{F}\}.$$

$\mathcal{S}$ is nonempty as $R \in \mathcal{S}$. Partially order $\mathcal{S}$ by inclusion. Clearly every totally ordered subset of $\mathcal{S}$ has a maximal element, its union. Thus, by Zorn's Lemma, $\mathcal{S}$ has a maximal element S. We claim that S is a transcendence basis for $\mathbf{E}$ over $\mathbf{F}$.

By definition, S is algebraically independent over $\mathbf{F}$, so we need only show that $\mathbf{E}$ is algebraic over $\mathbf{B} = \mathbf{F}(S)$. Suppose not and let $\alpha \in \mathbf{E}$ be transcendental over $\mathbf{B}$. Then, by Lemma 6.1.9, $\tilde{S} = S \cup \{\alpha\}$ is algebraically independent, contradicting the maximality of S. □

Corollary 6.1.13. *Let* **E** *be an extension of* **F**.

(1) **E** *has a transcendence basis over* **F**.

(2) If R is any subset of **E** *that is algebraically independent over* **F**, *then* **E** *has a transcendence basis over* **F** *that contains R.*

(3) If T is any subset of **E** *with* **E** *an algebraic extension of* $\mathbf{F}(T)$, *then* **E** *has a transcendence basis over* **F** *that is contained in T.*

Proof. For (1), let $R = \emptyset$ and $T = \mathbf{E}$ in Theorem 6.1.12. For (2), let $T = \mathbf{E}$ in Theorem 6.1.12. For (3), let $R = \emptyset$ in Theorem 6.1.12. □

The following "replacement lemma" is the key step in the proof of Theorem 6.1.15.

Lemma 6.1.14. *Let* **E** *be an extension of* **F** *and let S and T be any two transcendence bases of* **E** *over* **F**. *Let s be any element of S and let $S' = S - \{s\}$. Then there is an element t of T such that $S'' = S' \cup \{t\}$ is a transcendence basis of* **E** *over* **F**.

Proof. By Lemma 6.1.9 we see that s is transcendental over $\mathbf{F}(S')$. In particular, **E** is transcendental over $\mathbf{F}(S')$. This implies that some element t of T is transcendental over $\mathbf{F}(S')$, as if not, $\mathbf{F}(T)$ would be algebraic over $\mathbf{F}(S')$. By assumption, T is a transcendence basis for **E** over **F**, so in particular **E** is algebraic over $\mathbf{F}(T)$. But **E** algebraic over $\mathbf{F}(T)$ and $\mathbf{F}(T)$ algebraic over $\mathbf{F}(S')$ implies **E** algebraic over $\mathbf{F}(S')$, a contradiction.

Since t is transcendental over S', S'' is algebraically independent, again by Lemma 6.1.9. Furthermore, s is algebraic over $\mathbf{F}(S'')$ as otherwise $S \cup \{t\}$ would be algebraically independent, which it is not as **E** is algebraic over $\mathbf{F}(S)$. Thus $\mathbf{F}(S)$ is algebraic over $\mathbf{F}(S'')$. Again, **E** is algebraic over $\mathbf{F}(S)$, and so **E** is algebraic over $\mathbf{F}(S'')$. Hence S'' satisfies both conditions for a transcendence basis of **E** over **F**. □

In the following theorem, we do not distinguish between the cardinality of infinite sets.

Theorem 6.1.15. *Let* **E** *be an extension of* **F**. *Then any two transcendence bases of* **E** *over* **F** *have the same number of elements.*

Proof. To prove the theorem it suffices to show that if **E** has a transcendence basis S over **F** with a finite number k of elements, then any subset T of **E** with more than k elements (perhaps with infinitely many elements) cannot be algebraically independent. We prove this by contradiction.

Let $S = \{s_1, s_2, \ldots, s_k\}$ and suppose that T is a transcendence basis having more than k elements. Apply Lemma 6.1.14 k times in succession. First we obtain a transcendence basis $S_1 = \{t_1, s_2, s_3, \ldots, s_k\}$ where t_1 is some element of T. Next we obtain a transcendence basis $S_2 = \{t_1, t_2, s_3, \ldots, s_k\}$ where t_1 and t_2 are some elements of T. Finally we obtain a transcendence basis $S_k = \{t_1, t_2, t_3, \ldots, t_k\}$ where each t_i is an element of T. But T has more than k elements, so there is an element t_{k+1} of T that is not in S_k. But then, by Lemma 6.1.9, t_{k+1} is transcendental over $\mathbf{F}(S_k)$, contradicting the fact that S_k is a transcendence basis for $\mathbf{E}$ over $\mathbf{F}$. □

With this theorem in hand, we may make the following important definition.

Definition 6.1.16. Let $\mathbf{E}$ be an extension of $\mathbf{F}$. Then tr-deg($\mathbf{E}/\mathbf{F}$), the *transcendence degree* of $\mathbf{E}$ over $\mathbf{F}$, is equal to the number of elements in any transcendence basis S of $\mathbf{E}$ over $\mathbf{F}$, tr-deg$(\mathbf{E}/\mathbf{F}) \in \{0, 1, 2, \ldots\} \cup \{\infty\}$. ◇

Remark 6.1.17. If $\mathbf{E} = \mathbf{F}(X_1, X_2, \ldots)$ then $\{X_1, X_2, \ldots\}$ is a transcendence basis for $\mathbf{E}$ over $\mathbf{F}$ and so tr-deg($\mathbf{E}/\mathbf{F}$) is the number of elements of this set. Thus we see that $\mathbf{F}(X_1, X_2, \ldots, X_k)$ is not isomorphic to $\mathbf{F}(X_1, X_2, \ldots, X_\ell)$ if $k \neq \ell$. ◇

Theorem 6.1.18. *Let* $\mathbf{B}$ *be an extension of* $\mathbf{F}$ *with transcendence basis* S *and let* $\mathbf{E}$ *be an extension of* $\mathbf{B}$ *with transcendence basis* T. *Then* $\mathbf{E}$ *is an extension of* $\mathbf{F}$ *with transcendence basis* $S \cup T$. *In particular,*

$$\text{tr-deg}(\mathbf{E}/\mathbf{F}) = \text{tr-deg}(\mathbf{E}/\mathbf{B}) + \text{tr-deg}(\mathbf{B}/\mathbf{F}).$$

Proof. Let $\mathbf{B}_0 = \mathbf{F}(S)$, so $\mathbf{B}_0 \subseteq \mathbf{B}$. Let $\mathbf{E}_0 = \mathbf{B}_0(T)$ and let $\mathbf{E}_1 = \mathbf{B}(T)$, so $\mathbf{E}_0 \subseteq \mathbf{E}_1 \subseteq \mathbf{E}$.

Since T is algebraically independent over $\mathbf{B}$, it is certainly algebraically independent over $\mathbf{B}_0$. This implies that $S \cup T$ is algebraically independent over $\mathbf{B}$. For suppose there were a nonzero polynomial $f(X_1, \ldots, Y_1, \ldots)$ with coefficients in $\mathbf{B}$ with $f(s_1, \ldots, t_1, \ldots) = 0$. By the algebraic independence of T none of the variables Y_i can appear, so $f(s_1, \ldots, Y_1, \ldots) = g(X_1, \ldots)$ for some nonzero polynomial $g(X_1, \ldots)$ with coefficients in $\mathbf{B}$. But then $g(s_1, \ldots) = 0$, contradicting the algebraic independence of S.

Now $\mathbf{F}(S \cup T) = (\mathbf{F}(S))(T) = \mathbf{B}_0(T) = \mathbf{E}_0$. Let $\alpha \in \mathbf{E}_1 = \mathbf{B}(T)$. By definition, $\alpha = p(T_1, \ldots)/q(T_1, \ldots)$ for some polynomials with coefficients in $\mathbf{B}$. Let $\mathbf{B}_1$ be the extension of $\mathbf{B}_0$ obtained by adjoining the finitely many coefficients of these polynomials, all of which are algebraic over $\mathbf{B}_0 = \mathbf{F}(T)$ (as $\mathbf{B}$ is algebraic over $\mathbf{B}_0$, since S is a transcendence basis). Then $\alpha \in \mathbf{B}_1(T)$, which is a finite extension of $\mathbf{E}_0 = \mathbf{B}_0(T)$, and so α is algebraic over $\mathbf{E}_0$. Since $\alpha \in \mathbf{E}_1$ was arbitrary, this shows that $\mathbf{E}_1$ is algebraic over $\mathbf{E}_0$.

By the definition of a transcendence basis, $\mathbf{E}$ is algebraic over $\mathbf{E}_1$, and we have just shown that $\mathbf{E}_1$ is algebraic over $\mathbf{E}_0$, so $\mathbf{E}$ is algebraic over $\mathbf{E}_0 = \mathbf{F}(S \cup T)$. Hence $S \cup T$ satisfies both conditions for a transcendence basis of $\mathbf{E}$ over $\mathbf{F}$.

Note that S and T are disjoint as $S \subset \mathbf{B} - \mathbf{F}$ and $T \subset \mathbf{E} - \mathbf{B}$, so the second part of the theorem immediately follows. □

Corollary 6.1.19. *Let* $\mathbf{E}$ *be any extension of* $\mathbf{F}$. *Then there is a field* $\mathbf{B}$ *with* $\mathbf{F} \subseteq \mathbf{B} \subseteq \mathbf{E}$ *and with* $\mathbf{B}$ *purely transcendental over* $\mathbf{F}$ *and* $\mathbf{E}$ *algebraic over* $\mathbf{B}$.

Proof. Let S be a transcendence basis for $\mathbf{E}$ over $\mathbf{F}$ and let $\mathbf{B} = \mathbf{F}(S)$. □

Remark 6.1.20. Comparing Corollary 6.1.19 with Lemma 6.1.3, we see that in the situation of Corollary 6.1.19, the field $\mathbf{B}$ is not unique. ◇

Let $\mathbf{B}$ and $\mathbf{D}$ be two extensions of $\mathbf{F}$, with $\mathbf{B}$ algebraic and $\mathbf{D}$ purely transcendental. Then Lemma 6.1.5 shows that $\mathbf{B}$ and $\mathbf{D}$ are disjoint extensions of $\mathbf{F}$. In fact, more is true.

Theorem 6.1.21. *Let* $\mathbf{B}$ *and* $\mathbf{D}$ *be subfields of* $\mathbf{E}$ *containing* $\mathbf{F}$. *Suppose that* $\mathbf{B}$ *is an algebraic extension of* $\mathbf{F}$ *and that* $\mathbf{D}$ *is a purely transcendental extension of* $\mathbf{F}$. *Then* $\mathbf{B}$ *and* $\mathbf{D}$ *are linearly disjoint extensions of* $\mathbf{F}$.

Proof. We first prove this in the case that $k = \text{tr-deg}(\mathbf{D}/\mathbf{F})$ is finite. In this case let $\mathbf{D} = \mathbf{F}(X_1, \ldots, X_k)$. We prove the theorem by induction on k.

To begin the induction, we let $k = 1$. Let $\{Y_1, \ldots, Y_\ell\}$ be a $\mathbf{F}$-linearly independent set of elements in $\mathbf{F}(X_1)$, and suppose $\sum_{j=1}^{\ell} \beta_j Y_j = 0$ with each $\beta_j \in \mathbf{F}$, and not all of them 0. Clearing denominators, we may assume that each Y_j is a polynomial in X_1. Performing a nonsingular change of basis (reordering and adding multiples of each of these polynomials to the others), if necessary, we may obtain a (necessarily $\mathbf{F}$-linearly independent) set of nonzero polynomials $\{Y_1', \ldots, Y_\ell'\}$ with $\deg(Y_1') > \deg(Y_2') > \cdots > \deg(Y_\ell')$, and $0 = \sum_{j=1}^{\ell} \beta_j Y_j = \sum_{j=1}^{\ell} \beta_j' Y_j'$ for uniquely determined elements $\beta_j' \in \mathbf{F}$, not all of which are 0. But this is a nontrivial polynomial relationship $f(X_1) = 0$ with f a polynomial with coefficients in $\mathbf{B}$, and thus we see that X_1 is algebraic over $\mathbf{B}$. But $\mathbf{B}$ is algebraic over $\mathbf{F}$, so that implies that X_1 is algebraic over $\mathbf{F}$, a contradiction.

Now suppose the theorem is true for $k-1$ and consider $\mathbf{F}(X_1, \ldots, X_k)$. Let $\{Y_1, \ldots, Y_\ell\}$ be a $\mathbf{F}$-linearly independent set of elements in $\mathbf{F}(X_1, \ldots, X_k)$, and suppose $\sum_{j=1}^{\ell} \beta_j Y_j = 0$ with $\beta_j \in \mathbf{F}$ for each j, not all of which are 0. By the truth of the $k-1$ case, we cannot have $\{Y_1, \ldots, Y_\ell\} \subseteq \mathbf{F}(X_1, \ldots, X_{k-1})$. Then, proceeding as in the $k = 1$ case, we obtain $f(X_k) = 0$

for some nontrivial polynomial f with coefficients in $\mathbf{B}(X_1, \ldots, X_{k-1})$. Then X_k is algebraic over $\mathbf{B}(X_1, \ldots, X_{k-1})$, and $\mathbf{B}(X_1, \ldots, X_{k-1})$ is algebraic over $\mathbf{F}(X_1, \ldots, X_{k-1})$, so X_k is algebraic over $\mathbf{F}(X_1, \ldots, X_{k-1})$, a contradiction.

Finally, a linear dependence of a set of elements of $\mathbf{F}(X_1, X_2, \ldots)$ must be a linear dependence among elements of $\mathbf{F}(X_1, \ldots, X_k)$ for some k, so the finite case implies the general case. □

Theorem 6.1.22. *Let* **B** *and* **D** *be subfields of* **E** *containing* **F**. *Suppose that* **B** *and* **D** *are linearly disjoint extensions of* **F**, *and that* **D** *is a purely transcendental extension of* **F**. *Then* **BD** *is a purely transcendental extension of* **B** *and* tr-deg(**BD**/**B**) = tr-deg(**D**/**F**).

Proof. Let $\mathbf{D} = \mathbf{F}(X_1, X_2, \ldots)$. Clearly $\mathbf{BD} = \mathbf{B}(X_1, X_2, \ldots)$ so to prove the theorem we need only show that $\{X_1, X_2, \ldots\}$ is algebraically independent over **B**. Suppose the elements of this set satisfy some polynomial relationship. This can only involve finitely many variables, so we may assume it is $f(X_1, \ldots, X_k) = 0$ for some k, where f is a polynomial with coefficients in **B**. Write this polynomial as a sum of terms, each of which is a distinct monomial in $\{X_1, \ldots, X_k\}$ times a coefficient in **B**. Now the distinct monomials in $\{X_1, \ldots, X_k\}$ are linearly independent over **F**, as $\{X_1, \ldots, X_k\}$ is algebraically independent over **F**. By the definition of linear disjointness, each of the coefficients in f must be 0, and so $f(X_1, \ldots, X_k)$ is the 0 polynomial, as required. □

Here are a pair of examples that illustrate many of the concepts and results that we have developed in this section.

Example 6.1.23. Let $\mathbf{E} = \mathbf{C}(X)$. Note that $\{X\}$, $\{X^2\}$, and $\{X^3\}$ are all transcendence bases for **E** over **C**. Set $U = X^2$ and $V = X^3$. Let $\mathbf{B}_1 = \mathbf{C}(U)$ and $\mathbf{B}_2 = \mathbf{C}(V)$. Note that **E** is a Galois extension of $\mathbf{B}_1$ of degree 2 with Galois group $\mathrm{Gal}(\mathbf{E}/\mathbf{B}_1)$ generated by the automorphism of **E** determined by $\sigma_1(X) = -X$. Also note that **E** is a Galois extension of $\mathbf{B}_2$ of degree 3 with Galois group $\mathrm{Gal}(\mathbf{E}/\mathbf{B}_2)$ generated by the automorphism of **E** determined by $\sigma_2(X) = \omega X$. Clearly $\mathbf{B}_1\mathbf{B}_2 = \mathbf{E}$ as $X = V/U \in \mathbf{B}_1\mathbf{B}_2$. Note that σ_1 and σ_2 generate a group G of automorphisms of **E** of order 6. Indeed this is a cyclic group with generator $\sigma_0 = \sigma_1\sigma_2^{-1}$ where $\sigma_0(X) = \exp(2\pi i/6)X$. Then $\mathbf{B}_0 = \mathrm{Fix}(G) = \mathbf{B}_1 \cap \mathbf{B}_2 = \mathbf{C}(W)$ where $W = X^6$. In this case we have

$$\text{tr-deg}(\mathbf{B}_1/\mathbf{B}_0) = \text{tr-deg}(\mathbf{B}_2/\mathbf{B}_0) = 0, \quad \text{tr-deg}(\mathbf{E}/\mathbf{B}_1) = \text{tr-deg}(\mathbf{E}/\mathbf{B}_2) = 0.$$

Observe that $\mathbf{E} = \mathbf{B}_1(V)$ and V has minimal polynomial $m_V(Z) = Z^2 - U^3 \in \mathbf{B}_1[Z]$, and similarly that $\mathbf{E} = \mathbf{B}_2(U)$ and U has minimal polynomial $m_U(Z) = Z^3 - V^2 \in \mathbf{B}_2[Z]$.

Finally, observe that in this case $\mathbf{B}_1$ and $\mathbf{B}_2$ are not disjoint extensions of $\mathbf{C}$. ◇

Example 6.1.24. Let $\mathbf{E} = \mathbf{C}(X)$. Note that $\{X\}$, $\{X^2\}$, and $\{(X+1)^2\}$ are all transcendence bases for $\mathbf{E}$ over $\mathbf{C}$. Set $U = X^2$ and $V = (X+1)^2$. Let $\mathbf{B}_1 = \mathbf{C}(U)$ and $\mathbf{B}_2 = \mathbf{C}(V)$. Note that $\mathbf{E}$ is a Galois extension of $\mathbf{B}_1$ of degree 2 with Galois group $\mathrm{Gal}(\mathbf{E}/\mathbf{B}_1)$ generated by the automorphism of $\mathbf{E}$ determined by $\sigma_1(X) = -X$. Also note that $\mathbf{E}$ is a Galois extension of $\mathbf{B}_2$ of degree 2 with Galois group $\mathrm{Gal}(\mathbf{E}/\mathbf{B}_2)$ generated by the automorphism of $\mathbf{E}$ determined by $\sigma_2(X) = -X - 2$. Clearly $\mathbf{B}_1\mathbf{B}_2 = \mathbf{E}$ as $X = (V - U - 1)/2 \in \mathbf{B}_1\mathbf{B}_2$. The group G of automorphisms of $\mathbf{E}$ generated by σ_1 and σ_2 contains the element $\sigma_0 = \sigma_2\sigma_1$ and $\sigma_0(X) = X + 2$. Hence $\mathbf{B}_0 = \mathrm{Fix}(G) = \mathbf{B}_1 \cap \mathbf{B}_2 = \mathbf{C}$ as for no nonconstant rational function $f(X)$ do we have $f(X) = f(X+2)$. In this case we have

$$\text{tr-deg}(\mathbf{B}_1/\mathbf{B}_0) = \text{tr-deg}(\mathbf{B}_2/\mathbf{B}_0) = 1, \quad \text{tr-deg}(\mathbf{E}/\mathbf{B}_1) = \text{tr-deg}(\mathbf{E}/\mathbf{B}_2) = 0.$$

Observe that $\mathbf{E} = \mathbf{B}_1(V)$ and V has minimal polynomial $m_V(Z) = Z^2 - 2(U+1)Z + (U-1)^2 \in \mathbf{B}_1[Z]$, and similarly that $\mathbf{E} = \mathbf{B}_2(U)$ and U has minimal polynomial $m_U(Z) = Z^2 - 2(V+1)Z + (V-1)^2 \in \mathbf{B}_2[Z]$.

Finally, observe that in this case $\mathbf{B}_1$ and $\mathbf{B}_2$ are disjoint but not linearly disjoint extensions of $\mathbf{C}$. ◇

Let us now give an application of the theory we have just developed to the theory of symmetric functions. Compare the proof of Lemma 3.1.12 and Remark 3.1.13. We adopt the notation of Section 3.1 here.

Theorem 6.1.25. *Let* $\mathbf{D}$ *be an arbitrary field, let* $\mathbf{E} = \mathbf{D}(X_1, \ldots, X_d)$ *be the field of rational functions in the variables* $X_1, \ldots, X_d$ *and let* $\mathbf{F} \subseteq \mathbf{E}$ *be the field of symmetric functions in* $X_1, \ldots, X_d$*. Let* $s_1, \ldots, s_d$ *be the elementary symmetric polynomials. Then* $S = \{s_1, \ldots, s_d\}$ *is algebraically independent. Furthermore,* $\mathbf{F}$ *is a purely transcendental extension of* $\mathbf{D}$ *and* S *is a transcendence basis for* $\mathbf{F}$ *over* $\mathbf{D}$.

Proof. By definition, $\mathbf{F}$ is the fixed field of the symmetric group S_d acting on $\mathbf{E}$ by permuting the variables, so $\mathbf{E}$ is an algebraic extension of $\mathbf{F}$; indeed, $\mathbf{E}$ is a Galois extension of $\mathbf{F}$ with Galois group $\mathrm{Gal}(\mathbf{E}/\mathbf{F}) = S_d$. We showed in Lemma 3.1.4 that $\mathbf{F} = \mathbf{D}(s_1, \ldots, s_d)$. Thus it only remains to show that $S = \{s_1, \ldots, s_d\}$ is algebraically independent. Suppose not. By Corollary 6.1.13 (3), S contains a proper subset R of cardinality $c < d$ that is a transcendence basis for $\mathbf{F}$ over $\mathbf{D}$. But then

$$c = 0 + c = \text{tr-deg}(\mathbf{E}/\mathbf{F}) + \text{tr-deg}(\mathbf{F}/\mathbf{D}) = \text{tr-deg}(\mathbf{E}/\mathbf{D}) = d,$$

a contradiction. □

6.2 Simple Transcendental Extensions

Definition 6.2.1. A *simple transcendental extension* of the field $\mathbf{F}$ is a field $\mathbf{E}$ obtained by adjoining a single transcendental element X to $\mathbf{F}$, $\mathbf{E} = \mathbf{F}(X)$. $\diamond$

Throughout this section we let $\mathbf{E} = \mathbf{F}(X)$.

A simple transcendental extension $\mathbf{E}$ is easy to describe: it is a purely transcendental extension of transcendence degree 1. But there are interesting questions we can ask about $\mathbf{E}$.

Lemma 6.2.2. *Let $Y = p(X)/q(X) \in \mathbf{E}$, with $p(X)$ and $q(X)$ relatively prime polynomials in $\mathbf{F}[X]$ of degrees d_1 and d_2 respectively, and let $\mathbf{B} = \mathbf{F}(Y)$. Suppose that $d > 0$. Let $d = \max(d_1, d_2)$. Then $(\mathbf{E}/\mathbf{B}) = d$.*

Proof. (1) Since $\mathbf{E} = \mathbf{F}(X)$, certainly $\mathbf{E} = \mathbf{B}(X)$. We prove the lemma by showing that X satisfies an irreducible polynomial of degree d over $\mathbf{B}$.

Evidently X is a root of the polynomial $f(Y, Z) = Yq(Z) - p(Z) \in \mathbf{B}[Z]$. Thus we need only show that this polynomial is irreducible.

It is easy to see that $f(Y, Z)$ is irreducible in $(\mathbf{F}[Z])[Y]$: $f(Y, Z)$ is linear in Y so any factorization would have to be of the form $f(Y, Z) = a(Z)(b(Z)Y + c(Z))$. Then $q(Z) = a(Z)b(Z)$ and $-p(Z) = a(Z)c(Z)$, and hence $a(Z)$ is a common factor of $p(Z)$ and $q(Z)$. But $p(Z)$ and $q(Z)$ are assumed relatively prime so $a(Z)$ is a unit, i.e., a constant polynomial.

Now $(\mathbf{F}[Z])[Y] = \mathbf{F}[Z, Y] = \mathbf{F}[Y, Z] = (\mathbf{F}[Z])[Y]$ so $f(Y, Z)$ is irreducible in $(\mathbf{F}[Z])[Y]$. We need to show it is irreducible in $\mathbf{B}[Y] = (\mathbf{F}(Z))[Y]$. But that is an immediate consequence of Gauss's Lemma, Lemma 4.1.4. (We proved Gauss's Lemma for $\mathbb{Z}$ and $\mathbf{Q}$, but the exact same proof works for any unique factorization domain R and its quotient field.) $\square$

We now determine the automorphism group of a simple transcendental extension. Recall that $GL_2(\mathbf{F})$ denotes the group of invertible 2×2 matrices with coefficients in the field $\mathbf{F}$. It is easy to see that its center $Z = Z(GL_2(\mathbf{F}))$ is the subgroup consisting of all nonzero scalar multiples of the identity matrix (and so is naturally isomorphic to the multiplicative group $\mathbf{F}^*$). As usual, we let $PGL_2(\mathbf{F})$ be the quotient $PGL_2(\mathbf{F}) = GL_2(\mathbf{F})/Z$. $PGL_2(\mathbf{F})$ is known as the projective linear group of degree 2 over $\mathbf{F}$.

Theorem 6.2.3. *Let $\mathrm{Aut}_{\mathbf{F}}(\mathbf{E})$ be the group of automorphisms of $\mathbf{E}$ that fix $\mathbf{F}$.*

(1) For a matrix $M = \begin{bmatrix} a & b \\ c & d \end{bmatrix} \in GL_2(\mathbf{F})$, let $\varphi_M : \mathbf{E} \to \mathbf{E}$ be defined by

$$\varphi_M(f(X)) = f\left(\frac{aX + b}{cX + d}\right).$$

Then $\varphi_M \in \mathrm{Aut}_{\mathbf{F}}(\mathbf{E})$. *Furthermore, any* $\psi \in \mathrm{Aut}_{\mathbf{F}}(\mathbf{E})$ *is* $\psi = \varphi_M$ *for some* $M \in GL_2(\mathbf{F})$.

(2) The map $\tilde{\Phi} : GL_2(\mathbf{F}) \to \mathrm{Aut}_{\mathbf{F}}(\mathbf{E})$ *given by* $\tilde{\Phi}(M) = \varphi_{M^{-1}}$ *induces an isomorphism*

$$\Phi : PGL_2(\mathbf{F}) \to \mathrm{Aut}_{\mathbf{F}}(\mathbf{E}).$$

Proof. (1) Observe that the polynomials $aX + b$ and $cX + d$ are relatively prime if and only if $\det(M) = ad - bc \neq 0$. Assuming that, let $Y = \psi(X) = (aX + b)/(cX + d)$. Since $\mathbf{E} = \mathbf{F}[X]$, ψ extends to a unique homomorphism from $\mathbf{E}$ to itself, and that homomorphism is φ_M. By Lemma 6.2.2, $\mathbf{F}(Y) = \mathbf{E}$, so φ_M is surjective, and it is clearly injective, so it is an automorphism. Furthermore, again by Lemma 6.2.2, if ψ is any automorphism of $\mathbf{E}$, then $Y = \psi(X)$ must be of the above form, i.e., $Y = \varphi_M(X)$ for some invertible matrix M, in which case $\psi = \varphi_M$.

(2) By (1), $\tilde{\Phi}$ is a surjection of sets. We now check that $\tilde{\Phi}$ is a homomorphism, i.e., that $\tilde{\Phi}(M_1M_2) = \tilde{\Phi}(M_1)\tilde{\Phi}(M_2)$. This is a direct computation, using the fact that $\psi \in \mathrm{Aut}_{\mathbf{F}}(\mathbf{E})$ is determined by $\psi(X)$:

Let $M_i^{-1} = \begin{bmatrix} a_i & b_i \\ c_i & d_i \end{bmatrix}$ for $i = 1, 2$. First note that

$$(M_1M_2)^{-1} = M_2^{-1}M_1^{-1} = \begin{bmatrix} a_2 & b_2 \\ c_2 & d_2 \end{bmatrix}\begin{bmatrix} a_1 & b_1 \\ c_1 & d_1 \end{bmatrix} = \begin{bmatrix} a_2a_1 + b_2c_1 & a_2b_1 + b_2d_1 \\ c_2a_1 + d_2c_1 & c_1b_1 + d_2d_1 \end{bmatrix}.$$

Then, noting that for any automorphism ψ and any rational function $f(X)$, the value of ψ on $f(X) \in \mathbf{E}$ is $f(\psi(X))$, so that the order of composition is *reversed*,

$$\begin{aligned}
(\tilde{\Phi}(M_1)\tilde{\Phi}(M_2))(X) &= (\varphi_{M_1^{-1}}\varphi_{M_2^{-1}})(X) = \varphi_{M_2^{-1}}(\varphi_{M_1^{-1}}(X)) \\
&= \varphi_{M_2^{-1}}\left(\frac{a_1X + b_1}{c_1X + d_1}\right) \\
&= \frac{a_2\frac{a_1X+b_1}{c_1X+d_1} + b_2}{c_2\frac{a_1X+b_1}{c_1X+d_1} + d_2} \\
&= \frac{a_2(a_1X + b_1) + b_2(c_1X + d_1)}{c_2(a_1X + b_1) + d_2(c_1X + d_1)} \\
&= \frac{(a_2a_1 + b_2c_1)X + (a_2b_1 + b_2d_1)}{(c_2a_1 + d_2c_1)X + (c_2b_1 + d_2d_1)} \\
&= \varphi_{M_2^{-1}M_1^{-1}}(X) = \varphi_{(M_1M_2)^{-1}}(X) = \tilde{\Phi}(M_1M_2)(X)
\end{aligned}$$

so $\tilde{\Phi}$ is an epimorphism of groups, and hence induces an isomorphism of groups $\Phi : GL_2(\mathbf{F})/\mathrm{Ker}(\Phi) \longrightarrow \mathrm{Aut}_{\mathbf{F}}(\mathbf{E})$. Let M be a matrix with $M^{-1} =$

$\begin{bmatrix} a & b \\ c & d \end{bmatrix}$. $\tilde{\Phi}(M) = \varphi_{M^{-1}}$ is the identity on $\mathbf{E}$ if and only if $\varphi_{M^{-1}}(X) = X$, i.e., if and only if $(aX+b)/(cX+d) = X$, which occurs if and only if $a = d \neq 0$ and $b = c = 0$, i.e., if and only if M^{-1} is a nonzero scalar matrix, and this is true if and only if M is a nonzero scalar matrix. □

Remark 6.2.4. (1) If you are familiar with fractional linear transformations, you will have recognized them in the statement of Theorem 6.2.3. But you may have been puzzled by the appearance of the matrix inverse in that statement. The group $PGL_2(\mathbf{R})$ acts on the upper half-plane $\mathcal{H} = \{z \in \mathbf{C} \mid \mathrm{Im}(z) > 0\}$ by fractional linear transformations (or Möbius transformations), $\rho_M(z) = (az+b)/(cz+d)$, where the matrix M is as in the statement of Theorem 6.2.3. In that action we have $\rho_{M_1M_2} = \rho_{M_1}\rho_{M_2}$, where on the right-hand side of this equation multiplication is simply composition of maps from $\mathcal{H}$ to itself. But that action is *not* the action of $PGL_2(\mathbf{F})$ on $\mathbf{E}$; as we remarked in the proof of Theorem 6.2.3, in this action the order of composition is reversed. (The automorphism ψ does not act on the rational function $f(X)$ by acting on the value of the function f on the variable X, but rather by acting on the variable X which is the argument to the function f, and this accounts for the reversal of order.)

(2) Evidently ψ defined by $\psi(X) = aX + b$, $a \neq 0$, defines an automorphism of $\mathbf{E}$ fixing $\mathbf{F}$, as does $\psi(X) = 1/X$. Thus $\mathrm{Aut}_{\mathbf{F}}(\mathbf{E})$ must contain the group generated by compositions of these automorphisms as a subgroup. It is easy to check that this subgroup is precisely the group of all fractional linear transformations. Thus, Theorem 6.2.3 says that all automorphisms of $\mathbf{E}$ fixing $\mathbf{F}$ are obtained in this way.

Theorem 6.2.5 (Lüroth). *Let* $\mathbf{F} \subset \mathbf{B} \subseteq \mathbf{E}$. *Then* $\mathbf{B}$ *is a simple transcendental extension of* $\mathbf{F}$, *i.e.,* $\mathbf{B} = \mathbf{F}(Y)$ *for some* $Y \in \mathbf{E}$.

Proof. Let $W \in \mathbf{B}$, $W \notin \mathbf{F}$. Then $W = a(X)/b(X)$ for some relatively prime polynomials $a(X)$ and $b(X)$ not both of which are constant, so, by Lemma 6.2.2, the element X of $\mathbf{E}$ is algebraic over $\mathbf{F}(W)$. Since $\mathbf{F}(W) \subseteq \mathbf{B}$, X is algebraic over $\mathbf{B}$. (This fact is also a consequence of Lemma 6.1.5 and Theorem 6.1.17, but is just as easy to prove directly.) Let X have minimum polynomial $m_X(Z)$ over $\mathbf{B}$,

$$m_X(Z) = Z^n + Y_{n-1}Z^{n-1} + \cdots + Y_0 \quad \text{with each } Y_i \in \mathbf{B}.$$

Then $(\mathbf{E}/\mathbf{B}) = n$. Since X is not algebraic over $\mathbf{F}$ some coefficient $Y = Y_{i_0}$ of $m_X(Z)$ is not in $\mathbf{F}$. We claim that $\mathbf{B} = \mathbf{F}(Y)$.

Let $\mathbf{B}_0 = \mathbf{F}(Y)$. Since $Y \in \mathbf{B}$, $\mathbf{B}_0 \subseteq \mathbf{B}$. Write Y as a rational function $Y = p(X)/q(X)$ in lowest terms. Then, by Lemma 6.1.2, we have that $(\mathbf{E}/\mathbf{B}_0) = d$

where d is the maximum of the degrees of $p(X)$ and $q(X)$. Then

$$d = (\mathbf{E}/\mathbf{B}_0) = (\mathbf{E}/\mathbf{B})(\mathbf{B}/\mathbf{B}_0) = n(\mathbf{B}/\mathbf{B}_0)$$

so if we show that $n = d$, then $\mathbf{B} = \mathbf{B}_0 = \mathbf{F}(Y)$ as required.

Now X is a root of $f(Z) = Yq(Z) - p(Z)$, so $m_X(Z)$ divides $f(Z)$ in $\mathbf{B}[Z]$. Note that $f(Z)$ is not the 0 polynomial as if it were, we would have $Y = p(Z)/q(Z)$ in $\mathbf{B}(Z)$, which is impossible as $Y \in \mathbf{B}$ and $p(Z)/q(Z) \notin \mathbf{B}$. Let $f(Z) = m_X(Z)g(Z)$. We may "clear denominators" in $m_X(Z)$ to obtain a polynomial

$$\tilde{m}(X, Z) = c_n(X)Z^n + c_{n-1}(X)Z^{n-1} + \cdots + c_0(X)$$

with $\{c_n(X), \ldots, c_0(X)\}$ a relatively prime set of polynomials in $\mathbf{F}[X]$, so that $\tilde{m}(X, Z)$ is a primitive polynomial of degree n when regarded as a polynomial in Z. By the definition of Y, we see that $p(X)$ divides $c_{i_0}(X)$ and $q(X)$ divides $c_n(X)$, so, if $\tilde{m}(X, Z)$ is regarded as a polynomial in X, its degree is at least d. For clarity we will write $\deg_Z(\tilde{m}(X, Z)) = n$ and $\deg_X(\tilde{m}(X, Z)) \geq d$.

Now $f(Z) = m_X(Z)g(Z)$, i.e.,

$$Yq(Z) - p(Z) = m_X(Z)g(Z).$$

Multiplying by $c_n(X)$, we obtain the equation

$$c_n(X)(Yq(Z) - p(Z)) = \tilde{m}(X, Z)g(Z),$$

a polynomial identity in $\mathbf{B}[X, Z]$. Substituting $Y = p(X)/q(X)$ and multiplying by $q(X)$, we obtain the equation

$$c_n(X)(p(X)q(Z) - q(X)p(Z)) = \tilde{m}(X, Z)g(Z)q(X),$$

a polynomial identity in $\mathbf{F}[X, Z]$. Now $c_n(X)$ and $\tilde{m}(X, Z)$ are relatively prime, so $\tilde{m}(X, Z)$ must divide $p(X)q(Z) - q(X)p(Z)$, i.e.,

$$r(X, Z) = (p(X)q(Z) - q(X)p(Z)) = \tilde{m}(X, Z)s(X, Z)$$

for some polynomial $s(X, Z) \in \mathbf{F}[X, Z]$.

Now $\deg_X(r(X, Z)) = \deg_Z(r(X, Z)) \leq d$. But we observed above that $\deg_X(\tilde{m}(X, Z)) \geq d$, so we must have $\deg_X(\tilde{m}(X, Z)) = d$. Then $\deg_X(s(X, Z)) = 0$, so $s(X, Z) = t(Z)$ for some polynomial $t(Z) \in \mathbf{F}[Z]$. As we have observed, $\tilde{m}(X, Z)$ is primitive when regarded as a polynomial in Z, i.e., the coefficients of the different powers of Z are relatively prime polynomials in X. But then the same is true for $r(X, Z) = \tilde{m}(X, Z)t(Z)$. But by symmetry, $r(X, Z)$ is then also primitive when regarded as a polynomial in X, so the coefficients of the different powers of X are relatively prime polynomials in Z, and hence $t(Z)$ is a constant polynomial. Thus $r(X, Z) = a\tilde{m}(X, Z)$ for some $a \in \mathbf{F}$. We then conclude that

$$n = \deg_Z(\tilde{m}(X, Z)) = \deg_Z(r(X, Z)) = \deg_X(r(X, Z)) = d,$$

as required. □

6.3 Plane Curves

Throughout this section, except when explicitly stated otherwise, we let $\mathbf{F} = \mathbf{C}$, *the field of complex numbers.*

Hitherto we have only considered purely transcendental extensions. In this section we investigate extensions that are not (necessarily) purely transcendental. Our examples are chosen from elementary algebraic geometry. Clearly, they are just the tip of an iceberg, but it is beyond our scope here to investigate this subject more broadly or deeply.

Definition 6.3.1. Let $f(X, Y) \in \mathbf{F}[X, Y]$ be an irreducible polynomial. Let $V(f) = \{(z_1, z_2) \in \mathbf{F}^2 \mid f(z_1, z_2) = 0\}$. Then $C = V(f)$ is the *plane curve* associated to $f(X, Y)$. The quotient field of $\mathbf{F}[X, Y]/\langle f(X, Y)\rangle$ is the *function field* of C. ◊

Remark 6.3.2. (1) Strictly speaking, we should call C as in Definition 6.3.1 an irreducible affine plane curve, but we adopt the shorter language.

(2) In Definition 6.3.1 we are using some commutative ring theory. In this definition, $\langle f(X, Y)\rangle$ denotes the ideal in $\mathbf{F}[X, Y]$ generated by the polynomial $f(X, Y)$. Since $\mathbf{F}[X, Y]$ is a unique factorization domain, and we are assuming $f(X, Y)$ is irreducible, then $f(X, Y)$ is prime and $\langle f(X, Y)\rangle$ is a prime ideal. Thus $\mathbf{F}[X, Y]/\langle f(X, Y)\rangle$ is an integral domain and we may indeed take its quotient field.

(3) We are restricting our attention to $\mathbf{C}[X, Y]$, rather than, for example, $\mathbf{Q}[X, Y]$, for the following reasons: $f(X, Y) = X^2 + Y^2 + 1$ is an irreducible polynomial in $\mathbf{Q}[X, Y]$, but $V(f) = \emptyset$ when we consider it over $\mathbf{Q}$, and we don't want to regard that as a curve. Also, if $f(X, Y)$ and $g(X, Y)$ are regarded as polynomials over $\mathbf{C}$, then $V(f) = V(g)$ if and only if $f(X, Y)$ and $g(X, Y)$ are constant multiples of each other, but this is not true over $\mathbf{Q}$, as we see from the examples $f(X, Y) = X^2 + Y^4 + 1$ and $g(X, Y) = X^4 + Y^2 + 1$. However, there are some (very!) interesting questions about $V(f)$

when $f(X, Y)$ is regarded as a polynomial over $\mathbf{Q}$, or over some field intermediate between $\mathbf{Q}$ and $\mathbf{C}$, and we will return to this point in Remark 6.3.13 below.

(4) Much of the discussion here is valid with $\mathbf{C}$ replaced by an algebraically closed field of arbitrary characteristic. We are restricting our attention to characteristic 0 for simplicity. ◇

Notation 6.3.3. In the situation of Definition 6.3.1, we write $\mathbf{C}(x, y)$ for the quotient field of the curve C, where, in the natural map from $\mathbf{C}[X, Y]$ to this quotient field, the image of X is x and the image of Y is y. ◇

Remark 6.3.4. Our notation $\mathbf{C}(x, y)$ is perfectly consistent with our previous notation. $\mathbf{C}(x, y)$ is an extension field of $\mathbf{C}$ obtained by adjoining the elements x and y. Also, this field $\mathbf{C}(x, y)$ is an extension of $\mathbf{C}$ of transcendence degree 1. If we assume that $f(X, Y)$ involves both X and Y (i.e, that it is not a polynomial in X alone, or a polynomial in Y alone), then each of x and y is transcendental over $\mathbf{C}$; moreover, in this case, $\{x\}$ and $\{y\}$ are each transcendence bases for $\mathbf{C}(x, y)$ over $\mathbf{C}$. Furthermore, in this case, the natural maps from $\mathbf{C}(X)$ to $\mathbf{C}(x)$, and from $\mathbf{C}(Y)$ to $\mathbf{C}(y)$, are field isomorphisms. ◇

Definition 6.3.5. In the situation of Definition 6.3.1, the curve $V(f)$ is a *rational curve* if its function field $\mathbf{C}(x, y)$ is rational. ◇

We now present some interesting examples of plane curves that are or are not rational.

Theorem 6.3.6. *Fix a positive integer n and consider the Fermat curve C_n given by $X^n + Y^n = 1$.*

(1) For $n = 2$, C_n is rational.

(2) For $n \geq 3$, C_n is not rational.

Proof. (1) We begin by recalling the polynomial identity $(T^2-1)^2+(2T)^2 = (T^2+1)^2$ which immediately yields the identity among rational functions

$$\left(\frac{T^2-1}{T^2+1}\right)^2 + \left(\frac{2T}{T^2+1}\right)^2 = 1.$$

Guided by this identity, we construct an embedding $\varphi : \mathbf{F}(x, y) \longrightarrow \mathbf{F}(t)$ defined by

$$\varphi(x) = (t^2-1)/(t^2+1) \qquad \text{and} \qquad \varphi(y) = 2t/(t^2+1).$$

We thus see that $\mathbf{F}(x, y)$ is isomorphic to a subfield of $\mathbf{F}(t)$. In fact, this subfield is $\mathbf{F}(t)$ itself. To show this, we need only show that φ is an epimorphism. To see this, observe that $\varphi(x)/\varphi(y) = (t^2 - 1)/(2t)$. Cross-multiply, regard this as a quadratic equation in t, and solve by the quadratic formula, using the relation $\varphi(x)^2 + \varphi(y)^2 = 1$, to obtain $t = (\varphi(x)+1)/\varphi(y)$. (Of course, once we know that $\mathbf{F}(x, y)$ is isomorphic to a subfield of $\mathbf{F}(t)$, Lüroth's theorem (Theorem 6.2.5) tells us that $\mathbf{F}(x, y)$ is purely transcendental, but not only is the above argument simpler and more direct, it also gives us more information.)

(2) The isomorphism in part (1) was due to the polynomial identity among squares that we began with in that case. As we shall see, the fact that there is no isomorphism for $n \geq 3$ is due to the fact that there is no similar polynomial identity.

Suppose $\mathbf{F}(x, y)$ were purely transcendental. Then we would have an isomorphism $\varphi : \mathbf{F}(x, y) \longrightarrow \mathbf{F}(t)$. Then $\varphi(x) = f_1(t)/f_2(t)$ where $f_1(t)$ and $f_2(t)$ are relatively prime polynomials in $\mathbf{F}[t]$, and similarly $\varphi(y) = g_1(t)/g_2(t)$ where $g_1(t)$ and $g_2(t)$ are relatively prime polynomials in $\mathbf{F}[t]$, and furthermore

$$\left(\frac{f_1(t)}{f_2(t)}\right)^n + \left(\frac{g_1(t)}{g_2(t)}\right)^n = 1.$$

In the remainder of this proof we abbreviate $f_1(t)$ by f_1, etc. Multiplying through by $(f_2 g_2)^n$ we obtain the polynomial identity

$$a^n + b^n = c^n$$

where $a = f_1 g_2$, $b = g_1 f_2$, and $c = g_1 g_2$ are polynomials in $\mathbf{F}[t]$, not all of which are constant. We show this identity has no solutions.

We prove this by contradiction. Assume there is at least one solution, and choose a solution where $d_{\max} = \max(\deg a, \deg b, \deg c)$ is as small as possible. In this case a, b, and c are certainly pairwise relatively prime. Rewrite this identity as

$$a^n = c^n - b^n = \prod_{k=1}^{n}(c - \lambda^k b),$$

where $\lambda = \exp(2\pi i/n)$. Note that at most one of the factors on the right-hand side of this equation can be a constant polynomial. We now use the fact that $\mathbf{F}[t]$ is a unique factorization domain. These factors are pairwise relatively prime, as if $c - \lambda^i b$ and $c - \lambda^j b$ had a common factor for some $i \neq j$, then, taking linear combinations, a and b would have that common factor. Hence each of these factors is itself an nth power, $c - \lambda_i b = e_i^n$, $i = 1, \ldots, n$. In particular, since $n \geq 3$, this is true for $i = 1, 2, 3$. Then simple algebra shows

$$\begin{aligned} e_3^n &= c - \lambda^3 b \\ &= -\lambda(c - \lambda b) + (1 + \lambda)(c - \lambda^2 b) \\ &= -\lambda e_1^n + (1 + \lambda) e_2^n. \end{aligned}$$

Note that $1 + \lambda \neq 0$ as $n \neq 2$. Then, choosing complex numbers α and β with $\alpha^n = -\lambda$ and $\beta^n = 1 + \lambda$, and setting $a' = \alpha e_1$, $b' = \beta e_2$, and $c' = e_3$, we have

$$(a')^n + (b')^n = (c')^n$$

with $d'_{\max} = \max(\deg a', \deg b', \deg c') \leq d_{\max}/n$, contradicting the minimality of $d_{\max}$. □

Theorem 6.3.7. *Fix positive integers $n \geq 3$ and $m \geq 3$. Let $h(X)$ be any polynomial of degree m with distinct roots and let $\mathcal{C}$ be the curve defined by $Y^n = h(X)$. Then $\mathcal{C}$ is not rational.*

Proof. The proof of this is an adaptation of the proof of Theorem 6.3.6(2).

Let $h(X)$ have roots $\lambda_1, \ldots, \lambda_m$. Then $\mathcal{C}$ is given by

$$Y^p = c_q(X - \lambda_1) \cdots (X - \lambda_m)$$

Replacing Y by $Y {c_m}^{1/n}$ if necessary, we may assume $h(X)$ is monic, and we do so henceforth. Assume that $\mathcal{C}$ is rational. Then we have a pair $f_1(t)$ and $f_2(t)$ of relatively prime polynomials in $\mathbf{F}[t]$, and also a pair $g_1(t)$ and $g_2(t)$ of relatively prime polynomials in $\mathbf{F}[t]$, with

$$\left(\frac{g_1}{g_2}\right)^n = \left(\frac{f_1}{f_2} - \lambda_1\right) \cdots \left(\frac{f_1}{f_2} - \lambda_m\right).$$

Multiplying this equation by $g_2^n f_1^m$ we obtain the polynomial identity

$$g_1^n f_2^m = g_2^n (f_1 - \lambda_1 f_2) \cdots (f_1 - \lambda_m f_2).$$

Now g_2^n must divide f_2^m, as g_2 is relatively prime to g_1, and f_2^m must divide g_2^n as f_2 is relatively prime to f_1, which implies that f_2 is relatively prime to each term $(f_1 - \lambda_i f_2)$. Hence g_2^n and f_2^m must be equal up to a unit factor, which we may again assume is 1, by an appropriate linear change of variable if necessary. Thus we obtain the identity

$$g_1^n = (f_1 - \lambda_1 f_2) \cdots (f_1 - \lambda_m f_2).$$

The terms on the right-hand side are pairwise relatively prime so each must be an nth power. Let $(f_1 - \lambda_i f_2) = e_i^n$ for $i = 1, 2, 3$. The three polynomials $(f_1 - \lambda_1 f_2)$, $(f_1 - \lambda_2 f_2)$, and $(f_1 - \lambda_3 f_2)$ are linearly dependent, and no two are multiples of each other, so there is a linear dependence between them with each of the three coefficients nonzero. By an appropriate linear change of variable we may assume this linear dependence is $e_1^n + e_2^n - e_3^n = 0$, and we have seen in the proof of Theorem 6.3.6(2) that this is impossible. □

In order to handle the remaining cases we first prove a technical lemma.

Lemma 6.3.8. *There does not exist a pair of polynomials $(f(X), g(X))$ in $\mathbf{F}[X]$, not both of which are constant, such that $a_i f(X) + b_i g(X)$ is a square in $\mathbf{F}[X]$, $i = 1, \ldots, 4$, where (a_i, b_i) are pairs of elements in $\mathbf{F}$, not both 0, satisfying the condition that $(a_j, b_j) \neq (ea_i, eb_i)$ for any e in $\mathbf{F}$ and any $j \neq i$.*

Proof. Assume there exists such a pair $(f(X), g(X))$, which we abbreviate as (f, g). Choose such a pair with $\max(\deg(f), \deg(g))$ minimal. Then certainly f and g are relatively prime. This implies that $a_i f + b_i g$ and $a_j f + b_j g$ are relatively prime for $i \neq j$. Let $p = a_1 f + b_1 g$ and $q = a_2 f + b_2 g$. Then there are unique pairs (a_i', b_i'), with $a_i' p + b_i' q = a_i f + b_i g$, $i = 3, 4$. Now

$$\prod_{i=1}^{4} (a_i f + b_i g) = pq(a_3' p + b_3' q)(a_4' p + b_4' q).$$

The left-hand side of this equation is a product of squares, hence a square. Our conditions on (a_i, b_i) ensure that the terms on the right-hand side of this equation are pairwise relatively prime, and hence each of them must be a square. Let $p = r^2$ and $q = s^2$. Then, with any arbitrary but fixed choice of square roots,

$$a_i' p + b_i' q = a_i' r^2 + b_i' s^2 = \left(\sqrt{a_i'}\, r + \sqrt{-b_i'}\, s\right)\left(\sqrt{a_i'}\, r - \sqrt{-b_i'}\, s\right), \quad i = 3, 4.$$

Now each of these products is a square, and the factors are again relatively prime, so they must each be squares. Thus, replacing f by r, g by s, and $(a_1, b_1), \ldots, (a_4, b_4)$ with $(c_1, d_1), \ldots, (c_4, d_4)$ where

$$(c_1, d_1) = \left(\sqrt{a_3'}, \ \sqrt{-b_3'}\right),$$
$$(c_2, d_2) = \left(\sqrt{a_3'}, -\sqrt{-b_3'}\right),$$
$$(c_3, d_3) = \left(\sqrt{a_4'}, \ \sqrt{-b_4'}\right),$$
$$(c_4, d_4) = \left(\sqrt{a_4'}, -\sqrt{-b_4'}\right),$$

we have a pair of polynomials (r, s) satisfying the given conditions with $\max(\deg(r), \deg(s)) < \max(\deg(f), \deg(g))$, a contradiction. □

Remark 6.3.9. Lemma 6.3.8 is sharp. If $f(X) = X^4 + 1$ and $g(X) = X^2$, then $a_i f(X) + b_i g(X)$ is a square for $(a_1, b_1) = (1, 2)$, $(a_2, b_2) = (1, -2)$, and $(a_3, b_3) = (0, 1)$.

Theorem 6.3.10. *Fix a positive integer $m \geq 3$. Let $h(X)$ be any polynomial of degree m with distinct roots and let C be the curve defined by $Y^2 = h(X)$. Then C is not rational.*

Proof. We begin by following the proof of Theorem 6.3.7, whose notation we adopt. Exactly as in that proof, we obtain an identity

$$g_1^2 = (f_1 - \lambda_1 f_2) \cdots (f_1 - \lambda_m f_2),$$

with the terms on the right-hand side pairwise relatively prime, and hence each a perfect square. In case $m \geq 4$, this is impossible by Lemma 6.3.3. In case $m = 3$, we have only 3 perfect squares. But in this case, the equation $g_2^2 = f_2^3$ implies that f_2 is a perfect square as well, so we may again apply Lemma 6.3.8. □

There is one case left open, which we now dispose of.

Theorem 6.3.11. *Let $h(X)$ be any polynomial of degree 2 with distinct roots and let C be the curve defined by $Y^3 = h(X)$. Then C is not rational.*

Proof. Let $h(X) = aX^2 + bX + c$. Then $Y^3 = aX^2 + bX + c = a(X + b/(2a))^2 + (c - b^2/(4a))$ so $(X + b/(2a))^2 = Y^3/a - (c/a - b^2/(4a^2))$, and by making a linear change of variable and interchanging X and Y we are back in the case of Theorem 6.3.10. □

Remark 6.3.12. Note that the proof of Theorem 6.3.6(1) holds over $\mathbf{Q}$, from which we may conclude that the curve $X^2 + Y^2 = 1$ is rational over every field of characteristic 0. Also note that the proofs of Theorem 6.3.6(2), Theorem 6.3.7, Theorem 6.3.10, and Theorem 6.3.11 hold over every algebraically closed field of characteristic 0, from which we may conclude that these curves are not rational over any field of characteristic 0.

Remark 6.3.13. Now let $\mathbf{F}$ be a subfield of $\mathbf{C}$. For $f(X, Y) \in \mathbf{F}[X, Y]$, it makes sense to consider not only $V_{\mathbf{C}}(f) = \{(z_1, z_2) \in \mathbf{C}^2 \mid f(z_1, z_2) = 0\}$ but also $V_{\mathbf{F}}(f) = V_{\mathbf{C}}(f) \cap \mathbf{F}^2 = \{(z_1, z_2) \in \mathbf{F}^2 \mid f(z_1, z_2) = 0\}$. Questions about $V_{\mathbf{F}}(f)$ are questions of great arithmetic (i.e., number-theoretic) interest. For example, let $\mathbf{F} = \mathbf{Q}$ and consider $f(X, Y) = X^n + Y^n - 1$ as in Theorem 6.3.6.

By clearing denominators, any point $(a/c, b/c) \in V_{\mathbf{Q}}(f)$ gives us integers a, b, and c satisfying the equation $a^n + b^n = c^n$, and conversely.

If $n = 2$, then a, b, and c form a Pythagorean triple. Setting $t = j/k$ in the proof of Theorem 6.3.6 and clearing denominators, this gives us a parameterization of Pythagorean triples by $a = j^2 - k^2$, $b = 2jk$, $c = j^2 + k^2$. With the restrictions $j > k > 0$ and j and k relatively prime and not both odd, this yields all primitive Pythagorean triples, without duplication. (A Pythagorean triple is primitive if a, b, and c are relatively prime.) This can all be found in Euclid.

On the other hand, if $n \geq 3$, Fermat claimed that this equation has no integer solutions with a, b, and c all nonzero. This is the famous *Fermat's last theorem*. This claim was made by Fermat in the middle of the seventeenth century and remained open for over 300 years, until being proved by Wiles at the end of the twentieth century. It is because of this that the curves C_n are called Fermat curves. $\diamond$

6.4 Exercises

Exercise 6.4.1. Let x and y be algebraically independent elements of a transcendental extension $\mathbf{E}$ of $\mathbf{F}$. Show that $\mathbf{F}(x)$ and $\mathbf{F}(y)$ are linearly disjoint extensions of $\mathbf{F}$.

Exercise 6.4.2. Consider the polynomial $f(X, Y) = X^2 - 2XY + Y^2 - 2 \in \mathbf{Q}[X, Y]$. Show that $f(X, Y)$ is irreducible. Show that $\{(r_1, r_2) \in \mathbf{Q}^2 \mid f(r_1, r_2) = 0\} = \emptyset$. Let $\mathbf{E} = \mathbf{Q}(x, y)$ be the function field of the "curve" $f(X, Y) = 0$. Show that $\mathbf{E} = \mathbf{Q}(\sqrt{2})(x)$. In particular, $\mathbf{E}$ is a transcendental but not completely transcendental extension of $\mathbf{Q}$.

Exercise 6.4.3. Let $f(X, Y) \in \mathbf{F}[X, Y]$ be an irreducible polynomial in which both X and Y appear. Show that x and y are both transcendence bases for the function field of the curve $f(X, Y) = 0$. (Hence this field has transcendence degree 1 over $\mathbf{F}$.)

Exercise 6.4.4. (a) Let $\mathbf{E}$ be a purely transcendental extension of $\mathbf{F}$ and let $\mathbf{B}$ be an algebraic extension of $\mathbf{F}$. Show that $\mathbf{BE}$ is an algebraic extension of $\mathbf{E}$ and that $(\mathbf{BE}/\mathbf{E}) = (\mathbf{B}/\mathbf{F})$.

(b) Let $\mathbf{E}$ be a purely transcendental extension of $\mathbf{F}$ and let $\mathbf{B}$ be a Galois extension of $\mathbf{F}$. Show that $\mathbf{BE}$ is a Galois extension of $\mathbf{E}$ and that $\mathrm{Gal}(\mathbf{BE}/\mathbf{E}) = \mathrm{Gal}(\mathbf{B}/\mathbf{F})$.

Exercise 6.4.5. An extension $\mathbf{E}$ of $\mathbf{F}$ is a *finitely generated* extension of $\mathbf{F}$ if $\mathbf{E}$ is obtained by adjoining finitely many elements (algebraic or transcendental) to $\mathbf{F}$. Show that $\mathbf{E}$ is finitely generated over $\mathbf{F}$ if and only if $k = \text{tr-deg}(\mathbf{E}/\mathbf{F})$ is finite and for some (and hence for any) transcendence basis $X_1, \ldots, X_k$ of $\mathbf{E}$ over $\mathbf{F}$, $\mathbf{E}$ is a finite extension of $\mathbf{F}(X_1, \ldots, X_k)$.

Exercise 6.4.6. Let $\mathbf{F} \subseteq \mathbf{B} \subseteq \mathbf{E}$. Show that $\mathbf{E}$ is finitely generated over $\mathbf{F}$ if and only if $\mathbf{E}$ is finitely generated over $\mathbf{B}$ and $\mathbf{B}$ is finitely generated over $\mathbf{F}$.

Exercise 6.4.7. Let D_{2n}, the dihedral group of order $2n$, be given by $D_{2n} = \langle a, b \mid a^n = 1, b^2 = 1, ba = a^{-1}b \rangle$. Then D_{2n} acts on $\mathbf{C}(X)$ where the action is determined by $a(X) = \exp(2\pi i/n)X$ and $b(X) = 1/X$. Show that $\text{Fix}(D_{2n}) = \mathbf{C}(T)$ where $T = X^n + X^{-n}$.

Exercise 6.4.8. (a) Let $\sigma_1(X) = 1/X$ and let $\sigma_2(X) = 1 - X$. Show that σ_1 and σ_2 generate a subgroup G of $\text{Aut}(\mathbf{C}(X)/\mathbf{C})$ that is isomorphic to the dihedral group D_6. (Considering the matrices that give σ_1 and σ_2, as in Theorem 6.2.3, will make this computation easier.)
(b) Show that $\text{Fix}(G) = \mathbf{C}(T)$ where $T = (X^2 - X + 1)^3/(X^2(X-1)^2)$.

Exercise 6.4.9. (a) Let $\mathbf{B}_1$ and $\mathbf{B}_2$ be extensions of $\mathbf{F}$. Show that $\text{tr-deg}(\mathbf{B}_1\mathbf{B}_2/\mathbf{B}_1) \leq \text{tr-deg}(\mathbf{B}_2/\mathbf{F})$ and that $\text{tr-deg}(\mathbf{B}_1\mathbf{B}_2/\mathbf{F}) \leq \text{tr-deg}(\mathbf{B}_1/\mathbf{F}) + \text{tr-deg}(\mathbf{B}_2/\mathbf{F})$. (Note that Example 6.1.24 gives an example of disjoint extensions $\mathbf{B}_1$ and $\mathbf{B}_2$ where we have strict inequality.)
(b) Let $\mathbf{B}_1$ and $\mathbf{B}_2$ be linearly disjoint extensions of $\mathbf{F}$. Show that $\text{tr-deg}(\mathbf{B}_1\mathbf{B}_2/\mathbf{B}_1) = \text{tr-deg}(\mathbf{B}_2/\mathbf{F})$ and that $\text{tr-deg}(\mathbf{B}_1\mathbf{B}_2/\mathbf{F}) = \text{tr-deg}(\mathbf{B}_1/\mathbf{F}) + \text{tr-deg}(\mathbf{B}_2/\mathbf{F})$.

Exercise 6.4.10. (a) Let $\mathcal{C}$ be the curve defined by $Y^2 = X^2 + X^3$. Show that $\mathcal{C}$ is rational over $\mathbf{Q}$ (and hence over any field of characteristic 0).
(b) Let $\mathcal{C}$ be the curve defined by $Y^2 = X^2 - X^4$. Show that $\mathcal{C}$ is rational over $\mathbf{Q}$ (and hence over any field of characteristic 0).

Exercise 6.4.11. Let $f(X, Y) = aX^2 + bXY + cY^2 + dX + eY + f$ be an arbitrary irreducible polynomial, and let $\mathcal{C}$ be the curve defined by $f(X, Y) = 0$. Show that $\mathcal{C}$ is rational over $\mathbf{C}$.

Exercise 6.4.12. Show that Theorem 6.3.7 and Theorem 6.3.10 hold for any polynomial $h(X)$ with at least 3 roots of multiplicity prime to n, in the case of Theorem 6.3.7, or to 2, in the case of Theorem 6.3.10.

Exercise 6.4.13. Determine conditions on p so that each case of Theorem 6.3.6, Theorem 6.3.7, Theorem 6.3.10, Theorem 6.3.11, Exercise 6.4.10, and Exercise 6.4.11 holds when $\mathbf{F}$ is an algebraically closed field of characteristic p.

Exercise 6.4.14. Let $\mathbf{F}$ be a field of characteristic p, let $\mathbf{E} \supset \mathbf{F}$ be a purely transcendental extension of $\mathbf{F}$, and let $\mathbf{E}' \supseteq \mathbf{E}$ be a finite extension of $\mathbf{E}$. Show that $\mathbf{E}'$ is not a perfect field.

A

Some Results from Group Theory

A.1 Solvable Groups

Definition A.1.1. A *composition series* for a group G is a sequence of subgroups $G = G_0 \supseteq G_1 \supseteq \ldots G_k = \{1\}$ with G_i a normal subgroup of G_{i-1} for each i. A group G is *solvable* if it has a composition series with each quotient G_i/G_{i-1} abelian. ◇

Example A.1.2. (1) Every abelian group G is solvable (as we may choose $G_0 = G$ and $G_1 = \{1\}$).

(2) If G is the group of Lemma 4.6.1, $G = \langle \sigma, \tau \mid \sigma^p = 1, \tau^{p-1} = 1, \tau\sigma\tau^{-1} = \sigma^r \rangle$ where r is a primitive root mod p, then G is solvable: $G = G_0 \supset G_1 \supset G_2 = \{1\}$ with G_1 the cyclic subgroup generated by σ. Then G_1 is a normal subgroup of G with G/G_1 cyclic of order $p-1$ and G_1 cyclic of order p.

(3) If G is the group of Lemma 4.6.3, $G = \langle \sigma, \tau \mid \sigma^4 = 1, \tau^2 = 1, \tau\sigma\tau^{-1} = \sigma^3 \rangle$, then G is solvable: $G = G_0 \supset G_1 \supset G_2 = \{1\}$ with G_1 the cyclic subgroup generated by σ. Then G_1 is a normal subgroup of G with G/G_1 cyclic of order 2 and G_1 cyclic of order 4.

(4) If G is a p-group, then G is solvable. This is Lemma A.2.4 below.

(5) If $G = S_n$, the symmetric group on $\{1, \ldots, n\}$, then G is solvable for $n \leq 4$. S_1 is trivial. S_2 is abelian. S_3 has the composition series $S_3 \supset A_3 \supset \{1\}$. S_4 has the composition series $S_4 \supset A_4 \supset V_4 \supset \{1\}$ where $V_4 = \{1, (12)(34), (13)(24), (14)(23)\}$. (Here A_n denotes the alternating group.)

(6) If $G = S_n$ or $G = A_n$, then G is not solvable for $n \geq 5$. This is Corollary A.3.6 below. ◇

Lemma A.1.3. *A finite group G is solvable if and only if it has a composition series satisfying any one of the following conditions:*

(1) G_{i-1}/G_i *is solvable for each* i.
(2) G_{i-1}/G_i *is abelian for each* i.
(3) G_{i-1}/G_i *is cyclic for each* i.
(4) G_{i-1}/G_i *is cyclic of prime order for each* i.

Proof. If G_{i-1}/G_i is solvable then we may "refine" the original sequence to $G \supseteq \cdots \supseteq G_{i-1} = H_0 \supseteq H_1 \supseteq \cdots \supseteq H_m = G_i \supseteq \cdots \supseteq G_k = \{1\}$ with H_j normal in H_{j-1} and H_{j-1}/H_j abelian; similarly we may refine a sequence with abelian quotients to one with cyclic quotients and one with cyclic quotients to one with quotients cyclic of prime order. On the other hand, if G_{i-1}/G_i is cyclic of prime order it is certainly solvable. □

Lemma A.1.4. *(1) Let* G *be a solvable group and let* H *be a subgroup of* G. *Then* H *is a solvable group.*

(2) Let G *be a solvable group and let* N *be a normal subgroup of* G. *Then* G/N *is a solvable group.*

(3) Let G *be a group and let* N *be a normal subgroup of* G. *If* N *and* G/N *are solvable, then* G *is solvable.*

Proof. Consider $G = G_0 \supseteq G_1 \supseteq \cdots \supseteq G_k \subseteq \{1\}$. Then $H = H_0 \supseteq H_1 \supseteq \cdots \supseteq H_k \supseteq \{1\}$ where $H_i = H \cap G_i$. It is easy to check that H_i is a normal subgroup of H_{i-1}, and then by standard theorems of group theory

$$\begin{aligned} H_{i-1}/H_i = (H \cap G_{i-1})/H \cap G_i &= (H \cap G_{i-1})/(H \cap G_{i-1}) \cap G_i \\ &\cong (H \cap G_{i-1})G_i/G_i \subseteq G_{i-1}/G_i \end{aligned}$$

is isomorphic to a subgroup of an abelian group so it is abelian.

(2) Let $\pi : G \to G/N = Q$ be the quotient map and consider $Q = Q_0 \supseteq \cdots \supseteq Q_k = \{1\}$ where $Q_i = \pi(G_i)$. Then $Q_i = G_iN/N \cong G_i/G_i \cap N$, so

$$\begin{aligned} \pi(G_{i-1})/\pi(G_i) &= (G_{i-1}N/N)/(G_iN/N) \\ &\cong (G_{i-1}/G_{i-1} \cap N)/(G_i/G_i \cap N) \end{aligned}$$

is isomorphic to a quotient of the abelian group G_{i-1}/G_i so it is abelian.

(3) Let $\pi : G \to G/N = Q$. Let $N = N_0 \supseteq \cdots \supseteq N_j = \{1\}$ and $Q = Q_0 \supseteq \cdots \supseteq Q_k = \{1\}$ be as in Definition A.1.1. Then the sequence of subgroups

$$G = \pi^{-1}(Q_0) \supseteq \pi^{-1}(Q_1) \supseteq \cdots \supseteq \pi^{-1}(Q_k) = N_0 \supseteq N_1 \supseteq \cdots \supseteq N_j = \{1\}$$

shows that G is solvable. □

So far, we have considered abstract solvable groups. However, when we consider solvable subgroups of symmetric groups, we can get very precise descriptions.

Lemma A.1.5. *Let G be a subgroup of S_p of order divisible by p, p a prime. Then any nontrivial normal subgroup K of G has order divisible by p.*

Proof. Since G has order divisible by p, it must contain an element of order p, which must be a p-cycle.

Regard S_p as operating on $T = \{1, \ldots, p\}$. Write $T = T_1 \cup \cdots \cup T_q$, a decomposition onto orbits of K. In other words, $i, j, \in T_m$ for some m if and only if there is some $\tau \in K$ with $\tau(i) = j$. Now for any $i, j \in T$ there is a $\sigma \in G$ with $\sigma(i) = j$. This implies that the action of G permutes $\{T_1, \ldots, T_q\}$ so in particular each T_m has the same cardinality. This cardinality then divides p, so must be 1 or p, and cannot be 1 as K is nontrivial. Thus we see that K operates transitively on T, so has order divisible by p (as the subgroup of K fixing 1, say, has index p). □

Lemma A.1.6. *Let G be a solvable subgroup of S_p of order divisible by p, p a prime. Then G contains a unique subgroup N of order p.*

Proof. Let $G = G_0 \supset G_1 \supset \cdots \supset G_k = \{1\}$ as in Definition A.1.1. We prove the lemma by induction on k.

If $k = 1$, $G = G_0 \supset G_1 = \{1\}$ and G is solvable, so G is abelian, and hence has a unique p-Sylow subgroup N, of order p.

Suppose the lemma is true for $k - 1$. Then G_1 is a normal subgroup of G, so applying Lemma A.1.5 with $K = G_1$ we see that G_1 has order divisible by p, and then by the inductive hypothesis that G_1 contains a unique subgroup N of order p. Now let N' be any subgroup of G of order p. Then N' and N are both p-Sylow subgroups, so are conjugate. Let $N' = \sigma N \sigma^{-1}$. As $N \subseteq G_1$, $N' = \sigma N \sigma^{-1} \subseteq \sigma G_1 \sigma^{-1} = G_1$ as G_1 is normal in G, so $N' = N$, as required. □

Let a group G operate on a set T. This action is *effective* if the only element σ or G with $\sigma(t) = t$ for all $t \in T$ is $\sigma = \text{id}$.

Let a group G_1 operate on a set T_1, and a group G_2 operate on a set T_2. These operations are *permutation isomorphic* if there is an isomorphism of groups $\psi: G_1 \to G_2$ and an isomorphism of sets (i.e., a bijection) $\Psi: T_1 \to T_2$ with $\Psi(\sigma(t)) = \psi(\sigma)(\Psi(t))$ for every $\sigma \in G_1, t \in T_1$.

Proposition A.1.7. *Let G be a solvable group operating effectively and transitively on a set T of cardinality p, p a prime. Then there is a subgroup H of $\mathbf{F}_p^*$ such that this operation is permutation isomorphic to*

$$G_H = \left\{ \begin{bmatrix} h & n \\ 0 & 1 \end{bmatrix} \mid h \in H, n \in \mathbf{F}_p \right\}$$

operating on

$$T = \left\{ \begin{bmatrix} i \\ 1 \end{bmatrix} \mid i \in \mathbf{F}_p \right\}$$

by left multiplication.

Proof. As we have observed, in this situation G has order divisible by p. (Let $t \in T$ be arbitrary and consider $\{\sigma \in G \mid \sigma(t) = t\}$. Since G acts transitively on T, this subgroup of G has index p.)

Since G operates effectively on T, G is isomorphic to a subgroup of the symmetric group S_p, so, by Lemma A.1.6, G contains a unique subgroup N of order p, which is generated by a p-cycle. Reordering, if necessary, we have that N and T are permutation isomorphic to $\left\{ \begin{bmatrix} 1 & n \\ 0 & 1 \end{bmatrix} \mid n \in \mathbf{F}_p \right\}$ operating on $\left\{ \begin{bmatrix} i \\ 1 \end{bmatrix} \mid i \in \mathbf{F}_p \right\}$ by left multiplication, as a suitable generator ν of N takes $\begin{bmatrix} i \\ 1 \end{bmatrix}$ to $\begin{bmatrix} i+1 \\ 1 \end{bmatrix}$ mod p. For simplicity of notation, we simply identify N with this group and T with this set.

Since N is the unique subgroup of order p of G, it is certainly a normal subgroup of G. Let $\sigma \in G$ be arbitrary. Then $\sigma\nu\sigma^{-1} = \nu^h$ for some $h \neq 0$. Let $\sigma\left(\begin{bmatrix} 0 \\ 1 \end{bmatrix}\right) = \begin{bmatrix} n \\ 1 \end{bmatrix}$. Then, for any i,

$$\begin{bmatrix} n+ih \\ 1 \end{bmatrix} = \nu^{ih}\left(\begin{bmatrix} n \\ 1 \end{bmatrix}\right) = \sigma\nu^i\sigma^{-1}\left(\begin{bmatrix} n \\ 1 \end{bmatrix}\right) = \sigma\nu^i\left(\begin{bmatrix} 0 \\ 1 \end{bmatrix}\right) = \sigma\left(\begin{bmatrix} i \\ 1 \end{bmatrix}\right),$$

so

$$\sigma\left(\begin{bmatrix} i \\ 1 \end{bmatrix}\right) = \begin{bmatrix} h & n \\ 0 & 1 \end{bmatrix}\begin{bmatrix} i \\ 1 \end{bmatrix}$$

and hence

$$G \subseteq \left\{ \begin{bmatrix} h & n \\ 0 & 1 \end{bmatrix} \mid h \in \mathbf{F}_p^*, \ n \in \mathbf{F}_p \right\}.$$

Thus, if H is the subgroup of $\mathbf{F}_p^*$ given by

$$H = \left\{ h \in \mathbf{F}_p^* \mid \begin{bmatrix} h & 0 \\ 0 & 1 \end{bmatrix} \in G \right\},$$

then $G = G_H$ as claimed. □

Corollary A.1.8. *Let G be a subgroup of S_p, p a prime, of order divisible by p. The following are equivalent:*

(1) G is solvable.

(2) The order of G divides $p(p-1)$.

(3) The order of G is at most $p(p-1)$.

Proof. (1) implies (2): If G is solvable, then, by Proposition A.1.7, $G = G_H$ for some H and then $|G|$ divides $p(p-1)$.

(2) implies (3): Trivial.

(3) implies (1): G has a p-Sylow subgroup N, and N is unique, as if G had another p-Sylow subgroup N', then G would have order at least p^2. Then the last part of the proof of Proposition A.1.7 shows that $G = G_H$ for some H. But $G_H \supset N \supset \{1\}$ with N normal in G. N is abelian and G_H/N is isomorphic to H which is abelian, so G is solvable. □

A.2 p-Groups

Throughout this section, p denotes a prime.

Definition A.2.1. A group G is a *p-group* if $|G|$ is a power of p. ◇

The *center* $Z(G)$ of a group G is defined by $Z(G) = \{\sigma \in G \mid \sigma\tau = \tau\sigma$ for every $\tau \in G\}$.

Lemma A.2.2. *Let G be a p-group. Then G has a nontrivial center (i.e., $Z(G) \neq \{1\}$).*

Proof. Write $G = C_1 \cup \cdots \cup C_k$, a union of conjugacy classes. If $\sigma_i \in C_i$, then $|C_i| = [G : H_i]$ where $H_i = \{\tau \in G \mid \tau\sigma_i\tau^{-1} = \sigma_i\}$. If $\sigma_i \in Z(G)$, then $C_i = \{\sigma_i\}$ and $|C_i| = 1$ (and $H_i = G$). If $\sigma_i \notin Z(G)$, then H_i is a proper subgroup of G, so $|C_i| = [G_i : H_i]$ is a positive power of p. Of course $|G|$ is a positive power of p. But one of the conjugacy classes, say C_1, consists of {id} alone, and $|C_1| = 1$ is not divisible by p. Now $|G| = |C_1| + \cdots + |C_k|$. Hence there must exist other classes C_i with $|C_i|$ not divisible by p, and hence with $|C_i| = 1$, and if $C_i = \{\sigma_i\}$ then $\sigma_i \in Z(G)$. □

Corollary A.2.3. *Let G be a p-group, $|G| = p^n$. Then there is a sequence of subgroups $G = G_0 \supset G_1 \supset \cdots \supset G_n = \{1\}$ with G_i a normal subgroup of G of index p^i, for each $i = 1, \ldots, n$.*

Proof. By induction on n. If $n = 1$, the result is trivial.

Now suppose the result is true for all groups of order p^{n-1}, and let G be an arbitrary group of order p^n. By Lemma A.2.2, the center $Z(G)$ of G is nontrivial. Since $|Z(G)|$ divides $|G|$, $|Z(G)|$ is also a power of p, so, in particular, $Z(G)$ contains an element of order p. The cyclic subgroup H of G generated by that element is a subgroup of G of order p, and, since $H \subseteq Z(G)$, H is a normal subgroup of G. Let $Q = G/H$ be the quotient group, and let $\pi : G \to Q$ be the canonical projection.

Now Q is a group of order p^{n-1}, so by the inductive hypothesis there is a sequence of subgroups $Q = Q_0 \supset Q_1 \supset \cdots \supset Q_{n-1} = \{1\}$ with Q_i a normal subgroup of Q of index p^i for each $i = 1, \ldots, n-1$. Let $G_i = \pi^{-1}(Q_i)$ for $i = 1, \ldots, n-1$ and $G_n = \{1\}$. Then $G = G_0 \supset G_1 \supset \cdots \supset G_n = \{1\}$ form a sequence as claimed. (Clearly G_i has index p^i. Also, G_i is a normal subgroup of G as if $g_0 \in G_i$ and $g \in G$, then $gg_0g^{-1} \in G_i$ as $\pi(gg_0g^{-1}) = \pi(g)\pi(g_0)(\pi(g))^{-1} \in Q_i$ as Q_i is a normal subgroup of Q.) □

Lemma A.2.4. *Let G be a p-group. Then G is solvable.*

Proof. By Corollary A.2.3, there is a sequence of subgroups $G = G_0 \supset G_1 \supset \cdots \supset G_n = \{1\}$ with G_i a normal subgroup of G, and hence certainly a normal subgroup of G_{i-1}, for each i. Furthermore, since $[G : G_i] = p^i$ for each i, $[G_{i-1} : G_i] = p$ and hence $|G_{i-1}/G_i| = p$, and so G_{i-1}/G_i is cyclic of order p and, in particular, is abelian, for each i. Hence, by Definition A.1.1, G is solvable. □

A.3 Symmetric and Alternating Groups

In this section we prove several results about symmetric and alternating groups. We let S_n be the symmetric group on $\{1, \ldots, n\}$ and we let $A_n \subseteq S_n$ be the alternating group. We regard the elements of S_n as functions on this set and recall that functions are composed by applying the rightmost function first. Thus, for example, $(1\ 2)(1\ 3) = (1\ 3\ 2)$. We recall that every element of S_n can be written as a product of disjoint cycles, and that disjoint cycles commute.

The first two results we prove allow us to conclude that certain subgroups of S_n are in fact equal to S_n.

Lemma A.3.1. *Let p be a prime and let G be a transitive subgroup of S_p that contains a transposition. Then $G = S_p$.*

Proof. By renumbering if necessary, we may assume the transposition is $\tau = (1\ 2)$.

Since G acts transitively on $\{1, \ldots, p\}$, it has order divisible by p and hence it has an element σ_0 of order p. Since p is a prime, σ_0 is a p-cycle. Thus there is some power $\sigma = \sigma_0^k$ of σ_0 with $\sigma(1) = 2$. By renumbering if necessary, we may then assume $\sigma = (1\ 2\ \cdots\ p)$. Now direct calculation shows that

$$\sigma^j \tau \sigma^{-j} = (1\,2\ \cdots\ p)^j (1\,2)(1\,2\ \cdots\ p)^{-j} = ((j+1)\ (j+2))$$

for $j = 0, \ldots, p-2$. Direct calculation then shows that

$$\begin{aligned}(2\,3)(1\,2)(2\,3) &= (1\,3),\\ (3\,4)(1\,3)(3\,4) &= (1\,4),\\ &\ \vdots\\ ((p-1)\ p)(1\ (p-1))((p-1)\ p) &= (1\ p)\end{aligned}$$

and furthermore that

$$(1\ k)(1\ j)(1\ k) = (j\ k) \text{ for any } j \neq k.$$

Thus G contains every transposition. But S_p is generated by transpositions, so $G = S_p$. □

Lemma A.3.2. *Let G be a transitive subgroup of S_n that contains an $(n-1)$-cycle and a transposition. Then $G = S_n$.*

Proof. Let ρ be the $n-1$-cycle. By renumbering if necessary, we may assume that $\rho = (1\ 2\ \cdots\ (n-1))$. Then the transposition is $\tau = (i\ j)$ for some i and j. Since G operates transitively on $\{1, \ldots, n\}$, there is an element γ of G with $\gamma(j) = n$. Then $\gamma(i) = k$ for some k. Direct calculation shows that

$$\gamma\tau\gamma^{-1} = (k\ n) = \tau_k \text{ and then that } \rho^i \tau_k \rho^{-i} = ((k+i)\ n) = \tau_{k+i},$$

where $k + i$ is taken mod $n - 1$. Finally, direct calculation shows that

$$\tau_i \tau_j \tau_i^{-1} = (i\ j) \text{ for any } i \neq j.$$

Thus G contains every transposition. But S_n is generated by transpositions, so $G = S_n$. □

Next we show some results on the structure of A_n and S_n. We first need the following technical lemma.

Lemma A.3.3. *If G is a normal subgroup of A_n that contains a 3-cycle, then $G = A_n$.*

Proof. We shall first show that if G contains a single 3-cycle, then it contains every 3-cycle. Note that if G contains a 3-cycle $(i\ j\ k)$, it contains $(i\ j\ k)^{-1} = (k\ j\ i)$.

By renumbering if necessary, we may assume that G contains the 3-cycle $(1\ 2\ 3)$. Then for each $i > 3$ (observing that an element of order 2 is its own inverse) G also contains the elements

$$((1\ 2)(3\ i))(3\ 2\ 1)((1\ 2)(3\ i)) = (1\ 2\ i),$$
$$((1\ 3)(2\ i))(1\ 2\ 3)((1\ 3)(2\ i)) = (1\ 3\ i),$$
$$((2\ 3)(1\ i))(3\ 2\ 1)((2\ 3)(1\ i)) = (2\ 3\ i);$$

then for each distinct $i, j > 3$, G also contains the elements

$$(1\ 2\ j)(2\ i\ 1) = (1\ i\ j),$$
$$(2\ 3\ j)(3\ i\ 2) = (2\ i\ j),$$
$$(3\ 1\ j)(1\ i\ 3) = (3\ i\ j);$$

finally, for each distinct $i, j, k > 3$, G also contains the elements

$$(k\ i\ 1)(1\ i\ j) = (i\ j\ k),$$

so G contains all 3-cycles, as claimed.

But we now claim that any element of A_n can be written as a product of 3-cycles. Since any element of A_n can be written as a product of an even number of transpositions, to show this it suffices to show that the product of any two transpositions can be written as a product of 3-cycles, and we see this from

$$(i\ j)(i\ k) = (i\ k\ j),$$
$$(i\ j)(k\ \ell) = (i\ k\ j)(i\ k\ \ell),$$

completing the proof. □

Theorem A.3.4. *For $n \geq 5$, A_n is a simple group (i.e., it has no normal subgroups other than A_n and $\{1\}$).*

Proof. Let $n \geq 5$ and let $G \neq \{1\}$ be a normal subgroup of A_n. We will show $G = A_n$.

By Lemma A.3.3, it suffices to show that G contains a 3-cycle, so that is what we shall do.

Let $\sigma \in G, \sigma \neq 1$. Then σ has order $m > 1$. Let p be a prime dividing m and let $\gamma = \sigma^{m/p}$. Then γ has order p. Now γ can be written as a product of disjoint cycles, $\gamma = \gamma_1 \cdots \gamma_k$, and since γ has order p, each of $\gamma_1, \ldots, \gamma_k$ is a p-cycle. Now we must consider several cases:

Case I: $p \geq 5$: By renumbering if necessary, we may assume that $\gamma_1 = (1\ 2\ 3\ 4\ 5\ 6\ 7\ \cdots\ p)$. Let $\tau_1 = (1\ 2)(3\ 4)$. Then $\beta_1 = \tau_1 \gamma_1 \tau_1^{-1} = (1\ 4\ 3\ 5\ 6\ 7\ \cdots\ p\ 2)$ and $\gamma_1 \beta_1^{-1} = (1\ 3\ 5\ 4\ 2)$. Letting $\beta = \tau_1 \gamma \tau_1^{-1}$ and using the fact that disjoint cycles commute, it easily follows that $\gamma\beta^{-1} = (1\ 3\ 5\ 4\ 2)$. Let $\tau_2 = (1\ 3)(2\ 4)$. Then $\tau_2(1\ 3\ 5\ 4\ 2)\tau_2^{-1} = (1\ 5\ 2\ 4\ 3)$ and $(1\ 3\ 5\ 4\ 2)(1\ 5\ 2\ 4\ 3) = (1\ 4\ 5)$.

Case II: $p = 3$: If $k = 1$, i.e., there is only a single 3-cycle, there is nothing to do, so assume $k \geq 2$. By renumbering if necessary, we may assume that $\gamma_1 = (1\ 2\ 3)$ and $\gamma_2 = (4\ 5\ 6)$. Let $\gamma_{12} = \gamma_1\gamma_2 = (1\ 2\ 3)(4\ 5\ 6)$. Let $\tau_1 = (1\ 2)(3\ 4)$. Then $\delta_{12} = \tau_1 \gamma_{12} \tau_1^{-1} = (1\ 4\ 2)(3\ 5\ 6)$ and $\gamma_{12}\delta_{12}^{-1} = (1\ 3\ 4\ 2\ 5)$. Letting $\delta = \tau_1 \gamma \tau_1^{-1}$, it then follows, as in Case I, that $\gamma\delta^{-1} = (1\ 3\ 4\ 2\ 5)$. Thus G contains a 5-cycle and we are back in Case I.

Case III: $p = 2$: Since we are in A_n, k is even, so $k \geq 2$. By renumbering if necessary, we may assume that $\gamma_1 = (1\ 2)$ and $\gamma_2 = (3\ 4)$. Let $\gamma_{12} = \gamma_1\gamma_2 = (1\ 2)(3\ 4)$. Let $\rho = (1\ 2\ 3)$. Then $\zeta_{12} = \rho\gamma_{12}\rho^{-1} = (1\ 4)(2\ 3)$ and $\gamma_{12}\zeta_{12}^{-1} = (1\ 3)(2\ 4)$. Letting $\zeta = \rho\gamma\rho^{-1}$, it then follows, as in Case I, that $\gamma\zeta^{-1} = \theta = (1\ 3)(2\ 4)$. Let $\tau_3 = (1\ 3)(2\ 5)$. (Note here is where we need $n \geq 5$.) Then $\tau_3 \theta \tau_3^{-1} = \theta' = (1\ 3)(4\ 5)$. But then $\theta\theta' = (2\ 4\ 5)$. □

Theorem A.3.5. *For $n \geq 5$, the only normal subgroups of S_n are S_n, A_n, and $\{1\}$.*

Proof. Let $n \geq 5$ and let $G \neq \{1\}$ be a normal subgroup of S_n. We will show $G \supseteq A_n$. Since A_n is a subgroup of S_n of index 2, this shows that $G = A_n$ or S_n.

The proof of this is almost identical to the proof of Theorem A.3.4. We begin in the same way. Let $\sigma \in G, \sigma \neq 1$. Then σ has order $m > 1$. Let p be a prime dividing m and let $\gamma = \sigma^{m/p}$. Then γ has order p.

If we are in Case I, Case II, or Case III with $k \geq 2$ of the proof of Theorem A.3.4, we conclude in exactly the same way that $G \supseteq A_n$. This leaves only the case where $p = 2$ and $k = 1$, i.e., where γ is a single transposition. But all transpositions in S_n are conjugate, and S_n is generated by transpositions, so in this case $G = S_n$. □

Corollary A.3.6. *For $n \geq 5$, neither A_n nor S_n is solvable.*

Proof. Let $n \geq 5$. By the definition of a solvable group (Definition A.1.1), if $G = A_n$ or S_n were solvable, it would have a composition series with abelian quotients. But by Theorem A.3.4, the only possible composition series for A_n is $A_n \supset \{1\}$, and by Theorem A.3.5, the only possible composition series for S_n are $S_n \supset \{1\}$ and $S_n \supset A_n \supset \{1\}$, and A_n is not abelian. □

B

A Lemma on Constructing Fields

In Section 2.2 we showed: If $f(X) \in \mathbf{F}[X]$ is an irreducible polynomial, then $\mathbf{F}[X]/\langle f(X)\rangle$ is a field. We did so by using arguments particular to polynomials. In this appendix we shall show that this result follow from more general results in ring theory. In particular, in this section we show how to construct fields from integral domains.

Definition B.1.1. Let R be an integral domain.

(1) An ideal $I \subset R$ is a *maximal ideal* if $I \subseteq J \subseteq R$, J an ideal, implies $J = I$ or $J = R$.

(2) An ideal $I \subset R$ is a *prime ideal* if $ab \in I$, with $a, b \in R$, implies $a \in I$ or $b \in I$. ◇

Lemma B.1.2. *(1) Let R be an integral domain. Then every maximal ideal is prime.*

(2) Let R be a PID (principal ideal domain). Then every prime ideal is maximal.

Proof. (1) Let I be a maximal ideal and let $ab = i \in I$. If $a \in I$, there is nothing to prove. Otherwise, let $J = \langle a, I\rangle$ (the ideal generated by a and I). Then $I \subset J$ and I is maximal, so $J = R$. In particular, $1 \in J$. Then $xa + i' = 1$ for some $x \in R$ and $i' \in I$. Then

$$b = 1 \cdot b = (xa + i')b = X(ab) + i'b = iX + i'b \in I.$$

(2) Suppose I is a prime ideal and J is an ideal with $I \subseteq J$. Then $I = \langle p\rangle$ for some $p \in R$ and $J = \langle q\rangle$ for some $q \in R$. Since $I \subseteq J$, $p \in J$ so $p = qr$ for some $r \in R$. Since I is prime, we conclude that:

(a) $q \in I$ in which case $J = \langle q\rangle \subseteq I$ so $J = I$; or

(b) $r \in I$ so $r = pr'$ for some $r' \in R$ in which case $p = qr = qpr' = p(qr')$ so $1 = qr'$ and hence $1 \in J$, so $J = R$. □

Lemma B.1.3. *Let R be an integral domain and $I \subset R$ an ideal. Then I is maximal if and only if R/I is a field.*

Proof. Let $\pi : R \to R/I$ be the canonical projection. First suppose I is maximal. Let $\bar{a} \in R/I, \bar{a} \neq 0$. Then $\bar{a} = \pi(a)$ for some $a \in R, a \notin I$. Since I is maximal, $R = \langle a, I \rangle$, so $1 = xa + i$ for some $x \in R$ and some $i \in I$. Then, if $\bar{x} = \pi(x)$,

$$1 = \pi(1) = \pi(xa + i) = \bar{x}\bar{a} + 0 = \bar{x}\bar{a},$$

so $\bar{a}$ is invertible.

Now suppose I is not maximal, and let $I \subset J \subset R$. Let $j \in J, j \notin I$, and let $\bar{j} = \pi(j)$. We claim $\bar{j}$ is not invertible. For suppose $\bar{j}\bar{k} = 1, \bar{j}, \bar{k} \in R/I$, and let $\bar{k} = \pi(k)$. Then $jk = 1 + i$ for some $i \in I$, so $1 = jk - i$ and hence $1 \in J$, a contradiction. □

Corollary B.1.4. *Let R be a PID and let $I = \langle i \rangle$ be the ideal generated by the element i of R. Then R/I is a field if and only if i is irreducible.*

Proof. In a PID, an element i is irreducible if and only if i is prime, and an ideal $I = \langle i \rangle$ is a prime ideal if and only if the element i is prime, so the corollary follows directly from Lemma B.1.2 and Lemma B.1.3. □

Corollary B.1.5. *Let $f(X) \in \mathbf{F}[X]$ be an irreducible polynomial. Then $\mathbf{F}[X]/\langle f(X) \rangle$ is a field.*

Proof. Since $\mathbf{F}[X]$ is a PID, this is a special case of Corollary B.1.4. □

C

A Lemma from Elementary Number Theory

In this appendix, we prove a lemma that is used in the proof of Theorem 4.7.1.

Lemma C.1.1. *Let p be a prime and let t be a positive integer. Then there are infinitely many primes congruent to* $1 \pmod{p^t}$.

Proof. We begin by recalling the following definition and facts: Let q be a prime and let a be an integer relatively prime to q. Then $\mathrm{ord}_q(a)$, the *order* of $a \pmod q$, is defined to be the smallest positive integer x such that $a^x \equiv 1 \pmod q$. Then, for an arbitrary integer y, $a^y \equiv 1 \pmod q$ if and only if $\mathrm{ord}_q(a)$ divides y. Also, since, by Fermat's Little Theorem, $a^{q-1} \equiv 1 \pmod q$, we have that $\mathrm{ord}_q(a)$ divides $q - 1$.

We now proceed with the proof, which we divide into two cases: $p = 2$ and p odd.

Case I: $p = 2$. For a nonnegative integer k, let $F_k = 2^{2^k} + 1$. Note that, if $j < k$, $F_j = 2^{2^j} + 1$ divides $2^{2^k} - 1 = (2^{2^j})^{2^{k-j}} - 1 = F_k - 2$, so $\gcd(F_j, F_k) = \gcd(F_j, 2) = 1$, i.e., $\{F_1, F_2, \dots\}$ are pairwise relatively prime. Thus, if q_k is a prime factor of F_k, then $\{q_1, q_2, \dots\}$ are all distinct.

Now let $q = q_s$ for $s \geq t - 1$. By definition, q divides F_s, i.e., $2^{2^s} + 1 \equiv 0 \pmod q$, $2^{2^s} \equiv -1 \pmod q$, and hence $2^{2^{s+1}} \equiv 1 \pmod q$. Thus $\mathrm{ord}_q(2)$ divides 2^{s+1} but does not divide 2^s, so $\mathrm{ord}_q(2) = 2^{s+1}$. Then, as we have observed above, 2^{s+1} divides $q - 1$, so $q \equiv 1 \pmod{2^{s+1}}$ and hence $q \equiv 1 \pmod{2^t}$.

Thus we see that $\{q_{t-1}, q_t, \dots\}$ are distinct primes congruent to $1 \pmod{2^t}$.

Case II: p odd. In this case we proceed by contradiction. Assume there are only finitely many primes $\equiv 1 \pmod{p^t}$, and let these be $q_1, \dots, q_r$. Let

$$a = 2q_1 \cdots q_r, \qquad c = a^{p^{t-1}}, \qquad N = a^{p^t} - 1 = c^p - 1,$$

and write $N = (c-1)M$ where $M = c^{p-1} + \cdots + 1$.

Claim: $c - 1$ and M are relatively prime.

Proof of claim: $M = c^{p-1} + \cdots + 1 = (c^{p-1} - 1) + (c^{p-2} - 1) + \cdots + (c-1) + p$, and each term except possibly the last one is divisible by $c - 1$, so $\gcd(c-1, M) = \gcd(c-1, p) = 1$ or p. But $a \equiv 2 (\mathrm{mod}\, p)$ so $c \equiv 2^{p^{t-1}} (\mathrm{mod}\, p)$. By Fermat's Little Theorem, $2^{p-1} \equiv 1 (\mathrm{mod}\, p)$, and $p^{t-1} \equiv 1 (\mathrm{mod}\, p-1)$, so $c \equiv 2^1 \equiv 2 (\mathrm{mod}\, p)$ and so $c - 1 \equiv 1 (\mathrm{mod}\, p)$. Thus p does not divide $c - 1$ and so $\gcd(c-1, M) = \gcd(c-1, p) = 1$.

Now let q be any prime dividing M. Then q divides $N = c^p - 1$, so $c^p \equiv 1 (\mathrm{mod}\, q)$, i.e., $a^{p^t} \equiv 1 (\mathrm{mod}\, q)$. So $\mathrm{ord}_q(a)$ divides p^t, i.e., $\mathrm{ord}_q(a) = 1, p, \ldots, p^{t-1}$ or p^t. But if $\mathrm{ord}_q(a) = 1, p, \ldots,$ or p^{t-1}, then $a^{p^{t-1}} \equiv 1 (\mathrm{mod}\, q)$, i.e., $c \equiv 1 (\mathrm{mod}\, q)$, so q divides $c - 1$, contradicting the fact that $c - 1$ and M are relatively prime. Hence $\mathrm{ord}_q(a) = p^t$. Then, as we have observed above, p^t divides $q - 1$, i.e., $q \equiv 1 (\mathrm{mod}\, p^t)$.

But certainly q does not divides a, while each of $q_1, \ldots, q_r$ does divide a, so $q \neq q_1, \ldots, q_r$, contradicting the hypothesis that $q_1, \ldots, q_r$ are all the primes congruent to $1 (\mathrm{mod}\, p^t)$. Hence there must be infinitely many such primes. □

Index